AP

Advanced Placement

Chemistry

Donna Bassolino, PhD
Claudine Land, PhD

XAMonline, Inc.

XAMonline, Inc.
21 Orient Avenue
Melrose, MA 02176
Toll Free: 1-800-301-4647
Email: info@xamonline.com
Web: www.xamonline.com
Fax: 1-617-583-5552

Library of Congress Cataloging-in-Publication Data
Bassolino, Donna

AP Chemistry/ Donna Bassolino
ISBN: 978-1-60787-639-7

1. AP   2. Study Guides   3. Chemistry

Disclaimer:

The opinions expressed in this publication are the sole works of XAMonline and were created independently from the College Board, or other testing affiliates. Between the time of publication and printing, specific test standards as well as testing formats and website information may change that are not included in part or in whole within this product. XAMonline develops sample test questions, and they reflect similar content as on real tests; however, they are not former tests. XAMonline assembles content that aligns with test standards but makes no claims nor guarantees candidates a passing score.

Printed in the United States of America
AP Chemistry
ISBN: 978-1-60787-639-7

Contents

SECTION I:
About AP Chemistry

What is AP Chemistry?

The Advanced Placement® program is designed to offer college credit while still in high school. The more than 30 AP courses culminate in an intensive final exam given every year in May.

AP Chemistry is a college level chemistry course, but the exam is in many ways more difficult than most college general chemistry final exams. The questions are difficult, thinking questions that are designed to see whether you truly understand the concepts underlying chemistry. Therefore, while many students take this course to earn college credit, a good score on the AP exam can also be used to demonstrate to college admissions offices that you can do well in a rigorous college level course; that you are a deep thinker (not just a memorizer) and are ready to excel at the college level and beyond.

AP Chemistry follows the guidelines set forth by the College Board (http:www/college board.org). This book follows these same guidelines, paying close attention to what the makers of the AP exam think are important for you to know. The exam itself has been developed and will be scored by professional AP exam readers; either professors who teach the same course in a college/university or experienced high school AP exam teachers. Visit www.collegeboard.org/apcreditpolicy to view AP credit and placement policies at more than 1,000 colleges and universities.

The overarching idea of the makers of the AP Chemistry course is that it should provide you with a college-level foundation to support future advanced course work in chemistry. Therefore, the test is really oriented towards demonstrating that you have enough knowledge about chemistry to be able to complete further work in this field—problem solving, understanding of the meaning of data, and deeply understanding theoretical chemistry so that you will be able to advance your own research ideas in the future. Questions favor inquiry-based investigations and scenarios that explore topics including atomic structure, intermolecular forces and bonding, chemical reactions, kinetics, thermodynamics, and equilibrium. The key concepts and related content that define the AP Chemistry course and exam are organized around underlying principles called the Big Ideas. They encompass core scientific principles, theories, and processes that cut across traditional boundaries and provide a broad way of thinking about the particulate nature of matter underlying observations about the physical world. These Big Ideas form the framework for this review book.

The following are the **Big Ideas** for AP Chemistry:

- **Big Idea 1:** The chemical elements are fundamental building materials of matter, and all matter can be understood in terms of arrangements of atoms. These atoms retain their identity in chemical reactions.

- **Big Idea 2**: Chemical and physical properties of materials can be explained by the structure and the arrangement of atoms, ions, or molecules and the forces between them.

- **Big Idea 3:** Changes in matter involve the rearrangement and/or reorganization of atoms and/or the transfer of electrons.

- **Big Idea 4:** Rates of chemical reactions are determined by details of the molecular collisions.

- **Big Idea 5:** The laws of thermodynamics describe the essential role of energy and explain and predict the direction of changes in matter

- **Big Idea 6:** Any bond or intermolecular attraction that can be formed can be broken. These two processes are in a dynamic competition, sensitive to initial conditions and external perturbations.

More details on the six Big Ideas for AP Chemistry can be found at collegeboard.org.

The Makeup of the Exam

The AP Chemistry exam is administered every May. The exam will be 3 hours and 15 minutes long and consists of the following parts:

- a 90-minute multiple-choice section for which calculators are not permitted. (60 questions)

- a 105-minute free-response section during which calculators are permitted. (7 questions)

As mentioned, the new curriculum involves more science process skills and scientific thinking than the old.

How the Exam is Scored

The multiple choice part of the test is scored by machine and the free response portion is scored by hand (every summer hundreds of professors, content specialists, and AP Chemistry teachers meet to grade the 300,000+ exams that are taken). Once both scores have been tallied, they are combined and then scaled. This raw score is then changed into a composite score ranging from 1 − 5.

The College Board proposes the following qualifications for each of the potential scores:

Exam Score	Recommendation
5	Extremely Well Qualified
4	Well Qualified
3	Qualified
2	Possibly Qualified
1	No Recommendation

The minimum score required for college credit to be granted is a 3. As mentioned above, many schools require scores of 4 or 5 in order to grant credit.

For comparison, the College Board gives the following as college grade equivalents of the AP Exam scores:

AP Exam Score	Letter Grade Equivalent
5	A
4	A-, B+, B
3	B-, C+, C
2	None
1	None

SECTION II:
What To Expect
From This Book

Using this Study Guide

This study guide is organized according to the Big Ideas set forth by the college board. The content assumes that you know the basics from an introductory chemistry course.

The first sections will review the major chemistry concepts that you should know, or at least with which you should be familiar, and specifically relates them to the College Board concepts they want you to understand. Throughout the chapter, call-outs directly quote these concepts—the Big Ideas and their subsections-- in the exact words of the makers of the AP exam. At the beginning of each chapter, a few bullets will remind you of what you should already know before tackling this chapter. It will also outline some key points that the chapter will emphasize.

Each chapter is broken down by main topic from the Big Ideas set forth by the College Board. Each section will also include sample multiple choice questions for that topic. While these are not necessarily in the same format as those found on the AP Chemistry exam, they will help to assess your knowledge of the different concepts. At the end of each section there will be a sample Free-Response question to test what you have learned. Answers are given for each of the sample questions as well as detailed rationale, points to remember, and pitfalls to watch out for.

There are two additional sections, Appendix I and II, which are a review of basic math that is necessary for solving problems in chemistry, and a review of organic chemistry and nomenclature.

At the end of the study guide there are two sample tests. These are designed to give hands-on experience that simulates the actual exam. Each question on these tests has a detailed answer explaining why it is correct and why the incorrect answers are wrong. Use this information to help guide future studies and learning.

Start by taking the diagnostic test at the beginning of the book. This will assess your strong and weak points.

Study and Testing Tips

What to study in order to prepare for the subject assessments is the focus of this study guide but equally important is *how* you study.

You can increase your chances of truly mastering the information by taking some simple, but effective steps.

1. **Some foods aid the learning process.** Foods such as milk, nuts, seeds, rice, and oats help your study efforts by releasing natural memory enhancers called CCKs (cholecystokinin) composed of tryptophan, choline, and phenylalanine. All of these chemicals optimize brain processes associated with memory. Keep your blood sugar steady by avoiding sweets and concentrating on healthy foods with whole grains and protein, and plenty of vegetables. Before studying, a light, protein-rich meal will help keep your energy level up without making you sleepy

   Likewise, before you take a test, stick to a light snack of energy boosting and relaxing foods. A glass of milk, a piece of fruit, or some peanuts all help you maintain energy and help you to relax and focus on the subject at hand.

2. **Get the concept then the details.** Too often we focus on the details and don't gather an understanding of the concept. However, if you simply memorize only dates, places, or names, you may well miss the whole point of the subject.

   A key way to make sure you understand these review concepts is to rephrase them and put them in your own words. By doing this, you will be testing your own understanding of the concepts, and at the same time improving your chances of remembering them. You will remember your own thoughts and words much better than someone else's, and subconsciously tend to associate the important details with the core concepts. Relate the concepts to your own experience ("real life") and "ask why"—think about why a given concept makes sense. Make sure you understand the College Board quotes called out in the text.

3. **Read for reinforcement and future needs.** Even if you only have 10 minutes, open this book and go over a section, including the College Board quotes. If you have a small amount of time, pick a single section and make sure you understand the fundamental ideas the section puts forth.

4. **Relax to learn so go into exile.** Our bodies respond to an inner clock called biorhythms. Burning the midnight oil works well for some people, but not everyone. Be honest about your optimal learning time, and get plenty of sleep.

   If possible, set aside a particular place to study that is free of distractions. Shut off the television, cell phone, and pager and exile your friends and family during your study period.

   If you really are bothered by silence, try background music. Light classical music at a low volume has been shown to aid in concentration over other types. Music that evokes pleasant emotions without lyrics is highly suggested. Try just about anything by Mozart. It relaxes you.

5. **Budget your study time.** Although you shouldn't ignore any of the material, *allocate your available study time in the same ratio that topics may appear on the test.* When you start your AP exam preparation process, divide up the time you have by the topics you need to review, and make weekly and daily goals on that basis.

6. **Stop using a calculator in daily life.** The AP Chemistry exam does not allow calculators for the multiple choice section, so if you are rusty on your arithmetic, it can eat up valuable time on the test. Calculate tax, tips, and other daily calculations in your head. By the time of the exam, you should be able to do all basic arithmetic quickly and effortlessly, so if you have forgotten your multiplication tables after years of using a calculator, it's time to re-learn them!

Test Taking Tips for Multiple Choice Questions _____

1. **Get smart, play dumb. Don't read anything into the question.** Don't make an assumption that the test writer is looking for something other than what is asked. Stick to the question as written and don't read extra things into it.

2. **Read the question and all the choices twice before answering the question.** You may miss something by not carefully reading, and then re-reading, both the question and the answers.

   For questions you partly understand, cross off one or two answers that you can identify as wrong. The more wrong answers you can cross off, the higher the probability of getting your guess right. You can mark these questions as well, and come back to them later if you have time.

   If you really don't have a clue as to the right answer, leave it blank on the first time through. Go on to the other questions, as they may provide a clue as to how to answer the skipped questions.

   Go through all the ones you know and then go back and review the ones you have skipped. Even if you have not finished the test, leave 10 minutes on the clock to transfer answers from the test booklet to the scan sheet, and in the process, randomly mark an answer for any questions you did not answer. *There is no penalty for a wrong answer.* Pick one letter and use it for all of them, so you don't waste time choosing a letter. This gives you a 25% chance of getting it right. Remember, though, if you can cross off any of the answers as wrong, *you may be able to narrow down the choices and improve your odds.* Only one thing is certain; if you don't put anything down, you will get it wrong! Do not leave any questions blank!

3. **Turn the question into a statement.** Look at the way the questions are worded. The syntax of the question usually provides a clue. Does it seem more familiar as a statement rather than as a question? Does it sound strange?

   By turning a question into a statement, you may be able to spot if an answer sounds right, and it may also trigger memories of material you have read.

4. **Look for hidden clues.** It's actually very difficult to compose multiple-choice questions without giving away part of the answer in the options presented. If two answers mean the same thing, neither is the correct answer. Also, an answer that is totally different from the other three is often wrong—a filler to make up a set of four answers.

In most multiple-choice questions you can often readily eliminate one or two of the potential answers. This leaves you with only two real possibilities and automatically your odds go to fifty-fifty for very little work. You should try to answer ALL questions. Remember that there is **NO penalty** for guessing.

5. **Trust your instincts.** For every fact that you have read, you subconsciously retain something of that knowledge. On questions that you aren't really certain about, go with your basic instincts. **Your first impression on how to answer a question is usually correct.**

6. **Mark your answers directly on the test booklet.** Don't bother trying to fill in the optical scan sheet on the first pass through the test. Use the paper to visually present the problem, and always use units in your calculations—this will make many mistakes show up clearly, such as a wrong answer due to multiplying by an amount instead of dividing by it.

   Just be very careful not to mis-mark your answers when you eventually transcribe them to the scan sheet, and leave enough time at the end to fill in all answers.

7. **Remember you are NOT permitted to use a calculator on the multiple-choice section.** Therefore most of the problems will involve calculations that are easy to do "by hand". If you find your answer is requiring you to do complex math that you cannot do easily without a calculator this is a clue that perhaps you are handling the calculations wrong. Go back and quickly review the problem and see if there are any assumptions you can make to simplify the calculation.

8. **Watch the clock!** You have a set amount of time to answer the questions. Don't get bogged down trying to answer a single question at the expense of 10 questions you can more readily answer. Leave difficult questions until the end; if you have time, answer them then.

Test Taking Tips for the Free-Response Questions _____

Part II – Free Response (80 minutes + 10 minute preparation)

The free response questions are usually the items that give students the most difficulty. This is not because they do not know the answers. The problem usually results from not organizing one's thoughts sufficiently and then getting them down on paper fast enough. Organization is key—you will save time by not rushing and instead carefully reading the question and outlining the steps for a question that requires multiple steps to solve. There are 7 free-response questions in this section. You have 105 minutes to complete this section and it is worth 50% of the total grade.

Section II contains two types of free-response questions (short and long). The first three are longer and are worth 10 points each. The last four are much shorter and are worth 4 points each. This section of the exam will contain questions relating to data observation and analysis, experimental design, identifying patterns, explaining observations and following the logical pathway to a problem. These problems are much more in-depth than the multiple-choice problems.

Students will be allowed to use a scientific calculator on the entire free-response section of the exam. Additionally, students will be supplied with a **periodic table of the elements and a formula and constants chart** to use on both the multiple-choice and free-response sections of the exam.

In order to get full credit for each question, you must answer in sufficient detail for the reader to believe you have a complete understanding of the topic. Remember also, readers are human beings—make sure your writing is legible and not too small. You will not lose credit for presenting incorrect information. (However, you will lose time, so make sure to answer the questions you understand well before the ones that are more difficult for you.) Other things the readers do not care about include: spelling, grammar, and penmanship. Obviously, if a reader is unable to determine what it was you wrote, they cannot grade it, but they do their best to interpret a student's "chicken scratch." So, don't worry about the writing being pretty! Instead, focus on presenting your data and ideas in a logical clear way that the reader will be able to follow.

Finally, the biggest piece of advice for answering the free response questions is to show all work. Include all units wherever possible. Even if you do not get the correct answer or finish the problem you may get partial credit if you show them that you know what you are doing.

Your Study Plan

Your study plan for using this book should begin by taking the diagnostic test. Set aside time just as you would for a real test and take the whole test in one sitting. After taking the test, review your answers with the answer key. Make sure you understand why the answer was correct or incorrect. If you are not sure, then this is an area/chapter you will need to review in more detail. At this point the overall score isn't as important as finding your weak areas.

Once you have taken the test you should proceed to the content sections and review the content and the practice questions at the end of each section. This is especially true for your weak areas. You may have to go back to your chemistry textbook for more detailed information if you find you don't have a good understanding of basic concepts. This will help you improve your understanding of that concept for the next practice test.

After you have reviewed the chapters and the sample test questions at the end of each chapter, you should proceed to the Practice Tests. These are set up just like a real AP chemistry test. There are 60 multiple choice questions and 7 Free Response questions. The answers with rationale are provide. Each multiple-choice question is worth one point. There is no penalty for a wrong answer. Scores from both sections are combined to give a raw score. For more on converting a raw score to a AP Exam score go to collegeboard.org.

SECTION III:
Diagnostic Test

Diagnostic Test 1 _____

Multiple Choice Questions

Instructions: The Diagnostic Test consists of multiple Choice questions designed to test your knowledge of each of the important areas outlined by the Big Ideas of the College Board.

The numbers next to the question (1.A.1) refer to the Big Idea and the sub section which reviews the information in that question.

The test is not timed because it is only for you to learn which areas need review.

No calculators are permitted for multiple choice questions on the exam - so you should not need to use a calculator on this test either. (After the test you might want to review Appendix I to review some Math)

Q1. 1.A.1:

The terrestrial composition of an element is: 50.7% as a stable isotope with an atomic mass of 78.9 u and 49.3% as a stable isotope with an atomic mass of 80.9 u. The element is probably

A. Se

B. Br

C. Ca

D. K

Q2. 1.A.2:

Which of the following is a correct chemical formula:

A. $Ba(CO_3)^2$

B. Mg_3N_2

C. Na_2ClO_4

D. Ca_2O_3

Q3. 1.A.3:

At STP, 20. µL of O_2 contain 5.4×10^{16} molecules. According to Avogadro's hypothesis, how many molecules are in 20. µL of Ne at STP?

A. 5.4×10^{15}

B. 1.0×10^{16}

C. 2.7×10^{16}

D. 5.4×10^{16}

Q4. 1.B.1:

How many neutrons are there in $^{60}_{27}\text{Co}$?

A. 27

B. 33

C. 60

D. 87

Q5. 1.B.2:

In C_2H_2, each carbon atom contains the following valence orbitals:

A. p only

B. p and sp hybrids

C. p and sp^2 hybrids

D. sp^3 hybrids only

Q6. 1.B.2:

An atom with the following electronic configuration [Ar] $3d^{10}$ $4s^2$ $4p^3$ must be:

A. Ca

B. Se

C. As

D. Sb

Q7. 1.B.2:

The electronic configuration for Iron (Fe) is :

A. $1s^2$ $2s^2$ $2p^6$ $3s_2$ $3p^6$ $4s^2$ $3d^6$

B. $1s^2$ $2s^2$ $2p^6$ $3s^2$ $3p^6$ $3d^8$

C. $1s^2$ $2s^2$ $2p^6$ $3s^2$ $3p^6$ $4s^2$

D. $1s^2$ $2s^2$ $2p^6$ $3s^2$ $3p^6$ $4s^2$ $4p^6$

Q8. 1.C.1:

Moving down a column on the periodic table:

I.   atomic radius increases

II. Ionization energy increases

III. Protons are added

IV. metallic characteristics increase

A. only I is true

B. I and II are true

C. I, III and IV are true

D. I, II, and III are true.

Q9. 1.C.2:

Match the theory with the scientist who first proposed it:

I.   Electrons, atoms, and all objects with momentum also exist as waves

II. Electron density may be accurately described by a single mathematical equation.

III. There is an inherent indeterminacy in the position and momentum of particles.

IV. Radiant energy is transferred between particles in exact multiples of a discrete unit.

A. I - de Broglie, II - Planck, III - Schrödinger, IV - Thomson

B. I - Dalton, II - Bohr, III - Planck, IV - de Broglie

C. I - Henry, II - Bohr, III - Heisenberg, IV - Schrödinger

D. I - de Broglie, II - Schrödinger, III - Heisenberg, IV - Planck

Q10. 1.D.1:

Match the orbital diagram for the ground state of carbon with the rule/principle it violates:

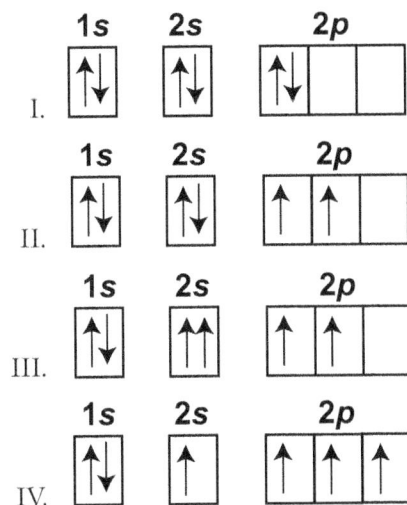

I.
1s	2s	2p
↑↓	↑↓	↑↓		

II.
1s	2s	2p
↑↓	↑↓	↑	↑	

III.
1s	2s	2p
↑↓	↑↑↑	↑	↑	

IV.
1s	2s	2p
↑↓	↑	↑	↑	↑

A. I - Pauli exclusion, II - Aufbau, III no violation, IV - Hund's
B. I - Aufbau, II - Pauli exclusion, III - no violation, IV Hund's
C. I - Hund's, II - no violation, III - Pauli exclusion, IV - Aufbau
D. I - Hund's, II - no violation, III Aufbau, IV Pauli exclusion

Q11. 1.D.2.:

Which of the following were used to study early models of the atom

A. oil solubility
B. gold foil experiment
C. x-rays
D. microwaves

Q12 1.D.3:

A photoelectron spectra shows 4 peaks. The atom is probably

A. Ne
B. Be
C. Na
D. Al

Q13. 1.E.1:

32.0 g of hydrogen and 32.0 grams of oxygen react to form water until the limiting reagent is consumed. What is present in the vessel after the reaction is complete?

A. 16.0 g O_2 and 48.0 g H_2O
B. 24.0 g H_2 and 40.0 g H_2O
C. 28.0 g H_2 and 36.0 g H_2O
D. 28.0 g H_2 and 34.0 g H_2O

Q14. 1.E.2:

Find the mass of CO_2 produced by the combustion of 15 kg of isopropyl alcohol in the reaction:

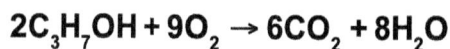

$$2C_3H_7OH + 9O_2 \rightarrow 6CO_2 + 8H_2O$$

A   33 kg
B. 44 kg
C. 50 kg
D. 60 kg

Q15. 2.A.1:

The normal boiling point of water on the Kelvin scale is closest to:
A. 112 K
B. 212 K
C. 273 K
D. 373 K

Q16. 2.A.2:

An ideal gas at 50.0° C and 3.00 atm is in a 300 cm^3 cylinder. The cylinder volume changes by moving a piston until the gas is at 50.0° C and 1.00 atm. What is the final volume?
A. 100. cm^3
B. 450. cm^3
C. 900. cm^3
D. 1.20 cm^3

Q17. 2.A.3:

Carbonated water is bottled at 25° C under pure CO_2 at 4.0 atm. Later the bottle is opened at 4° C under air at 1.0 atm that has a partial pressure of 3×10^{-4} atm CO_2. Why do CO_2 bubbles form when the bottle is opened?
A. CO_2 falls out of solution due to a drop in solubility at the lower air pressure.
B. CO_2 falls out of solution due to a drop in solubility at the lower CO_2 pressure.
C. CO_2 falls out of solution due to a drop in solubility at the lower temperature.
D. CO_2 is formed by the decomposition of carbonic acid.

Q18. 2.B.1:

Why does $CaCl_2$ have a higher normal melting point than NH_3?
A. London dispersion forces in $CaCl_2$ are stronger than covalent bonds in NH_3.
B. Covalent bonds in NH_3 are stronger than dipole-dipole bonds in $CaCl_2$.
C. Ionic bonds in $CaCl_2$ are stronger than London dispersion forces in NH_3.
D. Ionic bonds in $CaCl_2$ are stronger than hydrogen bonds in NH_3.

Q19. 2.B.2:

Which intermolecular attraction explains the following trend in straight-chain alkanes?

Condensed structural formula	Boiling point (° C)
CH_4	-161.5
CH_3CH_3	-88.6
$CH_3CH_2CH_3$	-42.1
$CH_3CH_2CH_2CH_3$	-0.5
$CH_3CH_2CH_2CH_2CH_3$	36.0
$CH_3CH_2CH_2CH_2CH_2CH_3$	68.7

A. London dispersion forces
B. Dipole-dipole interactions
C. Hydrogen bonding
D. Ion-induced dipole interactions

Q20. 2.B.3:

Which of the following regarding hydrogen bonding are true
I.   hydrogen-bonds are stronger than dipole-dople bonds
II. H_2O and NH_3 have higher melting points due to hydrogen bonding
III. Solid water is less dense than liquid water due to hydrogen bonds
IV. Hydrogen bonds occur when there is a large electronegativity difference between the
     two atoms.

A   only II is true
B.   II and III are true
C   I, II, III, IV are true
D.   I, II and IV are true

Q21 2.C.1:

In a covalent bond:
A. One atom loses its electrons to the other
B. the atoms for a 120 degree angle
C. two atoms share electrons
D. There is a large difference in the electronegativities of the atoms

Q22. 2.C.2:

Ionic Bonds result from :
A. electrons shared equally by the two ions
B. one atom looses electrons and the other atom gains electrons
C   removing a p shell electron from one of the atoms.
D. adding an electron to the s shell of an atom

Q23. 2.C.3:

What happens to the metallic bond as you go from potassium to calcium to scandium

A. The strength of the metallic bonds increases from potassium to calcium to scandium.

B. The strength of the metallic bonds decreases from potassium to calcium to scandium.

C. The sea of electrons increases the number of metallic bonds from potassium to calcium to scandium.

D. The number of metallic bonds decreases from potassium to calcium to scandium.

Q24. 2.C.4:

Which of the following is a proper Lewis dot structure of CHClO?

A.

B.

C.

D.

Q25. 2.D.1:

List the substances NH_3, PH_3, $MgCl_2$, Ne, and N_2 in order of increasing melting point.

A. $N_2 < Ne < PH_3 < NH_3 < MgCl_2$

B. $N_2 < NH_3 < Ne < MgCl_2 < PH_3$

C. $Ne < N_2 < NH_3 < PH_3 < MgCl_2$

D. $Ne < N_2 < PH_3 < NH_3 < MgCl_2$

Q26. 2.D.2:

Metallic solids:

A. are good conductors of heat and electricity,

B. all have very high melting points

C. have tightly bound electrons

D. include compounds such as NaCl and Fe_2O_3

Q27. 2.D.3:

Covalent Network Solids:

I. include substances like diamonds, graphite, and silicon dioxide

II. have very high melting points (almost 4000° C).

III. All the electrons are held tightly between the atoms, and aren't free to move and therefore don't conduct electricity.

IV. insoluble in water and organic solvents.

A. All are true

B. I and II

C. II and III

D. II and IV are true

Q28. 2.D.4:

Low weight molecular solids have the following properties

A. high melting points, good conductors

B. never exist in a molten states

C. may be polymerized into large molecules with commercial and biological applications

D. typically bond together to form colloids.

Q29. 3.A.1:

Balance the equation for the neutralization reaction between phosphoric acid and calcium hydroxide by filling in the blank stoichiometric coefficients.

$$\underline{\quad}H_3PO_4 + \underline{\quad}Ca(OH)_2 \rightarrow \underline{\quad}Ca_3(PO_4)_2 + \underline{\quad}H_2O$$

A. 4, 3, 1, 4

B. 2, 3, 1, 8

C. 2, 3, 1, 6

D. 2, 1, 1, 2

Q30. 3.A.2:

An experiment requires 100. mL of a 0.500 M solution of $MgBr_2$. How many grams of $MgBr_2$ will be present in this solution?

A. 9.21 g

B. 11.7 g

C. 12.4 g

D. 15.6 g

Q31. 3.B.1:

$Zn + 2HCl \rightarrow ZnCl + H_2$ is an example of

A. a neutralization reaction

B. a single replacement reaction

C. solution chemistry

D. ion exchange chemistry

Q32. 3.B.2a:

Which statement about acids and bases is NOT true ?

A. All strong acids ionize in water

B. All Lewis acids accept and electron pair

C. All Bronstead bases use OH^- as a proton acceptor

D. All Arrhenius acids form H^+ ions in water.

Q32B 3.B.2b:

NH₄F is dissolved in water. Which of the following are conjugate acid/base pairs present in the solution?

I. NH_4^+/NH_4OH

II. HF/F^-

III. H_3O+/H_2O

IV H_2O/OH^-

A. I, II, and III

B. I, III, and IV

C. II and IV

D. II, III, and IV

Q33. 3.B.3:

Which reaction is not a redox process?

A. Combustion of octane: $2C_8H_{18} + 25O_2 \rightarrow 16CO_2 + 18H_2O$

B. Depletion of a lithium battery: $Li + MnO_2 \rightarrow LiMnO_2$

C. Corrosion of aluminum by acid: $2Al + 6HCl \rightarrow 2AlCl_3 + 3H_2$

D. Taking an antacid for heartburn:

$$CaCo_3 + 2HCl \rightarrow CaCl_2 + H_2CO_3 \rightarrow CaCl_2 + CO_2 + H_2O$$

Q34. 3.C.1:

Which if the following is an example of a chemical change:

A. dissolving sugar in tea

B. rusting of a nail

C. getting a dent in your car.

D. melting of ice

Q35. 3.C.2:

Which statement about reactions is true?

A. All spontaneous reactions are exothermic and cause an increase in entropy.

B. An endothermic reaction that increases the order of the system cannot be spontaneous.

C. A reaction can be non-spontaneous in one direction and also non-spontaneous in the opposite direction.

D. Melting snow is an exothermic process.

Q36. 3.C.3:

Given $E° = -2.37$ V for Mg^{2+} $(aq) + 2e^- \rightarrow Mg$ (s) and $E° = 0.80$ V for Ag^+ $(aq) + e^- \rightarrow Ag$ (s), what is the standard potential of a voltaic cell composed of a piece of magnesium dipped in a 1 M Ag^+ solution and a piece of silver dipped in a 1 M Mg^{2+} solution?

A. 0.77 V

B. 1.57 V

C. 3.17 V

D. 3.97 V

Q37. 4.A.1a:

The exothermic reaction $2NO\ (g) + Br_2\ (g) \leftrightarrow 2NOBr\ (g)$ is at equilibrium.

According to Le Chatelier's principle:

A. Adding Br_2 will increase [NO].

B. An increase in container volume (with T constant) will increase [NOBr].

C. An increase in pressure (with T constant) will increase [NOBr].

D. An increase in temperature (with P constant) will increase [NOBr].

Q38. 4.A.2:

If a reaction does not depend on the concentration of the reactants, the reaction:

A. is always spontaneous

B. has a zero order rate law

C. has a first order rate law

D. must be using a catalyst

Q39. 4.A.1b:

At a certain temperature, T, the equilibrium constant for the reaction $2NO\ (g) \leftrightarrow N_2\ (g) + O_2\ (g)$ is $K_{eq} = 2 \times 10^3$. If a 1.0 L container at this temperature contains 90 mM N_2, 20 mM O_2, and 5 mM NO, what will occur?

A. The reaction will make more N_2 and O_2.

B. The reaction is at equilibrium.

C. The reaction will make more NO.

D. The temperature, T, is required to solve this problem.

Q40. 4.B.1:

Which of the following are true for uni-molecular reactions ?

I. unimolecular reaction occurs when a molecule rearranges itself to produce one or more products

II. Radioactive decay is a umimolecular reaction

III. The rate at which a substance decomposes is independent of its concentration

IV. The reaction involves the collision of two molecules

A. only I is true

B. I and II are true

C. II and III are true

D. I, II, and III are true

Q41. 4.B.2

Write the equilibrium expression Keq for the reaction:

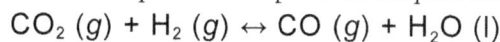

$CO_2 (g) + H_2 (g) \leftrightarrow CO (g) + H_2O (l)$

A. $\dfrac{[CO][H_2O]}{[CO_2][H_2]^2}$

B. $\dfrac{[CO_2][H_2]}{[CO][H_2O]}$

C. $\dfrac{[CO][H_2O]}{[CO_2][H_2]}$

D. $\dfrac{[CO]}{[CO_2][H_2]}$

Q42. 4.B.3:

Collision Theory involves:

A. activation energy, velocity and work

B. potential energy, entropy, and frequency

C. activation energy, temperature, and concentration

D. potential energy, temperature, and entropy

Q43. 4.C.1:

A multistep reaction:

A. consists of a series of small reactions that add up to the overall reaction

B. is dependent only on the concentration of the reactants

C. is independent of a catalyst

D. always has a higher order rate constant K.

Q44. 4.C.2:

The reaction: $(CH_3)_3CBr(aq) + OH^-(aq) \rightarrow (CH_3)_3COH(aq) + Br^-(aq)$ occurs in three elementary steps:

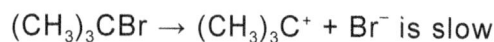

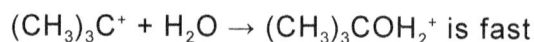

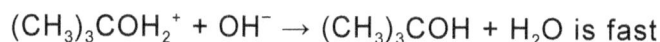

$(CH_3)_3CBr \rightarrow (CH_3)_3C^+ + Br^-$ is slow

$(CH_3)_3C^+ + H_2O \rightarrow (CH_3)_3COH_2^+$ is fast

$(CH_3)_3COH_2^+ + OH^- \rightarrow (CH_3)_3COH + H_2O$ is fast

What is the rate law for this reaction?

A. $Rate = k\left[(CH_3)_3CBr\right]$

B. $Rate = k\left[OH^-\right]$

C. $Rate = k\left[(CH_3)_3CBr\right]\left[OH^-\right]$

D. $Rate = k\left[(CH_3)_3CBr\right]^2$

Q45. 4.C.3:

A reaction between NO and H_2 occurs in the following three-step process:

$$NO + NO \rightarrow N_2O_2 (fast)$$
$$N_2O_2 + H_2 \rightarrow N_2O + H_2O (slow)$$
$$N_2O + H_2 \rightarrow N_2 + H_2O (fast)$$

The reaction intermediates are:

A. H_2O and N_2O

B. N_2O_2 and N_2O.

C. H_2 and N_2O

D. H_2 and H_2O

Q46. 4.D.1:

Catalysts function by lowering the activation energy of an elementary step in a reaction mechanism, and by providing a new and faster reaction mechanism.

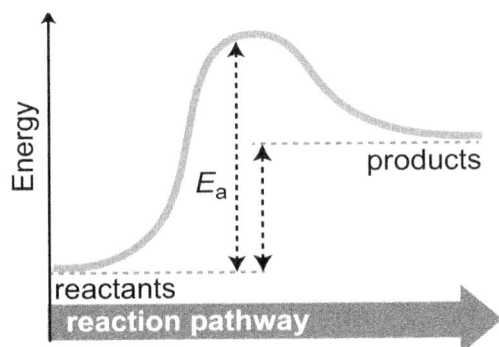

What could cause this change in the energy diagram of a reaction?

A. Adding a catalyst to an endothermic reaction

B. Removing a catalyst from an endothermic reaction

C. Adding a catalyst to an exothermic reaction

D. Removing a catalyst from an exothermic reaction

Q47. 4.D.2:

Which statements about reaction rates are true?

I. A catalyst will shift an equilibrium to favor product formation.

II. Catalysts increase the rate of forward and reverse reactions.

III. A greater temperature increases the chance that a molecular collision will overcome a reaction's activation energy.

IV. A catalytic converter contains a homogeneous catalyst.

A. I and II

B. II and III

C. II, and IV

D. I, III, and IV

Q48. 5.A.1:

Heat is added to a pure solid at its melting point until it all becomes liquid at it's freezing point. Which of the following occur(s)?

A. Intermolecular attractions are weakened.

B. The kinetic energy of the molecules does not change.

C. The freedom of the molecules to move about increases.

D. All of the above.

Q49. 5.A.2:

Which of the following statements are true ?

A. One cannot convert heat completely into work

B. Every isolated system become more disordered over time

C. heat flows spontaneously from a hot body to a cold one.

D. All of the above.

Q50. 5.B.1:

10.0 kJ of heat are added to one kilogram of Iron at $10.^\circ$ C. What is its final temperature? The specific heat of iron is 0.45 J/g$^\circ$ C.

A. 22° C

B. 27° C

C. 32° C

D. 37° C

Q51. 5.B.2:

What is the standard heat of combustion of CH_4 (g)? Use the following data:

Standard heats of formation

$CH_4(g)$	−74.8 kJ/mol
$CO_2(g)$	−393.5 kJ/mol
$H_2O(l)$	−285.8 kJ/mol

A. −890.3 kJ/mol

B. −604.5 kJ/mol

C. −252.9 kJ/mol

D. −182.5 kJ/mol

Q52. 5.B.3:

Which of the following occurs when NaCl dissolves in water?

A. Heat is required to break bonds in the NaCl crystal lattice.

B. Heat is released when hydrogen bonds in water are broken.

C. Heat is required to form bonds of hydration.

D. The hydrogen end of the water molecule is attracted to the Cl^- ion.

Q53. 5.B.4:

The following procedure was developed to find the specific heat capacity of metals:

1. Place pieces of the metals in an ice-water bath so their initial temperature is $0°$ C.
2. Weigh a Styrofoam cup.
3. Add water at room temperature to the cup and weigh it again.
4. Add a cold metal from the bath to the cup and weigh the cup a third time.
5. Monitor the temperature drop of the water until a final temperature at thermal equilibrium is found.

_____ is also required as additional information in order to obtain heat capacities for the metals. The best control would be to follow the same protocol except to use _____ in Step 4 instead of a cold metal.

A. The heat capacity of water / a metal at $100°$ C
B. The heat of formation of water / ice from the $0°$ C bath
C. The heat of capacity of ice / glass at $0°$ C
D. The heat capacity of water / water from the $0°$ C bath

Q54. 5.C.1:

Potential energy is associated with a particular geometric arrangement of atoms or ions and the electrostatic interactions between them.

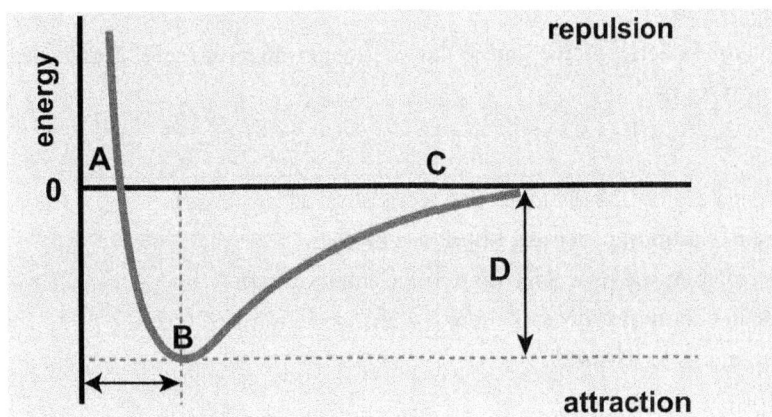

In the energy diagram shown above, which point represents the equilibrium bond distance between the two atoms.

A. A.
B. B.
C. C.
D. D.

Q55. 5.C.2:

Given the following heats of reaction:

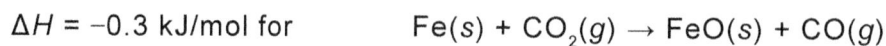

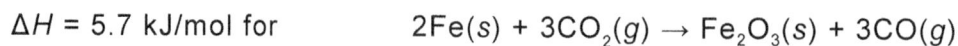

$\Delta H = -0.3$ kJ/mol for $\qquad$ $Fe(s) + CO_2(g) \rightarrow FeO(s) + CO(g)$

$\Delta H = 5.7$ kJ/mol for $\qquad$ $2Fe(s) + 3CO_2(g) \rightarrow Fe_2O_3(s) + 3CO(g)$

and $\Delta H = -5.7$ kJ/mol for $\qquad$ $3FeO(s) + CO_2(g) \rightarrow Fe_3O_4(s) + CO(g)$

use Hess' Law to determine the heat of reaction for:

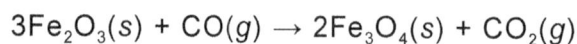

$\qquad$ $3Fe_2O_3(s) + CO(g) \rightarrow 2Fe_3O_4(s) + CO_2(g)$

A. -10.8 kJ/mol

B. -9.9 kJ/mol

C. -9.0 kJ/mol

D. -8.1 kJ/mol

Q56. 5.D.1:

What happens to the potential energy of the dipole-dipole interaction as temperature increases?

A. The potential energy of the dipole-dipole interaction decreases as T increases

B. The potential energy of the dipole-dipole interaction increases as T increases

C. The potential energy of the dipole-dipole interaction remains the same regardless of temperatures

D. The potential energy of the dipole-dipole interaction varies with T differently for different systems.

Q57. 5.D.2:

Chemical processes can be distinguished from physical processes

A. Because in chemical processes bonds are broken

B. Because in chemical processes there is a change of state

C. Because in chemical processes there is a change in temperature

D. Because energy is released.

Q58. 5.D.3:

Noncovalent and intermolecular interactions are important in

A. Acid Neutralization reactions

B. Subsitution Reactions

C. Protein Folding process

D. Passive Transport process

Q59. 5.E.1:

Which statement about thermochemistry is true?

A. Particles in a system move about less freely at high entropy.

B. Water at 100° C has the same internal energy as water vapor at 100° C.

C. A decrease in the order of a system corresponds to an increase in entropy.

D. At its sublimation temperature, dry ice has higher entropy than gaseous CO_2.

Q60 5.E.2:

It is observed that the a certain reaction starting from its elements goes essentially to completion. Which of the following is a true statement about the thermodynamic favorability of the reaction?

A. The reaction is favorable and driven by an enthalpy change only.

B. The reaction is unfavorable and driven by an entropy change only.

C. The reaction is favorable and driven by both enthalpy and entropy changes.

D. The reaction is unfavorable due to both enthalpy and entropy changes

Q61. 5.E.3:

Water boills at 100° C with a molar heat of vaporization of -43.9kJ. At 100 °C what is the entropy change of the system ?

A. -439 J/K

B. 43.9 J/K

C. -118 J/K

D. Cannot be determined with out the value for ΔG

Q62 5.E.4:

Which of the following can change the value of ΔG for a chemical reaction

A. Change in the total pressure of the systematic

B. Change in the concentration of the products

C. Change in the Temperature of the system

D. Change in the volume of the system

Q63 5.E.5:

For an equilibrium reaction such as

$$A + B \leftrightarrow C + D$$

A. At equilibrium, $\Delta G = 0$

B. At equilibrium, Q=K.

C. When K_{eq} is large, almost all reactants are converted to products.

D. All of the above

Q64. 6.A.1:

At equilibrium, _____.

A. all chemical reactions have ceased

B. the rate constants of the forward and reverse reactions are equal

C. the rates of the forward and reverse reactions are equal

D. the value of the equilibrium constant is 1

Q65. 6.A.2:

The current state of a system undergoing a reversible reaction can be characterized by the extent to which reactants have been converted to products. The relative quantities of reaction components are quantitatively described by the reaction quotient, Q

For a given reaction, if Q is larger than K, in what direction will the reaction proceed?

A. Forward, until Q = K

B. Reverse, until Q = K

C. Forward and reverse reactions will be equal, since the reaction is at equilibrium

D. Reverse if K is small and forward if K is large

Q66. 6.A.3:

When a system is at equilibrium, all macroscopic variables, such as concentrations, partial pressures, and temperature, do not change over time. Equilibrium results from an equality between the rates of the forward and reverse reactions, at which point Q=K.

As a reaction proceeds towards equilibrium, the rate of the reverse reaction:

A. stays constant before and after equilibrium has been reached

B. decreases until it becomes the same as the forward reaction

C. increases to become a constant, nonzero rate at equilibrium

D. decreases to become zero at equilibrium

Q67. 6.A.4:

Which of the following equilibrium reactions is favored to produce reactants at 298K

A. $COCl_2 \rightarrow CO(g) + Cl(g)$ $K_{eq} = 2.2 \times 10^{-10}$

B. $2NO_2 \rightarrow N_2O_4(g)$ $K_{ew} = 2.15 \times 10^2$

C. $H_2(g) + Cl_2(g) \leftrightarrow 2HCl(g)$ $K_{eq} = 4 \times 10^{31}$ at 300 K

D. $2NO(g) \leftrightarrow N_2(g) + O_2(g)$ is $K_{eq} = 2 \times 10^3$

Q68. 6.B.1:

What would happen to the reaction below if the temperature of the system is increased from the equilibrium temp of 300 to 500

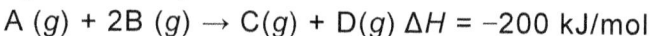

$A\ (g) + 2B\ (g) \rightarrow C(g) + D(g)\ \Delta H = -200$ kJ/mol

A. There will be an increase the production of all products

B. There will be an increase the production of all reactants

C. There will be no change to the system

D. The prediction cannot be made without additional information

Q69 6.B.2:

Predict the direction in which the reaction $H_2(g) + I_2(g) \leftrightarrow 2HI(g)$ will proceed if initial concentrations are 0.004 mM H_2, 0.006 mM I_2, and 0.011 mM HI given $K_{eq}=48$.

A. The reaction will proceed to the left

B. The reaction will proceed to the right

C. The reaction is in equilibrium

D. Cannot be determined without knowing the temperature

Q70 6.C.1:

A student wishes to make a buffer solution with a pH = 8. Which of the following would be the best choice ?

A. oxalic acid $\times$ 10^{-2}

B. hypochlorous 3.5×10^{-8}

C. acetic acid 1.8×10^{-5}

D. uric acid 1.3×10^{-4}

Q71 6.C.2:

What is the pH of a buffer made of 0.128 M sodium formate (HCOONa) and 0.072 M formic acid (HCOOH)? The pK_a of formic acid is 3.75.

A. 2.0

B. 3.0

C. 4.0

D. 5.0

Q72 6.C.3:

The solubility of $CoCl_2$ is 54 g per 100 g of ethanol. Three flasks each contain 100 g of ethanol. Flask #1 also contains 40 g $CoCl_2$ in solution. Flask #2 contains 56 g $CoCl_2$ in solution. Flask #3 contains 5 g of solid CoCl2 in equilibrium with 54 g $CoCl_2$ in solution. Which of the following describes the solutions present in the liquid phase of the flasks?

A. #1 - saturated, #2 - supersaturated, #3 - unsaturated.

B. #1 - unsaturated, #2 - miscible, #3 - saturated.

C. #1 - unsaturated, #2 - supersaturated, #3 - saturated.

D. #1 - unsaturated, #2 - not at equilibrium, #3 - miscible.

Q73 6.D.1:

A certain reaction has a ΔH = 2.98 kJ, ΔS = 10.0 J/K. State whether the reaction is spontaneous at 200K, 298K and 300K.

A. the reaction is spontaneous at all temperatures

B. the reaction is not spontaneous at 200, at equilibrium at 298K and sp at 300K

C. the reaction is spontaneous at 200K, but not spontaneous at 298 or 300K.

D. cannot determine with the given information.

Diagnostic Test Answer Key _____

1.	B	26.	A	50.	C
2.	B	27.	A	51.	A
3.	D	28.	C	52.	D
4.	B	29.	C	53.	D
5.	B	30.	A	54.	B
6.	C	31.	B	55.	B
7.	A	32a.	C	56.	A.
8.	C	32b.	D	57.	A
9.	D	33.	D	58.	C
10.	C	34.	B	59.	C
11.	B	35.	B	60.	A
12.	C	36.	C	61.	C
13.	C	37.	C	62.	C
14.	A	38.	B	63.	D
15.	D	39.	A	64.	C
16.	C	40.	B	65.	B
17.	B	41.	D	66.	C
18.	D	42.	C	67.	A
19.	A	43.	A	68.	B
20.	C	44.	A	69.	B
21.	C	45.	B	70.	B
22.	B	46.	B	71.	C
23.	A	47.	B	72.	C
24.	C	48.	D	73.	B
25.	D	49.	D.		

SECTION IV:
Content Review
For AP Exam

Chapter 1: Big Idea #1

Atoms, Elements and the Building Blocks of Matter

Big Idea 1: The chemical elements are fundamental building materials of matter, and all matter can be understood in terms of arrangements of atoms. These atoms retain their identity in chemical reactions.

What you should already know:

- the basics of the periodic table
- the basic components of an atom (proton, neutron, electron)
- the basic concept of the mole

What you will learn from this chapter:

- trends of the periodic table
- electronic configurations
- how to work with moles and mass
- spectral analysis
- how models in chemistry have been refined using experimental data

---

Chapter 1.1 Matter and Atoms

The word "matter" describes everything that has physical existence, defined as anything that has mass and takes up space. The Law of Conservation of Matter (also known as the Law of Conservation of Mass), developed by Antoine Lavoisier in 1774 states that all matter is made up of atoms, which are never created or destroyed in chemical reactions, only rearranged. This atomic theory also successfully predicted the behavior of matter in chemical reactions that had not been studied at the time. As a result, the atomic theory has stood for 200 years with only minor modifications.

The atom-based make-up of matter allows it to be separated into categories. The two main classes of matter are pure substances and mixtures. Each of these classes can also be divided into smaller categories such as elements, compounds, homogeneous mixtures, and heterogeneous mixtures, based on composition.

Pure Substances:

A pure substance is a form of matter with a definite composition and distinct properties that are exactly the same throughout the whole quantity of substance. This type of matter can not be separated by ordinary processes like filtering, centrifuging, boiling or melting.

"All matter is made of atoms. There are a limited number of types of atoms; these are the elements."

Pure substances are divided into elements and compounds.

Elements: A single type of matter, containing one and only one specific type of atom, is called an element. Elements can not be broken down any farther by ordinary chemical processes. Examples are hydrogen and oxygen. Two or more elements can join together chemically into a **molecule**.

Compounds: When two or more different elements chemically combine, they become a **compound**. A compound may be broken down into its component elements by processes that cause chemical change, such as electrolysis. Large compounds can also be broken down into smaller compounds. Compounds have a uniform composition regardless of the sample size or source of the sample.

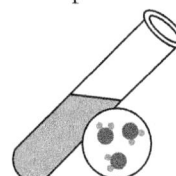

Examples are table salt (made up of sodium and chloride atoms) and pure water (made up of oxygen and hydrogen atoms).

Note: A molecule is made up of two or more atoms joined together chemically. A compound is a molecule that contains at least two different elements. Therefore, all compounds are molecules but not all molecules are compounds. Molecules can differ from each other because they are made up of different elements, but they can also differ even if they are made up of the same elements. For example, oxygen, O_2, and ozone, O_3, have very different properties, even though they are composed of the same element.

Because the fundamental nature of compounds is discrete-- combinations of individual atoms rather than "amounts of element"—mass ratios of the elements making up the compounds are seen in whole-number ratios of the component elements. These mass ratios can be expressed as an empirical formula, which represents the lowest whole number ratio of atoms in a compound. Unlike structural formulas, which reflect the pattern of connections between atoms in a compound, empirical formulas will be the same for two molecules of the same elements with identical mass percent of their constituent atoms, even if the arrangement of their atoms is different.

Mixtures:

Two or more substances that are not chemically bound together into a single compound. Mixtures may be made up of any proportions of the components and can be physically, or mechanically, separated by processes like filtering, centrifuging, boiling, or melting. The components do not form new chemical bonds with each other, and separated componenets can be mixed back together again to easily recreate the original mixture.

"Molecules are composed of specific combinations of atoms; different molecules are composed of combinations of different elements and of combinations of the same elements in differing amounts and proportions."

"Atoms are conserved in physical and chemical processes."

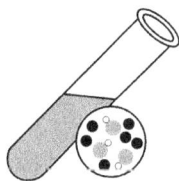

Homogenous and heterogenous mixtures:

Homogeneous mixtures (also known as solutions) have the same composition and properties throughout the mixture, with a uniform color and distribution of solute and solvent particles. Examples of homogenous mixtures are the gases in the air in a still room and the liquid in a glass of wine or saltwater.

In heterogenous mixtures, the components are not distributed uniformly. Examples include vegetable soup or carbonated soda.

An easy way to remember this is through the example of oil-and-vinegar salad dressing, which is a heterogenous mixture that includes two homogenous mixtures. Vinegar is a homogenous mixture of the compounds acetic acid and water, with tiny amounts of other compounds that give rise to flavor. Vegetable oil is a homogenous mixture made up of triglycerides and other compounds. When mixed together, along with some spices, they form a heterogenous mixture of bubbles and solid particles that are not distributed uniformly. The heterogenous mixture-- the oil, vinegar, and solid spices-- can be separated by centrifugation, or simply letting the mixture sit for awhile, and can be easily recombined into salad dressing by stirring or shaking. The homogenous mixtures— the oil and the vinegar-- can each, in turn, be separated into their component compounds, also through physical means, such as distillation. However, the component compounds making up these homogenous mixtures— such as acetic acid or water-- cannot be separated further by ordinary physical means. To break these compounds down into their component elements or smaller compounds, a chemical reaction needs to take place, and at that point their fundamental nature would be changed.

The purity of a mixture can be tested by looking at the mass ratio of the components. A pure mixture will be expected to have a certain mass ratio based on the atoms that make it up; any deviation from this expected value indicates an impurity. Impurities can also be intentionally added to a pure mixture; a small quantity of such an intentionally added impurity is called a dopant.

Atoms are conserved in physical and chemical processes

In any transformation of matter, whether physical or chemical, mass is conserved because the number of atoms is conserved. Therefore, it is possible to calculate product masses if you know reactant masses, and reactant masses if you know product masses. These types of transformations can be depicted symbolically, and in such depictions, the number of atoms much be conserved (more on this in Chapter 3). With the exception of radioactive atoms, no atoms are lost or gained, but instead cycle through water, land, the atmosphere, and living organisms.

Chapter 1.2 The Elements

The identity of an **element** depends on the **number of protons** in the nucleus of the atom. This value is called the **atomic number** and it is sometimes written as a subscript before the symbol for the corresponding element. Atoms and ions of a given element that differ in the number of neutrons have a different mass and are called isotopes.

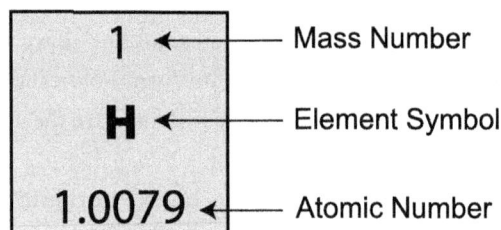

1 ←	—— Mass Number
H ←	—— Element Symbol
1.0079 ←	—— Atomic Number

The **mass number** of an atom is the sum of the number of protons and neutrons in the nucleus. This is used in chemistry as the mass of an atom because electrons are so tiny that their mass is considered as approximately zero. It may be written as a superscript before the atomic symbol. Since an element can include isotopes with different numbers of neutrons, the mass number of each element in the periodic table represents the **average** number of nucleons, and is therefore generally not a whole number. For example, in the periodic table, the mass number of hydrogen is 1.0079. This is because most hydrogen atoms contain one proton and no neutrons, but a small fraction of hydrogen atoms, known as deuterium or heavy hydrogen, contain one neutron and one proton. These hydrogen atoms have a mass number of 2, but there are very few of them, so when all of the heavy hydrogen atoms are averaged in with all of the regular hydrogen atoms, the average number of nucleons is 1.0079.

A carbon atom with 6 protons and 8 neutrons would be represented by

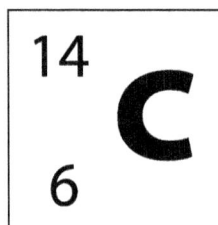

$$^{14}_{6}\text{C}$$

The number of neutrons may be found by subtracting the atomic number from the mass number. For example, uranium-235 has $235 - 92 = 143$ neutrons because it has 235 nucleons and 92 protons.

While different isotopes have different natural abundances and have different nuclear properties, an atom's chemical properties are almost entirely due to electrons. In other words, less common isotopes, such as carbon 14, depicted above, will react in the same way as more common isotopes, such as carbon 12, which has 6 neutrons. As far as their chemical "behavior" is concerned, it's the electrons and protons that count, not the neutrons.

Protons have a positive charge, **neutrons** have no charge, and **electrons** have a negative charge. Atoms have no net charge because they have an equal number of protons and electrons. **Anions** are negative ions and contain more electrons than protons. **Cations** are positive ions and contain more protons than electrons. Electrons are prevented from flying away from the nucleus by the attraction that exists between opposite electrical charges—the positive charge of the nucleus and the negative charge of the electrons. This force is known as **electrostatic** or **coulombic** attraction.

Chapter 1.3 Early Models of the Atom

Prior to the late 1800s, atoms were thought to be small, spherical and indivisible particles that made up matter. However, with the discovery of electricity and the investigations that followed, this view of the atom changed to the one we are familiar with today—that, while atoms are fundamental building blocks of matter that cannot be

Dalton's 1808 AD Elements			
	wt.	wt.	wt.	wt.
Hydrogen 1	Sulphur 13	Strontium 46	Lead 95
Azote 5	Magnesia 20	Barytes 68	Silver 100
Carbon 5	Lime 23	Iron 38	Gold 100
Oxygen 7	Soda 28	Zinc 56	Platina 140
Phosphorus 9	Potash 42	Copper 56	Mercury 167

"Atoms are so small that they are difficult to study directly; atomic models are constructed to explain experimental data on collections of atoms."

split apart by normal means, they are made up of dynamic, interacting particles.

The ideas that pure materials called elements existed and that these elements were composed of fundamentally indivisible units called atoms were proposed by ancient philosophers, even though they had little evidence. Modern atomic theory is credited to the work of **John Dalton** published in 1803-1807. Observations made by him and others about the composition, properties, and reactions of many compounds led him to develop the following postulates:

1. Each element is composed of small particles called atoms.

2. All atoms of a given element are identical in mass and other properties.

3. Atoms of different elements have different masses and differ in other properties.

4. Atoms of an element are not created, destroyed, or changed into a different type of atom by chemical reactions.

5. Compounds form when atoms of more than one element combine.

6. In a given compound, the relative numbers and kinds of atoms are constant.

Dalton determined and published the known relative masses of a number of different atoms. He also formulated the law of partial pressures. Dalton's work focused on the ability of atoms to arrange themselves into molecules and to rearrange themselves via chemical reactions, but he did not investigate the composition of atoms themselves. **Dalton's model of the atom** was a tiny, indivisible, indestructible **particle** of a certain mass, size, and chemical behavior, but Dalton did not deny the possibility that atoms might have a substructure.

While Dalton understood the discrete nature of atoms as the basis of all substance, he did not understand the nature of atoms themselves. Fundamentally, the basis for the structure of the atom is the attraction between positive and negative charges. If two charges are opposite, they attract each other; if both are positive or both are negative, they repel each other. Coulomb's law states that the force between two charged particles is proportional to the magnitude of the charges and is inversely proportional to the square of the distance between them. In other words, potential energy is proportional to (charge 1 * charge 2)/(distance between charge 1 and charge 2). This understanding of the atom evolved over time, though, with many scientists refining Dalton's initial model into a model that more accurately reflects the true nature of the atom.

J. J. Thomson, was the first to examine the atom's substructure, and to realize that electrical charges were fundamental to the nature of atoms. In the mid-1800s, scientists had studied a form of radiation called "cathode rays" or "electrons" that originated from the negative electrode (cathode) when electrical current was forced through an evacuated tube. Thomson determined in 1897 that **electrons have mass**, and because many different cathode materials release electrons, Thomson proposed that the **electron is a subatomic particle**. **Thomson's model of the atom** was a uniformly positive particle with electrons contained in the interior. This has been called the "plum-pudding" model of the atom where the pudding represents the uniform sphere of positive electricity and the bits of plum represent electrons.

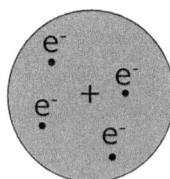

Ernest Rutherford studied atomic structure in 1910-1911 by firing a beam of alpha particles at thin layers of gold leaf, called the gold-foil experiment. Thomson's model envisioned the positively charged part of the atom as extending throughout its volume, forming a diffuse positive background in which the electrons "floated". If this were the case, the path of an alpha particle should be deflected only slightly if it struck an atom, akin to light shining through water. However, instead, Rutherford observed that most alpha particles went straight forward, but some alpha particles bounced almost backwards, like they had been reflected from something small and very dense. This suggested that **nearly the entire mass of an atom is contained in a small positively charged nucleus.**

Rutherford's model of the atom was an analogy to the sun and the planets. A small positively charged nucleus is surrounded by circling electrons and mostly by empty space.

Max Planck had determined in 1900 that **energy is transferred by radiation in**

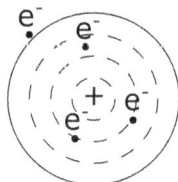

exact multiples of a discrete unit of energy called a quantum. Quanta of energy are extremely small, and may be found from the frequency of the radiation, hv, using the equation:

$$E = h\,v$$

where E is a quantum of energy measured in Joules, h is Planck's constant, and v is the frequency of the radiation. An amount of light that contains one quantum of energy is a photon. Planck's constant is approximately equal to 6.626×10^{-34} Joule • seconds.

Niels Bohr incorporated Planck's quantum concept into Rutherford's model of the atom in 1913 to explain the **discrete frequencies of radiation emitted and absorbed by atoms/ions with one electron** (H, He$^+$, and Li^{2+}). This electron is attracted to the positive nucleus and is closest to the nucleus at the **ground state** of the atom.

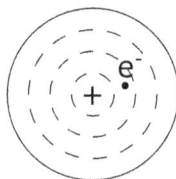

When the electron absorbs enough energy, it moves into an orbit further from the nucleus and the atom is said to be in an electronically **excited state**. The energy needed to move an electron from one orbit to another is a discrete quantity; that is, when a certain amount is absorbed, a threshold is passed and the electron moves. When the electron moves back, it emits the same discrete amount of radiation energy that was required to change its orbit. If sufficient energy is absorbed, the electron separates from the nucleus entirely, and the atom is ionized.

The minimum energy required to remove the least tightly held electron from an atom or ion is called the atom's **first ionization energy**. The ionization energy of any electron in an atom or ion is the minimum energy needed to remove that electron from its atom or ion. This can be estimated via the principles in Coulomb's law—the farther an electron is from the nucleus, the lower its ionization energy, and the higher the nuclear charge, the higher the ionization energy. The discrete frequencies of radiation emitted and absorbed by the atom correspond (using Planck's constant) to discrete energies and in turn to discrete distances from the nucleus. **Bohr's model of the atom** was a small positively charged nucleus surrounded mostly by empty space and by electrons orbiting at certain discrete distances ("shells") corresponding to discrete energy levels.

"As is the case with all scientific models, any model of the atom is subject to refinement and change in response to new experimental results. In that sense, an atomic model is not regarded as an exact description of the atom, but rather a theoretical construct that fits a set of experimental data."

Photoelectron spectroscopy (PES) provides direct evidence for the shell model of the atom. Light consists of photons, discrete "packets" of radiation with energy defined by E=hv, where h is a constant (Planck's constant and v is the frequency of the light. The photoelectric effect, the phenomenon of light ejecting electrons from a material, requires the photon to have enough energy to eject an electron. PES determines the energy needed for this ejection of an electron from a material. From this energy, the shell structure of an atom can be deduced—the intensity of the photoelectron signal can be used to calculate the number of electrons in that energy level.

Going beyond the work of Niels Bohr, PES can be used to infer the structure of atoms with multiple electrons. In helium, both electrons have the same energy—they are identical, and roughly the same distance from the nucleus as the electron in hydrogen. In contrast, Lithium has an outermost electron that is further from the nucleus than the electron of hydrogen. This electron is said to be in a different shell.

Outer electrons, called valence electrons, require much less energy to leave the atom, and therefore are the ones that are mainly involved in chemical reactions. Inner electrons, called core electrons, are close to the nucleus and serve to "shield" the valence electrons from the electrostatic attraction of the nucleus. This makes it easier for the valence electrons to leave, or interact with other atoms.

Chapter 1.4 Arrangement of Electrons ─────────────

It is important to remember that a scientific model is simply a construction to explain experimental data about things that cannot be observed directly, such as atoms. When a model cannot adequately account for new data, the model is modified or refined. A model of an atom is not an exact description of an atom, but rather a theoretical construct that fits a set of experimental data.

So, while a mental picture of electrons whizzing around a central core may be useful, it is important to realize what it is and what it is not: It is a helpful, accurate representation that fits the experimental data; it is not the "true picture" of an atom. (The formula H2O is similar—certainly, no one thinks that the nature of water is tiny letter Hs and Os, but this is an accurate and useful way to think about water when doing chemistry.) Werner Heisenberg highlighted another issue: that of uncertainty. The realization that both matter and radiation interact as waves led **Werner Heisenberg** to the conclusion in 1927 that the act of observation and measurement of subatomic particles requires the interaction of one wave with another. For example, if you are looking at an object, the information you are receiving is actually a wave of light that has bounced off the object. This is, for practical purposes, irrelevant when you are observing or measuring large objects, but it becomes important when observing subatomic particles, which act as waves themselves. This interaction results in an **inherent uncertainty** in the location and momentum of particles observed. Specifically, the more precisely you know the location of a subatomic particle, the less precisely you will know its momentum, and vice versa. This limitation in the measurement of phenomena at the subatomic level is known as the **Heisenberg uncertainty principle**, and it applies to the location and momentum of electrons in an atom.

"The atom is composed of negatively charged electrons, which can leave the atom, and a positively charged nucleus that is made of protons and neutrons. The attraction of the electrons to the nucleus is the basis of the structure of the atom."

The Quantum Mechanical model

The Bohr model explained the hydrogen atom well, but not larger atoms. A robust model must explain results over a wide range of experimental circumstances. In light of a large body of data collected after Bohr developed his model, the new model that best explains the atom is the **quantum mechanical model**.

When **Erwin Schrödinger** studied the atom in 1925, he replaced the idea of precise orbits with regions in space called **orbitals** where electrons were likely to be found. Differences in electron-electron repulsion are responsible for the differences in energy between electrons in different orbitals in the same shell. This is seen as differences in orbital shape. **Schrödinger's model of the atom** is a mathematical formulation of quantum mechanics that describes the electron density of orbitals. **The Schrödinger equation** describes the probability that an electron will be in a given region of space, a quantity known as **electron density**.

This is the atomic model that has been in use from shortly after it was introduced up to the present.

This model explains the movement of electrons to higher energy levels when exposed to radiation energy. It also explains the emission of electromagnetic radiation (e.g., light) accompanying the movement of electrons to lower energy levels when the source of energy has disappeared. Using Plank's equation, the frequency of this emitted EMR can be determined.

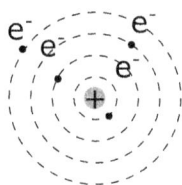

The shells of an atom contain subshells made up of orbitals that have different shapes. Each s subshell consists of one spherical orbital, known as an s orbital, that may contain up to 2 electrons. Each p subshell consists of up to three roughly figure-8 shaped orbitals called p orbitals, each of which may contain up to 2 electrons, for a total of 6 electrons in the p subshell. They are usually referred to as x, y, and z to denote their 3-dimensional orientation.

The diagrams below represent surfaces of constant electron density in the $1s$, $2p$, and $3d$ orbitals.

Each orbital of any type may hold up to two electrons, one spinning in a clockwise direction and the other in a counterclockwise direction. **Wolfgang Pauli** helped develop quantum mechanics in the 1920s by developing this concept of **spin** and the **Pauli exclusion principle**, which states that if two electrons occupy the same orbital, they must have different spin (intrinsic angular momentum). This principle has been generalized to other quantum particles.

"The currently accepted best model of the atom is based on the quantum mechanical model."

"The atoms of each element have unique structures arising from interactions between electrons and nuclei."

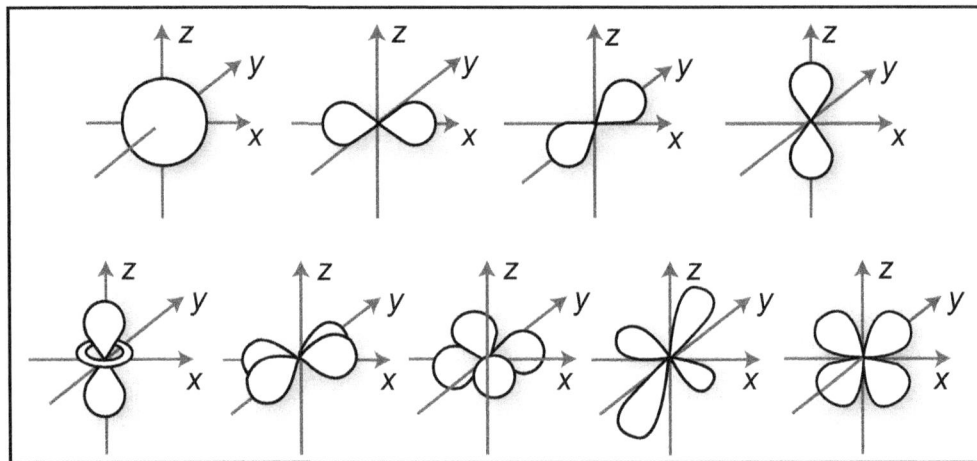

Subshell energy levels

Electrons naturally occupy the positions with the lowest possible energy levels; atoms with this configuration are said to be in the ground state. The **Aufbau principle** or **building-up rule** states that **electrons at ground state fill orbitals starting at the lowest available energy levels**. It can be useful to think of this as electrons filling slots, with two slots per orbital, although we know from Schrödinger that these are not static places but rather are probability "clouds" where the electron is likely to be at any given moment. The first of these "slots" to be filled will be in the 1s orbital, which has a roughly spherical shape. After two electrons have filled the 1s orbital, the next lowest energy space is the 2s orbital, which, like 1s, is roughly spherical in shape. If two electrons have filled this orbital, the next lowest subshell is 2p, with its three infinity-symbol shaped orbitals. Within a p subshell, electrons will occupy one orbital each until all three orbitals are filled. (Because of Coulomb's law and electron repulsion, electrons behave much like strangers boarding an airplane or a bus; you can think of a p subshell like a mini-bus with three 2-person seats being filled. If three strangers get on the bus, they will tend to each take their own seat, rather than two of them sharing one seat and leaving one seat empty. Only after all three seats have a passenger in them will the next stranger sit next to someone to fill the seat; likewise, only after all three orbitals in the p subshell have an electron in them will the next electron share an orbital to fill it.) Each subshell, however, will be filled (with two electrons per orbital) before electrons occupy the next-higher energy subshell. So, if you have an atom with 4 electrons in shell 2, two of the electrons will be in 2s, and one each will occupy their own 2p orbital; the electrons will only occupy the p orbitals when there is no more space in the s subshell. Careful, though— a d subshell in a lower shell can have a higher energy level than an s subshell in a higher shell. For example, the 4s orbital has a lower energy level than the 3d orbitals.

Here is a good way to visualize the electron filling order:

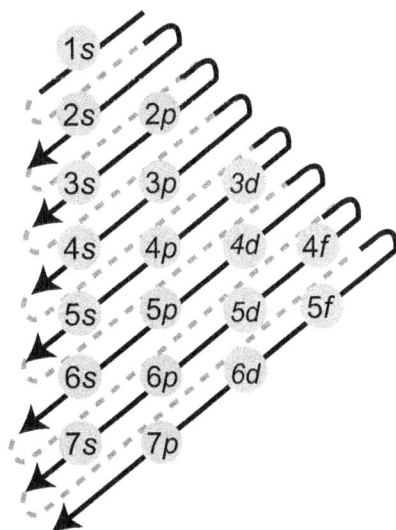

The following list orders subshells by increasing energy level:

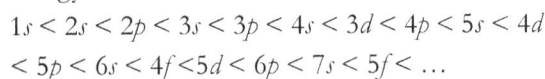

$$1s < 2s < 2p < 3s < 3p < 4s < 3d < 4p < 5s < 4d < 5p < 6s < 4f < 5d < 6p < 7s < 5f < \ldots$$

The reason for this pattern is electron repulsion. For single-electron atoms (H, He$^+$, and Li^{2+}), when the electron is above the ground state (e.g., excited by light), subshells within a shell are all at the same energy level, and an orbital's energy level is only determined by the shell number. However, in all other atoms, multiple electrons repel each other. Electrons in orbitals closer to the nucleus create a screening or **shielding effect** on electrons further away from the nucleus, preventing them from receiving the full attractive force of the nucleus.

Drawing electron arrangements

Electron arrangements (also called electron shell structures) in an atom may be represented using three methods: an **electron configuration**, an **orbital diagram**, or an **energy level diagram**. All three methods require knowledge of the subshells occupied by electrons in a certain atom. .

The first way of representing electrons in their orbitals is an **electron configuration**. An electron configuration is a **list of subshells** with superscripts representing the **number of electrons** in each subshell. For example, a boron atom has 5 electrons. According to the Aufbau principle, two will fill the $1s$ subshell, two will fill the higher energy $2s$ subshell, and one will occupy the $2p$ subshell, which has an even higher energy. The electron configuration of boron is $1s^2 2s^2 2p^1$. Similarly, the electron configuration of a vanadium atom with 23 electrons is:

$$1s^2 2s^2 2p^6 3s^2 3p^6 4s^2 3d^3.$$

Configurations are also written with their shell numbers together instead of ordering subshells by their energy level:

$$1s^2 2s^2 2p^6 3s^2 3p^6 3d^3 4s^2.$$

Electron configurations are often written to emphasize the outermost electrons. This is done by writing the symbol in brackets for the element with a full p subshell from the previous shell and adding the **outer electron configuration** onto that configuration. The element with the last full p subshell will always be a noble gas from the right-most column of the periodic table. For the vanadium example, the element with the last full p subshell has the configuration 1s22s22p63s23p6. This is 18Ar. The configuration of vanadium may then be written as [Ar]4s23d3 where 4s23d3 is the outer electron configuration.

Orbital diagrams assign electrons to individual orbitals so the energy state of individual electrons may be found. This way of depicting electron arrangements is most like the bus-boarding analogy above, and shows how electrons occupy orbitals within a subshell. **Friedrich Hund**, in the 1920s, determined a set of **rules to determine the ground state** of a multi-electron atom. **Hund's rule** states that **before any two electrons occupy the same orbital, other orbitals in that subshell must first contain one electron each with parallel spins**. Electrons with up and down spins are shown by half-arrows, and these are placed in lines of orbitals (represented as boxes or dashes) according to Hund's rule, the Aufbau principle, and the Pauli exclusion principle. Below is the orbital diagram for vanadium:

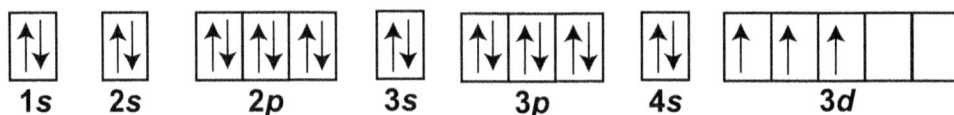

Another way to visualize electrons in the shells of an atom is an **energy level diagram**, an orbital diagram that shows subshells with higher energy levels higher up on the page. The energy level diagram of vanadium is:

Valence shell electrons and the periodic table

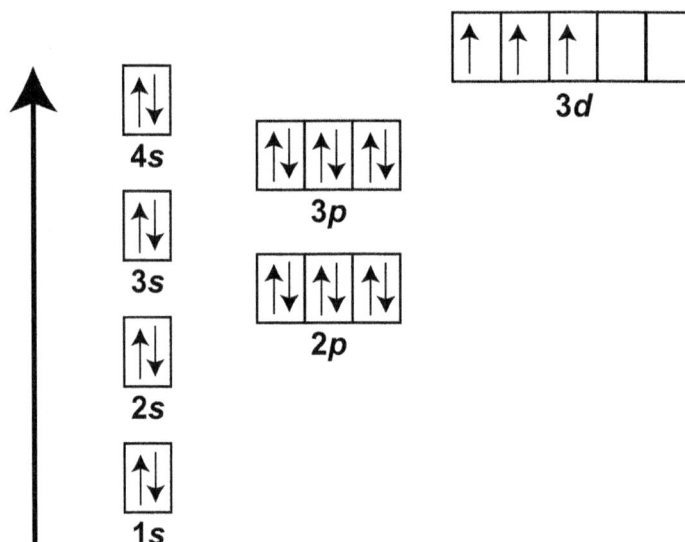

The valence shell electrons are the most important electrons for determining chemical properties. For example, the electron configuration of Se is $[Ar]4s^2 3d^{10} 4p^4$, and its valence shell electron configuration is $4s^2 4p^4$. (Even though the $3d$ subshell fills after the $4s$ subshell, the third shell is considered a closed shell at the point when the $4p$ subshell begins to fill. The $3d$ electrons are therefore not valence electrons.) Valence shell electrons give rise to the chemical character of their elements. For example, elements with d subshell valence electrons have metallic properties. This can be seen in the periodic table below. Elements with valence electrons in s shells behave similarly to each other, as do those with their valence electrons in p shells, d shells, or f shells.

The table is divided up into **blocks corresponding to the subshell** designation of the most recent orbital to be filled by the Aufbau principle. Elements in the *s*- and *p*-blocks are known as **main-group elements**. The *d*-block elements are called **transition metals**. The *f*-block elements are called **inner transition metals**.

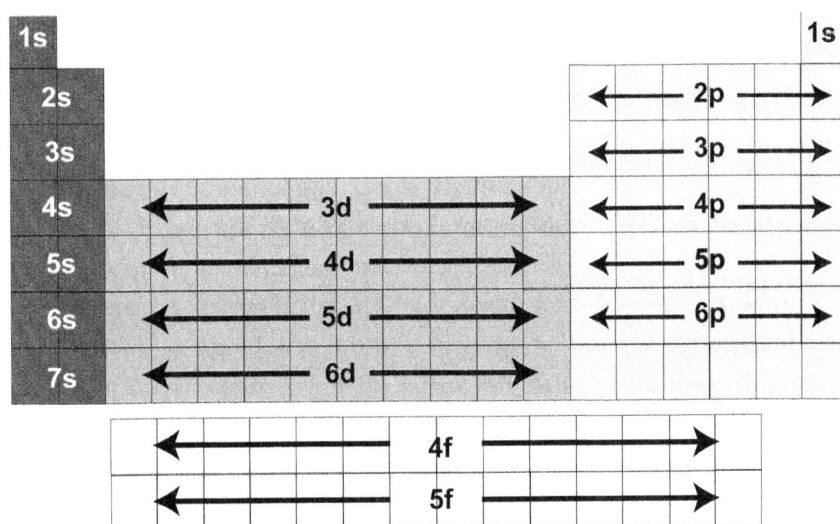

The maximum number of electrons in each subshell (2, 6, 10, or 14) determines the number of elements in each block, and the order of energy levels for subshells create the pattern of blocks. This has important consequences for the physical and chemical properties of the elements. The outermost shell or valence shell number (for example, 4 for Se) is also the period number for the element in the table.

Atoms in the d- and f-blocks often have unexpected electron shell structures that cannot be explained using simple rules. Some heavy atoms have unknown electron configurations because the number of different frequencies of radiation emitted and absorbed by these atoms is very large.

Melting point, density, and properties of compounds formed are all related to valence electrons. Elements with valence electrons in s orbitals tend to donate their electrons when forming compounds, while elements with *p* valence electrons tend to accept electrons. Oxides, hydrides, and halides are compounds with O, H, and halogens, respectively—elements with valence electrons in *p* subshells. Measures of intermolecular attractions other than melting point are also higher for metal oxides, hydrides, and halides than for the nonmetal compounds.

Chapter 1.5 Trends in the Periodic Table

The first periodic table was developed in 1869 by Dmitri **Mendeleev** several decades before the basic structure of atoms was understood. Mendeleev arranged the elements in order of increasing atomic mass into **columns of similar physical and chemical properties**. He then boldly **predicted the existence and the properties of undiscovered elements** to fill the gaps in his table. These interpolations were initially treated with skepticism until three of Mendeleev's theoretical elements were discovered

"Elements display periodicity in their properties when the elements are organized according to increasing atomic number. This periodicity can be explained by the regular variations that occur in the electronic structures of atoms. Periodicity is a useful principle for understanding properties and predicting trends in properties. Its modern-day uses range from examining the composition of materials to generating ideas for designing new materials."

and were found to have the properties he predicted. It is the correlation with properties—not with electron arrangements—that have placed the periodic table at the beginning of most chemistry texts.

Today, the 115 named elements are organized into the Periodic Table. The design of this table provides much information about a particular element or group of elements. Also, while Mendeleev created the table without knowledge about atomic substructure, modern chemists can use it to quickly and easily obtain information about the protons, neutrons, and electrons of any atom.

In the modern periodic table, **elements are arranged in numerical order by atomic number**. The elements in a **column are known as a group**, and groups are numbered from 1 to 18. Older numbering styles used roman numerals and letters. **Elements with similar properties are called a family** or a **chemical series**. For example, all of the elements in Group 18 are non-reactive gases, known as noble gases, whereas the metals in Group 1 are so highly reactive that they are only found in compounds, except under highly protective laboratory conditions. The modern table, like Mendeleev's, places elements with similar properties into columns. **A row of the periodic table is known as a period**. Periods of the known elements are numbered from 1 to 7.

Some groups in the periodic table are identified with specific group names: Group 1 elements are known as alkali metals, Group 2 as alkaline earth metals, Group 17 as halogens, and Group 18 as the noble gases. Other families in the periodic table encompass more than one group, such as the transition metals (Groups 3-12). Groups 13-16 contain post-transition metals, metalloids, and non-halogen nonmetals such as oxygen, carbon, and sulfur. Families tend to behave in the same way. For example, halogens tend to react strongly to form salts with alkali or alkaline earth metals (their name means salt-generating), or compounds with other nonmetals like carbon. Transition metals tend to be relatively non-reactive, highly malleable (can be hammered or pressed into different shapes), and good conductors of electricity—these properties make many of them useful as materials such as jewelry, tools, building materials, and electrical wires.

Seven elements are found in nature as **diatomic molecules: (H_2, N_2, O_2, and the halogens: F_2, Cl_2, Br_2, and I_2)**. Two mnemonic devices to remember the diatomic elements are: "$Br_2I_2N_2Cl_2H_2O_2F_2$" (pronounced "Brinklehof") and "**H**ave **N**o **F**ear **O**f **I**ce **C**old **B**eer." These molecules are attracted to one another through **London dispersion forces**, weak attractive forces resulting when the electrons in the molecules are momentarily unevenly distributed, forming temporary dipoles.

Note that **hydrogen** is *not* an alkali metal. Hydrogen is a colorless gas and is the most abundant element in the universe, but H_2 is very rare in the atmosphere because it is light enough to escape gravity and reach outer space. Hydrogen atoms form more compounds than any other element.

	alkali metals		halogens
	alkaline earth metals		noble gases

Alkali metals (group 1 elements) are shiny, soft, metallic solids. They have **low melting points and low densities** compared with other metals because they have a weaker metallic bond. In the alkali metals, **measures of intermolecular attraction, including melting points, decrease further down the periodic table, due to weaker metallic bonds** as the size of these atoms increases. Alkali metals are highly reactive, and will explosively form compounds if exposed to water or other polar compounds or elements, such as halogens.

Alkaline earth metals (group 2 elements) are grey, metallic solids. They are harder, denser, and have a higher melting point than the alkali metals, but values for these properties are still low compared to most of the transition metals, and they are highly reactive. Measures of metallic bond strength, such as melting points, do not follow a simple trend down the periodic table for these metals.

Halogens (group 17 elements) have an irritating odor, and are highly toxic. All elemental halogens are found as diatomic molecules, and these molecules are attracted to each other by London forces. London forces increase in strength further down the periodic table, and their melting points reflect this. London forces make Br_2 a liquid and I_2 a solid at 25 °C. The lighter halogens are gases.

Noble gases (group 18 elements) have no color or odor and exist as **individual gas atoms** attracted to each other only by London forces. These attractions also increase with period number.

Periodicity can be used to design new materials with certain predicted properties, by considering that elements within the same group may behave in similar ways. For example, Si and Ge are both in Group 14. Since SiO_2 can be a ceramic, perhaps GeO_2 may also be a ceramic.

"Many properties of atoms exhibit periodic trends that are reflective of the periodicity of electronic structure."

Intermolecular forces contribute to density by bringing nuclei closer to each other, so the periodicity is similar to trends for melting point. These group-to-group differences are superimposed on a general trend for density to increase with period number because heavier nuclei make the material denser.

Trends among properties of compounds may often be deduced from trends among their atoms, but caution must be used. For example, the densities of three potassium halides are:

2.0 g/cm^3 for KCl

2.7 g/cm^3 for KBr

3.1 g/cm^3 for KI.

We would expect this trend for increasing atomic mass within a group. We might also expect the density of KF to be less than 2.0 g/cm^3, but it is actually 2.5 g/cm^3 due to a change in crystal lattice structure.

Atomic radius

The size of an atom is not an exact value (due to of the probabilistic nature of electron density), but we may compare radii among different atoms using some standard. In the periodic table, the sizes of neutral atoms increase with period number and decrease with group number.

Atomic radius decreases left to right due to the electrons being pulled in tighter to the nucleus with the addition of each electron and proton. The atomic radius increases going down the periodic table due to the addition of another level of electrons with each row. The smallest atom, at the top right of the periodic table, is helium.

Note that many atomic properties and trends within the periodic table can be understood using three principles: Coulomb's law, the shell model, and the concept of shielding/effective nuclear charge. Four such properties you will need to understand, in terms of these three principles, are:

1. First ionization energy

2   Atomic and ionic radii

3. Electronegativity

4. Typical ionic charges

Chapter 1.6 The Mole and Mass Conversions

The periodic table lists the relative atomic mass of each element. This mass has no units. It is the mass of one mole of that substance. The mole is an important unit of measurement in chemistry. It is defined as 6.02×10^{23} units. Much like a dozen refers to 12 of anything, a mole can refer to 6.02x1023 of anything. This is known as **Avogadro's number** and it useful for converting between units of measure in chemical calculations. The equation **Moles = gr/ molar mass** is used to convert between grams and moles. For example 1 mole of carbon (C) = 12g, 1 mole of methane (CH$_4$) = 16g.

Atomic mass is a mass relative to carbon-12, and ^{12}C has an assigned value of exactly

12 u. This means that **the mass number of an atom is usually a good guess at the atom's atomic mass**, and for ^{12}C it is an exact value and not a guess at all. For example, uranium-238 has an atomic mass of 238.05 u.

The mass of a compound is the sum of the masses of its constituent atoms. The atoms in a compound are listed in an **empirical formula**. This type of formula simply lists the atoms; it does not reflect the structure of the molecule, so two molecules with different structures but the same numbers of the same elements will have the same empirical formula.

In a solution, the amount of solute dissolved in a given amount of solvent is called concentration. Solutions that cannot dissolve any additional solute are referred to as saturated solutions. Qualitative terms such as concentrated and dilute are used to describe solutions with relativity high or low solute concentrations, respectively. However, these expressions give very little information about the actual quantities present in the solution.

Expressions of percent by mass or volume, molarity, molality, or parts per million give more information about the quantity of solute and solvent present in a solution.

Percent by mass or volume expresses the amount of solute as a percentage of the total solution, either by mass or by volume.

$$\% \text{ by mass} = \frac{\text{mass of solute}}{\text{mass of solute} + \text{mass of solvent}} \times 100\%$$

Molarity, M, is an expression of concentration that compares the number of moles of solute in the solution to the total volume in liters of the solution.

Molarity is the most frequently used concentration unit in chemical reactions because it reflects the number of solute moles available. By using Avogadro's number, the number of molecules in a flask--a difficult image to conceptualize in the lab--is expressed in terms of the volume of liquid in the flask—a straightforward image to visualize and actually manipulate.

Example: What is the molarity of a 5.00 liter solution that was made with 10.0 moles of $CuCl_2$?

Solution: We can use the original formula. Note that in this particular example, where the number of moles of solute is given, the identity of the solute ($CaCl_2$) has nothing to do with solving the problem.

$$\text{Molarity} = \frac{\text{\# of moles of solute}}{\text{Liters of solution}}$$

Given: # of moles of solute = 10.0 moles
Liters of solution = 5.00 liters

Molarity $= \dfrac{10.0 \text{ moles of } CaCl_2}{5.00 \text{ Liters of solution}} = 2.00 \text{ M}$

Answer = 2.00 M

A **mole fraction** is used to represent a component in a solution as a portion of the entire number of moles present. If you were able to pick out a molecule at random from a solution, the mole fraction of a component represents the probability that the molecule you picked would be that particular component. The mole fractions for all components must add up to one and mole fractions have no units.

Molality is the moles of a solute per mass of the solution. Typical units for molality are mol/kg.

Chapter 1.7 Spectroscopy

Spectroscopy can be broadly defined as the use of the interaction between light and matter as a tool to yield information about a substance.

Relationships among electron energy levels, photons, and atomic spectra

The quantum structure of the atom describes electrons in discrete energy levels surrounding the nucleus. When an electron moves from a high energy orbital to a lower energy orbital, a quantum of electromagnetic radiation is emitted, and for an electron to move from a low energy to a higher energy level, a quantum of radiation must be absorbed. The particles that carry this electromagnetic force are called **photons**. Photons of light have energy levels that differ with frequency; the energy level of a photon can be calculated by Planck's equation ($E = h\nu$), where ν is the frequency of the light and h is **Planck's constant** ($h = 6.63 \times 10^{-34}$ J$\times$s). When a molecule absorbs a photon, the energy in the molecule increases by the amount of energy in one photon. Conversely, when a molecule emits a photon, the energy in the molecule is decreased by the amount in one photon. The quantum structure of the atom predicts that only photons corresponding to certain wavelengths of light will be emitted or absorbed by each atom. These distinct wavelengths are measured by **atomic spectroscopy**.

In **atomic absorption spectroscopy,** a continuous spectrum (light consisting of all wavelengths) is passed through the element. The frequencies of absorbed photons are then determined as the electrons increase in energy. **An absorption spectrum** in the visible region usually appears as a rainbow of color stretching from red to violet interrupted by a few black lines corresponding to distinct wavelengths of absorption.

In atomic emission spectroscopy, the electrons of an element are excited by heating or by an electric discharge. The frequencies of emitted photons are then determined as the electrons release energy. In the case of visible light, these frequencies are seen as colors. Emission spectroscopy uses the energy given off *after* absorption to establish molecular structure.

An **emission spectrum** in the visible region typically consists of lines of light at certain colors corresponding to distinct wavelengths of emission. The bands of emitted or absorbed light at these wavelengths are called **spectral lines**. **Each element has a unique line spectrum**. Light from a star (including the sun) may be analyzed to determine what elements are present.

Uses of spectroscopy in chemistry

Looking at the interaction of matter and the component colors of light (which are the energy components of light) is a practical way to glean information about various physical properties of the matter, such as temperature, mass, luminosity and composition. It is a tool used by all types of chemists in the identification of substances based on the spectrum that results from the interaction between molecules and light. There are three different types of light/matter interactions, detailed further below: scattering spectroscopy, absorption spectroscopy, and emission spectroscopy.

- **Scattering spectroscopy:** Light-scattering spectroscopy is primarily used for particle size measurements. It relies on the amount of light scattered by a substance and is dependent on wavelength.

- **Absorption spectroscopy:** As the name implies, absorption spectroscopy is a measurement of the light absorbed by a substance. The specific wavelengths of light absorbed by a substance can tell us about the nasture of that substance, such as the types and amounts of chemical constituents in a mixture. When a molecule absorbs light, the absorption, A, can be ascertained from the **Beer-Lambert Law: A =εbc, where** ε = constant of absorptivity, b = path length and c = concentration. Since different molecules absorb light of different wavelengths, the absorption spectrum will show different bands corresponding to structural groups within the molecule. There are several different types of absorption spectroscopy that are commonly used, including UV/Vis spectroscopy and IR spectroscopy.

  UV/Vis Spectroscopy: This type of absorption spectroscopy uses a combination of ultraviolet (UV) and visible light wavelengths to help ascertain the molecular structure of a substance. The absorption of light in this region corresponds to the excitation of outer shell electrons which occurs as ground state electrons (those in the lowest energy level) are promoted to excited states.

  IR Spectroscopy: Similar in methodology to UV/Vis, IR spectroscopy uses infrared (IR) light (400 – 14,000 nm) to determine molecular structure based on the types of bonds present, rather than the specific atomic structure. This is possible due to the vibrational frequencies generated by the bonded atoms when light is absorbed. To measure a sample, a beam of infrared light is passed through the sample and the amount of energy absorbed at each wavelength is recorded. From this, an absorbance spectrum may be plotted. This shows at which wavelengths the sample absorbs the IR light and allows an interpretation of which bonds are present.

- **Emission Spectroscopy:** In contrast to absorption spectroscopy, where the properties of a substance can be determined from the light the molecule absorbs, emission spectroscopy uses the energy given off *after* absorption to establish molecular structure. The molecule first absorbs energy, elevating electrons to higher energies. As these electrons return to their ground state, the substance radiates this energy as characteristic emissions that are directly related to the decrease in electron energy as it drops in orbitals.

- **Fluorescence Spectroscopy:** Fluorescence spectroscopy uses high energy photons to excite a sample, which will then emit photons of another energy. In order for this to occur, a fluorophore is a necessary component of the molecule being tested. A fluorophore is a functional group which will absorb energy at one wavelength and emit it at another. The strength of this emission is measured as a function of wavelength.

Other: Not all types of spectroscopy fit into the three above categories. Here are two examples.

- **NMR spectroscopy:** Nuclear magnetic resonance (NMR) spectroscopy exploits the magnetic properties of atomic nuclei to obtain molecular structure. Although many different nuclei are NMR active, by far the most common NMR probes are ^{1}H and ^{13}C. NMR allows information to be garnered from a single atom and details the environment of that atom for structure determination. When placed in a magnetic field, NMR-active nuclei resonate at different, and specific, frequencies. Since the resonance frequency is dependent on the strength of the magnetic field, it is reported as a chemical shift. Magnetic Resonance Imaging (MRI), originally known as NMRI, uses NMR to create images of internal structures in the body.

- **Mass spectrometry:** Mass spectrometry (MS) is a technique used to measure mass-to-charge ratios of ions generated in a mass spectrometer, based on the simple idea that different compounds have different masses. In the instrument, a sample is ionized, or broken down into the simplest ions of the substance. The spectrometer measures the relative abundances of each of the ions present and reports the abundance as a function of mass. The resulting spectrum is representative of the masses of the components of the substance and can be used to identify the parent compound.

Mass spectrometry and isotopes

Mass spectrometry is the technique that provided some of the most important evidence that Dalton's model of the atom—that all atoms of a certain element are exactly the same—was incorrect. Mass spectrometry led to the discovery of isotopes-- atoms of the same element with different mass. This difference in mass is due to different numbers of neutrons. For example, most carbon atoms have 6 protons and 6 neutrons, for a total atomic mass of 12. However, a small minority of carbon atoms have 7 neutrons, for a total mass of 13, or 8 neutrons, for a total atomic mass of 14. These differences do not affect the chemical behavior of the atoms in obvious ways, like differences in protons and electrons do, although they do have small effects. For example, differences in ratio are

"Chemical analysis provides a method for determining the relative number of atoms in a substance, which can be used to identify the substance or determine its purity."

seen in C_3 and C_4 plants, allowing scientists to analyze the diet of an animal based on the carbon isotope content in their collagen. Carbon 14 is radioactive and decays over time, providing paleontologists with a strong tool for dating very old once-living matter. The atomic mass given in the periodic table for each element (for example, 12.011 for carbon) is the average mass of the atoms of that element. This value can be determined from mass spectra.

Consider the mass spectrum of boron:

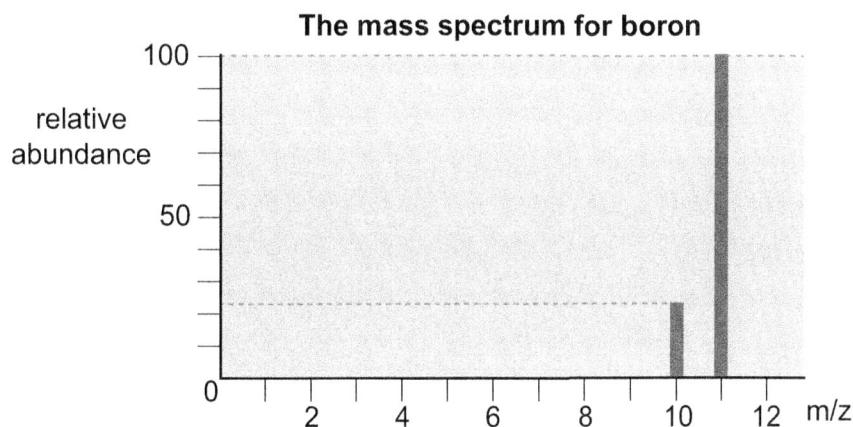

The mass spectrum for boron

relative abundance

100

50

0

2   4   6   8   10   12   m/z

This spectrum shows that there are two isotopes of boron, one with an atomic mass of 10 and one with an atomic mass of 11. The heights of the peaks tells us how abundant each isotope is. (This is a relative number, not a measurement with units.) In the case of boron, of 123 isotopes, 23 have an atomic mass of 10 and 100 have an atomic mass of 11. The total mass of these would be (23 x 10) + (100 x 11) = 1330. Dividing by 123, we obtain 10.8, the relative atomic mass we see in the Periodic Table.

Elements with no stable isotopes are often listed in tables of atomic masses with a number in brackets. This value is the mass number of the isotope with the longest half-life. A list of isotopic compositions and atomic masses for all natural isotopes is at http://physics.nist.gov/cgi–bin/Compositions/stand_alone.pl.

"An early model of the atom stated that all atoms of an element are identical. Mass spectrometry data demonstrate evidence that contradicts this early model."

Sample Test Questions (Big Idea #1) _____

Multiple Choice Questions

Instructions: This section consists of some practice multiple choice questions as well as one Free Response question related to this chapter. You may NOT use a calculator for the Multiple Choice. However calculators are permitted for the Free Response.

1. The photoelectron spectra to the right depict the energy required to remove a $1s$ electron from helium and hydrogen. Why is the peak for helium to the left of the peak for hydrogen?

   A. There is greater electron repulsion in helium than hydrogen.

   B. The s subshell in hydrogen is only half filled.

   C. The nuclear charge is greater in helium than in hydrogen.

   D. Helium has a larger volume than hydrogen.

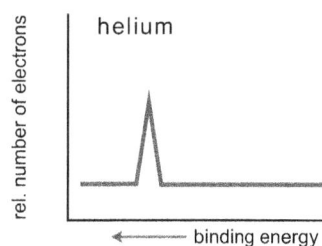

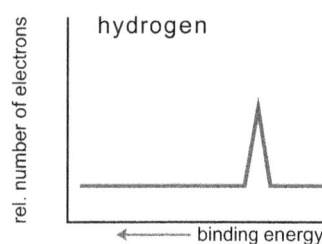

2. Based on trends in the periodic table, which of the following statements are true?

   A. Fluorine is larger in volume than oxygen because it has more protons and neutrons

   B Fluorine is smaller in volume than oxygen because it has a greater nuclear charge

   C. Fluorine is smaller in volume than oxygen because atomic mass is inversely proportional to volume

   D. Fluorine is larger in volume than oxygen because it has more valence electrons

3. Based on trends in the periodic table, calcium will be most similar, in terms of chemical properties, to:

   A. Potassium because it is similar in atomic mass

   B. Potassium because it is in the same period

   C. Strontium because it is has a similar nuclear charge

   D. Strontium because it has the same valence electron configuration

4. The following is a depiction of the mass spectrum of carbon dioxide:

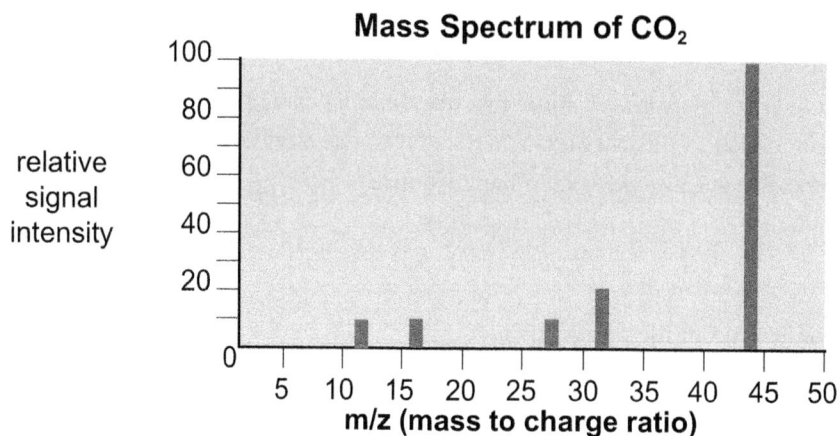

Mass Spectrum of CO_2

relative signal intensity

m/z (mass to charge ratio)

Which of the following elements and molecules are present in the sample?

A. C, O_2, CO_2, and Ru

B. C, O_2, CO_2, and CO

C. C, O, CO, and CO_2

D. CO_2 only

5. If a mass spectrum showed a line at 16, what might the substance be?

A. CH_4

B. O_2

C. Li and Be

D. A or B

6. A spectrum shows a rainbow of color interrupted by black lines. What type of spectrum does this represent?

A. An absorption spectrum

B. An emission spectrum

C. A scattering spectrum

D. Nuclear magnetic resonance

7. How many valence electrons are in titanium (Ti) and what subshell are they in? What is titanium's outermost shell?

A. 2 electrons in the 3d subshell; shell 3 is the outermost shell

B. 2 electrons in the 4d subshell; shell 4 is the outermost shell

C. 2 electrons in the 3d subshell; shell 4 is the outermost shell

D. 4 electrons in the 4d subshell; shell 4 is the outermost shell

8. How many orbitals exist within the 4f subshell?

   A. 4
   B. 5
   C. 6
   D. 7

9. One morning you ate a bowl of cereal that contained all your daily vitamins and minerals. One of these minerals was Iron. If your bowl of cheerios had 0.55g Iron, how many atoms of Iron did you eat ?

   A. 55 atoms
   B. 6.02×10^{23} atoms
   C. 6.02×10^{21} atoms
   D. 3.11×10^{25} atoms

10. Which of these sets of elements will have similar chemical and physical properties?

   A. Ir,Pt,Au
   B. C,Si,Ge
   C. Ag,Au,Pt
   D. Al,Si,P

11. Magnesium is one of the least dense elements (d = 1.74 g/cm3). What is the volume of a 10.0 g sample of the metal?

   A. 5.75 cm³
   B. 17.4 cm³
   C. 57.5 cm³
   D. 174 cm³

12. A sample gives the following spectra :

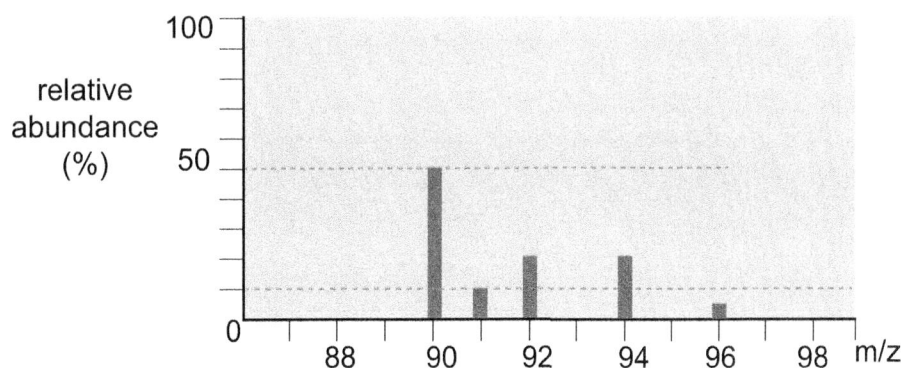

   The sample is probably:
   A. Zr
   B. Y
   C. Mo
   D. Sc

13 What is the electron configuration for a calcium ion, Ca^{2+}?

A.

1s	2s	2p			3s	3p			4s	3d				
↑↓	↑↓	↑↓	↑↓	↑↓	↑↓	↑↓	↑↓	↑↓	↑↓	↑↓	↑↓	↑↓		

B.

1s	2s	2p			3s	3p			4s	3d				
↑↓	↑↓	↑↓	↑↓	↑↓	↑↓	↑↓	↑↓	↑↓	↑↓	↑↓	↑↓	↑	↑	

C.

1s	2s	2p			3s	3p			4s	3d				
↑↓	↑↓	↑↓	↑↓	↑↓	↑↓	↑↓	↑↓	↑↓	↑↓	↑↓	↑	↑	↑	↑

D.

1s	2s	2p			3s	3p			4s	3d				
↑↓	↑↓	↑↓	↑↓	↑↓	↑↓	↑↓	↑↓	↑↓	↑↓	↑↓		↑↓		↑↓

14. Which of the following represents iron (Fe)?

A. $1s^2\ 2s^2\ 2p^6\ 3s^2\ 3p^6$

B. $1s^2\ 2s^2\ 2p^6\ 3s^2\ 3p^6\ 4s1$

C. $1s^2\ 2s^2\ 2p^6\ 3s^2\ 3p^6\ 4s^2$

D. $1s^2\ 2s^2\ 2p^6\ 3s^2\ 3p^6\ 4s2\ 3d^2$

15. Which energy sublevel is being filled by the elements Ga through Kr?

A.   3d

B.   4s

C.   4p

D.   4d

16. Phenacyl chloride is a riot control gas having the formula C_7H_7OCl. What is the percentage of carbon in the compound.

A.   4.91%

B.   8.42%

C.   24.9%

D.   58.9%

Free Response Questions (Big Idea #1)

Sample Test Questions

1. A mass spectrum such as the one shown below is given to a student in a chemistry lab.

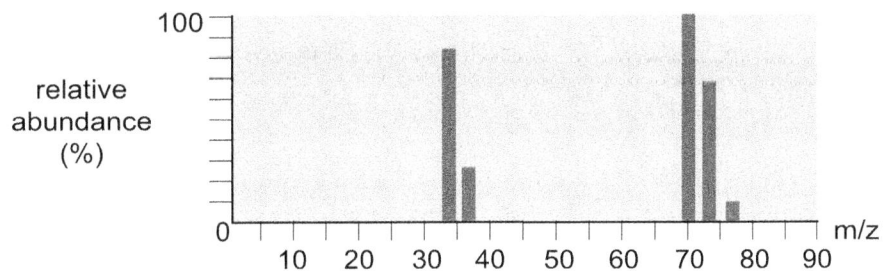

A. Which atom do you think this is? Explain why you think this and explain the various peaks in the spectra.

B. Write out the electronic configuration of this element.

C. Draw an electron diagram showing the arrangement of the electrons.

1. **Answer C:**

   Helium has two protons while hydrogen has one. The greater nuclear charge from these protons serves to hold the electrons in more tightly—they have greater binding energy and therefore it is more difficult to remove them.

2. **Answer B:**

   Fluorine is smaller in volume because it has more protons, creating a greater nuclear charge. The electrons are held in more tightly and closely by this greater nuclear charge. This is why, as you move to the right in the periodic table, the volume of the atoms gets smaller even as they are getting larger in terms of mass.

3. **Answer D:**

   Elements in the same group in the periodic table behave similarly, in terms of chemical properties, when they have the same valence electron configuration

4. **Answer C:**

   The masses shown are 12, 16, 28, and 44. These correspond to C, O, CO, and CO_2. O_2 is not in the sample; it would have shown up as a line at 32, so A and B are incorrect. D is incorrect because it does not account for the first three lines.

5. **Answer A:**

   CH4 has an atomic mass of 16. O_2 has an atomic mass of 32. Li and Be would have separate mass lines at 7 and 9, not a single line at 16.

6. **Answer A:**

   This is the appearance of an absorption spectrum in the visible region.

7. **Answer C:**

   Subshell 4s is filled; therefore, shell 4 is the outermost shell. However, the 3d subshell fills after the 4s subshell, so the outermost electrons are in 3d. Having valence electrons in d subshells is a characteristic of the transition metals, and leads to metallic properties such as electrical conductivity.

8. **Answer D:**

   There are $2l+1$ orbitals in each subshell. Thus the s subshell has only one orbital, the p subshell (2+1) has three orbitals, the d subshell has 5 orbitals (4+1) and the f subshell has 7 (6+1).

9. **Answer C:**

   This can be calculated by simple unit analysis. From periodic table Iron has a molecular weight of 55.3g.

   0.55g X   1mol/ 55.3g Fe x 6.02×10 atoms/mo = ~ 6.02×10

10. **Answer B:**

The chemical properties depend on the number of valence shell electrons since these electrons participate in chemical bond formation with other elements. Therefore the chemical properties of the elements present in the same group/column are similar.

11. **Answer A:**

Density = M/V if D= 1.74g/cm3 and the mass is 10g substitute in the numbers

D = M /V

1.74 g/cm = 10g/Vol

Vol =   10g/ 1.74g/cm = 5.74 cm

12. **Answer A:**

The sample is Zirconium. It has a molecular weight of 91 which is close to the largest peak. The 5 peaks in the mass spectrum shows that there are 5 isotopes of zirconium - with relative isotopic masses of 90 (at about 50% relative abundance), 91, 92, 94 and 96.

Y has a molecular weight of 88 which is not found in the spectra. Sc has a molecular weight of 45, also not found in the spectra. Mo has a molecular weight of 96, which while found in the spectra is a very small peak.

13. **Answer A:**

Calcium has 20 electrons and has an electron configuration of

A. $1s^2\ 2s^2\ 2p^6\ 3s^2\ 3p^6 4s^2$

Ca2+ is missing the two valance electrons and therefore has the electron configurations

A. $1s^2\ 2s^2\ 2p^6\ 3s2\ 3p^6$

14. **Answer C:**

Iron has 26 electrons. All the answers show 26 electrons, however only C does not violate any of the filling rules.

15. **Answer C:**

These elements are in the 4th row. Therefore as you go across the periodic table the 4s shell is filled, followed by the 3d shell for the transition metals, followed by the 4p shell for elements Ga through Kr?.

16. **Answer D:**

The % carbon is found by finding the mass of carbon divided by the total mass.

The total mass is found by

7C $\times$ 12 = 84

7H $\times 1$ = 7

1O $\times$ 16 = 16

1Cl $\times$ 35.0 = 35

Total Molecular weight = 142

%Carbon = 84/142 = 59%

Answers to Free Response (Big Idea #1)

1. A. This spectrum is probably for chlorine. Chlorine has two isotopes, ^{35}Cl and ^{37}Cl,. However chlorine rarely exists very long as a single atom. It normally produces the dimer C. The second set of lines represent the Cl molecules. There are three peaks because you can have two ^{35}Cl molecules or two Cl377 or $^{35}Cl/^{37}Cl$.

    B. $1s^2 2s^2 2p^6 3s^2 3p^5$

    C.

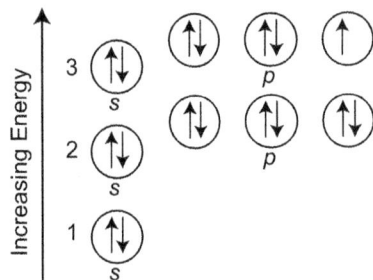

Electron Configuration for Chlorine (Cl)

Chapter 2: Big Idea #2
Bonding and Phases

Big Idea 2:
Chemical and physical properties of materials can be explained by the structure and the arrangement of atoms, ions, or molecules and the forces between them.

What you should already know:

- properties of atoms
- trends in the periodic table
- electron configurations
- moles, mass, volume, and pressure
- math as reviewed in this book

What you will learn:

- physical and chemical properties of matter
- intermolecular attractive forces
- molecular and energetic nature of solids, liquids, and gases
- solutions, alloys, and separation methods
- kinetic molecular theory- ideal gas laws
- intramolecular forces (chemical bonds)
- the VSEPR model, Lewis diagrams, and molecular orbital theory

---

Chapter 2.1 : Physical and Chemical Properties

Physical properties

A physical property of matter is a property that can be determined without inducing a chemical change. Melting point, boiling point, specific heat, hardness, density, and conductivity are all examples of physical properties. Any given element or chemical compound will have a unique set of physical properties and can be identified through these properties.

While matter may go through physical changes, these changes do not alter the fundamental nature of the element or compound; there is no chemical modification of the elements or compounds that make up the matter. For example, ice may be broken into smaller pieces but this does not change the nature of the ice. Matter may also go undergo phase changes among gas, liquid, and solid forms without changing its fundamental

"Matter can be described by its physical properties. The physical properties of a substance generally depend on the spacing between the particles (atoms, molecules, ions) that make up the substance and the forces of attraction among them."

nature. For example, ice can melt, then re-freeze. Whether broken, melted, or boiled, the substance, HO, remains the same.

Chemical properties

Chemical properties of a substance are properties affecting the substance's ability and tendency to change. The chemical properties of an element or compound are related to its interaction with other elements and compounds. They determine its ability to undergo a chemical reaction. For example, these include electronegativity, oxidation state, ionization potential, chemical structure, and types of chemical bonds.

Unlike a phase change, which is a change between two physical states, chemical change rearranges atoms to form a new molecule. The idea that chemical changes are due to the rearrangement of atoms in molecules formed the basis of Dalton's atomic theory.

Matter and Energy

Molecules have kinetic energy (they move around), and they also have intermolecular attractive forces (they stick to each other). Molecules can undergo three different types of motion—vibration, rotation, and translation (the ability to change location). The relationship between kinetic energy, encompassing all three types of motion, and intermolecular attractive forces determine whether a collection of molecules will be a gas, liquid, or solid. While physical properties, including intermolecular attractive forces, are intrinsic to a substance, the kinetic energy of the substance changes as energy is added to or taken from a system. It is the interaction between the two that determines the physical state of a substance. For example, the same substance, water, can be solid, liquid, or gas depending on the kinetic energy in the system. At a given level of kinetic energy, though, water will be liquid while iron will be solid. This is because of the different intermolecular attractive forces of water vs. iron. These attractive forces are a physical property of the substance.

Similarly, chemical properties like electronegativity are intrinsic to an element or compound, but whether it will undergo a chemical reaction will depend on the kinetic energy in the system.

Chapter 2.2 The different properties of gases, liquids, and solids

A **gas** has an indefinite shape and an indefinite volume. The kinetic model for a gas is a collection of widely separated molecules, each moving in a random and free fashion, with negligible attractive or repulsive forces between them. Gases will expand to occupy a larger container so there is more space between the molecules. Gases can also be compressed to fit into a small container so the molecules are less separated. Gases undergo all three types of motion—vibration, rotation, and translation—but translation is the

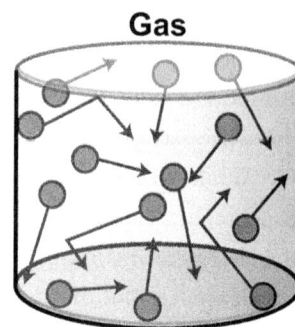

Gas

most important. Since the intermolecular forces are negligible in a gas, gas molecules can move rapidly through space. **Diffusion** occurs when one material spreads into or through another. Gases diffuse rapidly and easily move from one place to another.

A **liquid** assumes the shape of the portion of any container that it occupies and has a specific volume. The kinetic model for a liquid is a collection of molecules attracted to each other with sufficient strength to keep them close to each other but with insufficient strength to prevent them from moving around randomly. Liquids undergo all three types of motion—vibration, rotation, and translation. Liquids have a higher density and are much less compressible than gases because the molecules in a liquid are closer together. Diffusion occurs more slowly in liquids than in gases because the molecules in a liquid stick to each other and are not completely free to move.

Liquid

Solid

A **solid** has a definite volume and definite shape. The kinetic model for a solid is a collection of molecules attracted to each other with sufficient strength to essentially lock them in place. Each molecule may vibrate, but it has an average position relative to its neighbors. If these positions form an ordered pattern, the solid is called crystalline. Otherwise, it is called amorphous. Solids have a high density and are almost incompressible because the molecules are close together. Molecules in a solid cannot undergo rotation or translation. Therefore, diffusion occurs extremely slowly, with molecules rarely altering their positions.

"The different properties of solids and liquids can be explained by differences in their structures, both at the particulate level and in their supramolecular structures."

When thinking of these properties, it is useful to imagine analogous groups of people. Whereas a gas is like strangers moving at random, getting close only if they randomly move near each other or if they are compressed in a tight space, a liquid is like a group of friends walking through a park-- they move around, with the shape of the group changing as they do, but they maintain contact with other members of the group so that the group moves as a whole even as individuals are constantly changing who they are talking to or walking next to. Whereas gases and liquids are like random strangers in a space or a group of close friends, respectively, solids are like people seated in a classroom—they can move their bodies, but they don't switch places. Solids and liquids, like a seated group and a group of close friends, are similar in their molar volume—though their arrangements and motion are different, the distance between individuals is similar so they take up about the same amount of space. If you don't know the answer to a question involving physical states of matter, think of the molecules as tiny people who are either strangers, close friends, or a formal seated group, and you will be in a good position to make an educated guess.

As described in Chapter 1, **mixtures** are composed of two or more pure substances that are not chemically combined. Mixtures may be of any proportion and can be physically separated by processes such as filtering, centrifuging, or distillation. In homogenous mixtures (solutions), the macroscopic characteristics are the same throughout the mixture.

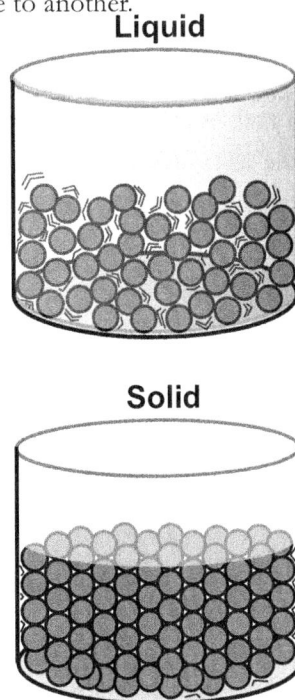

In contrast, heterogenous mixtures have different compositions in different locations, and can include matter in different phases. For example, ice water is a mixture of solid H_2O and liquid H_2O. Carbonated soda is a mixture of liquid and carbon dioxide gas. Gases and solids can also form heterogeneous mixtures—smoke is a common example of this.

Whether a mixture is homogenous or heterogenous depends on scale. For example, a colloid like milk can be homogenous on a scale of centimeters but heterogenous on the scale of micrometers. That is, each cubic centimeter of homogenized milk may have the same composition as each other cubic centimeter. However, focusing in more closely, one cubic micrometer of milk may differ in composition from another cubic micrometer. By convention, when a mixture is heterogenous enough to stop light (light cannot shine through it, as is seen with colloids like milk), it is considered heterogenous.

Mixtures can also change from heterogenous to homogenous and vice-versa over time. For example, when dye is added to water, it forms a heterogenous solution, with the colored dye cloud moving slowly through the colorless water. After several hours of diffusion, though, the mixture may become homogenous, with the dye molecules uniformly distributed in the water. A glass of sugar water is a homogenous mixture, but if it is put in a freezer for 30 minutes, it will be a heterogenous mixture of bits of frozen and liquid sugar water with different concentrations of sugar in each.

Chapter 2.3 Forces of attraction between particles _____

2.3.1 Intermolecular Forces

Intermolecular forces are the forces that hold matter together or push it apart, and include the various types of attraction **between molecules** that are responsible for the cohesion of liquids and solids. In addition to determining physical properties such as viscosity and hardness, intermolecular forces dictate what state the molecule will be in under certain conditions of kinetic energy. Intermolecular forces are intrinsic to molecules; they do not change as molecules undergo phase changes. In other words, while a substance's phase at any given time is subject to how much heat is in the system, the specific heat needed to cause a substance to change phase is a property of that substance, and is independent of the amount of heat in a system.

While these forces are generally much weaker than intramolecular forces (chemical bonds), they are still very important in explaining the behavior of matter at different temperatures and pressures, as well as the interaction between different types of matter. One important arena for intermolecular forces is solutions. When one substance is interspersed uniformly throughout another, forming a homogenous mixture, the mixture is called a solution. Any phase of matter (solid, liquid, or gas) can be dissolved in any other phase of matter. The dissolved matter is the solute and the matter in which the solute is dissolved is the solvent. The concentration of the solute is one important factor in determining the physical properties of the solution. The other is solubility, which, along with factors like temperature, helps determine how concentrated the solution can be. Solubility properties of substances are due to intermolecular forces.

"Forces of attraction between particles (including the noble gases and also different parts of some large molecules) are important in determining many macroscopic properties of a substance, including how the observable physical state changes with temperature."

"Solutions are homogenous mixtures in which the physical properties are dependent on the concentration of the solute and the strengths of all interactions among the particles of the solutes and solvent."

Liquid solutions, and especially aqueous ones, are a particularly important type of solution. Since they are a homogenous mixture of a solute in liquid, they can be expressed in terms of molarity (moles of solute per liters of solution). Liquid solutions are characterized by three general properties:

1. The components cannot be separated by using filter paper (all components are small enough to pass through).

2. They are translucent (no components are large enough to scatter visible light).

3. The components can be separated using processes that disrupt the intermolecular interactions that hold the solute and solvent molecules together.

Ion-dipole interactions

Salts tend to separate into their component ions as they dissolve in polar solvents such as water. An ion with a full charge in a polar solvent will **orient nearby solvent molecules** so that their opposite partial charges are pointing towards the ion. The presence of ions in water can be detected through tests of electrical conductivity. In aqueous solution, certain salts react to form solid **precipitates** if a combination of their ions is insoluble.

For example, table salt is NaCl. In water, it ionizes to Na^+ and Cl^-. The Na^+ ions will tend to be attracted to the O end of the H_2O molecule, while the Cl^- ions will tend to orient toward the H atoms in the H_2O molecule. The polarity of the water molecule makes it easier for the ionic compound to dissolve, as it helps pull the ions apart. The bonds are not intramolecular bonds, and the dissolving of NaCl in water is not a chemical reaction. The salt water is a homogenous solution that can be separated into its components (NaCl and H_2O) by the physical process of evaporation.

Dipole-dipole interactions

The intermolecular forces between polar molecules are known as dipole-dipole interactions. The partial positive charge of one molecule is attracted to the partial negative charge of its neighbor.

Hydrogen bonds

Hydrogen bonds are particularly **strong dipole-dipole interactions** that form between the **H atom** of one molecule and an **F, O, or N atom** of an adjacent molecule. The partial positive charge on the hydrogen atom is attracted to the partial negative charge on the electron pair of the other atom. The hydrogen bond between two water molecules is shown as the dashed line below:

Hydrogen bonds are usually not a bonding mechanism within a single molecule. However, very large molecules, such as proteins and RNA, are so big that they fold around and form hydrogen bonds with themselves. These within-molecule intermolecular bonds allow them to fold into 3-dimensional shapes that provide structure or play functional roles in living organisms. One of the most famous hydrogen bond structures is DNA. DNA is composed of two very long strands that are bonded together by hydrogen bonds that link the two strands together like rungs on a ladder.

"Dipole forces result from the attraction among the positive ends and negative ends of polar molecules. Hydrogen bonding is a strong type of dipole-dipole force that exists when very electronegative atoms (N, O, and F) are involved."

Even when intermolecular bonds are present within a single molecule, they are still intermolecular, and can be broken by physical means. For example, when you boil proteins, the 3-dimensional structures based on hydrogen and other intermolecular bonds will break, and the proteins will turn into long molecules without their functional shape. Double-stranded DNA can also be separated into two single strands by boiling, a process known as denaturation.

Intermolecular forces are also sometimes described as "secondary molecular bonds".

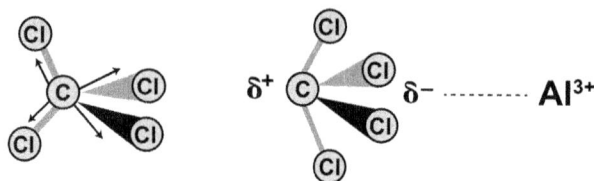

Ion-induced dipole

When a nonpolar molecule (or a noble gas atom) encounters an ion, its **electron density is temporarily distorted** resulting in an **ion-induced dipole**, and will be attracted to the ion.

For example, carbon tetrachloride, CCl_4, has polar bonds but is a nonpolar molecule due to the tetrahedral symmetry of those bonds. An aluminum cation will draw the non-bonded electrons of the chlorine atom towards it, distorting the molecule (this distortion has been exaggerated in the figure) and creating an attractive force as shown by the dashed line below.

Dipole-induced dipole

The partial charge of **a permanent dipole may also induce a dipole in a nonpolar molecule**, resulting in an attraction similar to but weaker than that created by an ion.

London dispersion force: induced dipole – induced dipole

Dipoles can also be induced in the absence of ions or polar molecules. Intermolecular attractions due to induced dipoles in a nonpolar molecule are known as London forces or Van der Waals interactions. These are very weak intermolecular forces. These forces occur because at any given moment, electrons are located within a certain region of the molecule, and **the instantaneous location of electrons will induce a temporary dipole** on neighboring molecules. For example, an isolated helium atom consists of a nucleus with a 2+ charge and two electrons in a spherical electron density cloud. An attraction between He atoms due to London dispersion forces occurs when the electrons happen to be distributed unevenly on one atom, inducing a dipole on its neighbor. This dipole is due

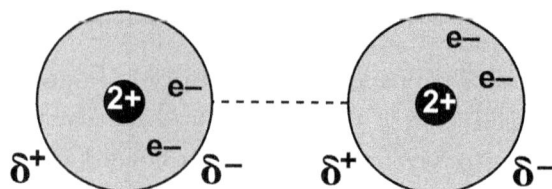

to intermolecular repulsion of electrons and the attraction of electrons to neighboring nuclei.

The strength of London dispersion forces **increases for larger molecules** because a larger electron cloud is more easily polarized. These forces are also enhanced by the presence of pi bonds. The strength of London dispersion forces also **increases for molecules with a larger surface area** because there is greater opportunity for electrons to influence neighboring molecules if there is more potential contact between the molecules. Paraffin in candles is an example of a solid held together by weak London forces between large molecules. These materials are soft.

Chapter 2.3.2 Impact on physical properties ⎯⎯⎯⎯⎯⎯

Solutions

Substances with similar intermolecular forces mix well together, and are said to be **miscible** or soluble in one another. The formation of solutions can be endothermic (for example, for sugar to dissolve well in water, heat is required) or exothermic (for example, ethanol mixes easily with an aqueous solution, even on ice).

Solutions can be separated into their components only by using methods that disrupt intermolecular forces. Examples of these include:

1. Chromatography: When a solution is allowed to flow through a substance, such as juice being wicked up by filter paper, the different components will separate on the basis of the strength of the intermolecular attraction to the solid substrate (e.g. filter paper) relative to the solvent (water). There are several types of chromatography, each with different principles of separation. Paper chromatography is used to identify chemicals like coloring agents in foods. Liquid chromatography uses liquid as the mobile phase to separate, identify, and quantify each component in a mixture. Gas chromatography utilizes a gas as the mobile phase to move the mixture of molecules through a column.

2. Evaporation: In both evaporation and distillation (below), intermolecular forces are disrupted by boiling. Substances with weaker intermolecular forces will boil at lower temperatures than substances with stronger intermolecular forces. If the goal is to retain the molecule with the stronger intermolecular forces, the evaporated molecule is simply allowed to dissipate, leaving the desired molecule behind. An example of this is obtaining salt from saltwater. In solutions like saltwater, where the boiling points of the components are extremely different, evaporation can be accomplished without adding energy to boil the solution. This type of room-temperature evaporation is used to separate sea salt from seawater in evaporating ponds.

3. Distillation: If the goal is to obtain the molecule with the weaker intermolecular bonds, this can be achieved by distillation. Distillation also begins with boiling, but the evaporated liquid is collected by condensation. For example, to separate

"London dispersion forces are attractive forces present between all atoms and molecules. London dispersion forces are often the strongest net intermolecular force between large molecules."

ethanol from wine, the wine is heated enough to allow ethanol to vaporize. The ethanol is allowed to condense on a cold surface above the boiling wine and the surface is slanted so that the condensed ethanol flows into a different container. (The name for this beverage, brandy, is related to the chemistry used to create it. The older, pre-shortened English term, brandywine, is derived from the Dutch brandewijn, meaning "burned wine".)

Phase changes

If two substances are being compared, the material with the **greater intermolecular attractive forces** (i.e. the stronger intermolecular bond) will require more energy to pull apart the molecules. Substances with greater intermolecular forces will have the following properties:

For solids:

Higher melting point
Higher enthalpy of fusion
Greater hardness
Lower vapor pressure

For liquids:

Higher boiling point
Higher critical temperature
Higher critical pressure
Higher enthalpy of vaporization
Higher viscosity
Higher surface tension
Lower vapor pressure

Capillary action is an important quality of liquids that is dependent on two intermolecular interactions: the intermolecular attraction between the molecules of the liquid and that between the molecules of the liquid and those of the container, called the adhesive force. Consider two liquids, water and mercury. Water has a stronger force of adhesion to glass than mercury, so in a capillary tube, the meniscus will be concave, with the water touching the sides of the tube higher than the meniscus. With mercury, the meniscus is convex, with the mercury adjacent to the sides lower than the meniscus. This is because the cohesive force for mercury is stronger than the adhesive force between mercury and glass.

For gases:

Intermolecular attractive forces are neglected for ideal gases, as they seldom have observable effects. When they do have observable effects, such as at very high pressure or near the condensation point, the gases will no longer exhibit ideal gas behavior.

For example, H_2O and NH_3 are liquids at room temperature because they contain hydrogen bonds. These bonds are of intermediate strength, so the melting point of these compounds is lower than room temperature and their boiling point is higher than room temperature. H_2S contains weaker dipole-dipole interactions than H_2O because the sulfur atoms do not form hydrogen bonds. Therefore, H_2S is a gas at room temperature due to its low boiling point. Small non-polar molecules such as CO_2, N_2, or atoms such as He are gases at room temperature due to very weak London forces, but larger non-polar molecules such as octane or CCl_4 may be liquids, and very large non-polar molecules such as paraffin will be soft solids.

Phase changes occur when the relative importance of kinetic energy and intermolecular forces is altered sufficiently for a substance to change its state. While the strength and nature of the intermolecular force is an unchanging property of the molecules themselves, the kinetic energy within the molecule changes as heat is added or removed.

In a solid, the energy of the intermolecular attractive forces (such as ionic or covalent bonds) is much stronger than the kinetic energy of the molecules (the vibrational energy within the molecules themselves). As temperature increases in a solid, the vibrations of individual molecules grow more intense and the molecules spread slightly further apart, decreasing the density of the solid.

In a liquid, the energy of the intermolecular attractive forces (such as dipole-dipole and London dispersion forces) is about as strong as the kinetic energy of the molecules. Therefore, both play a role in the properties of liquids. Liquids will be discussed in detail later.

In a gas, the energy of intermolecular forces is much weaker than the kinetic energy of the molecules. Kinetic molecular theory is usually applied to gases.

The **normal melting point** (T_m) and **normal boiling point** (T_b) of a substance are defined at 1 atm. Note that freezing point and melting point refer to the same temperature approached from different directions.

Substances with strong intermolecular attractive forces will remain solid even when kinetic energy is high, whereas those with weak intermolecular attractive forces will be gases even when kinetic energy is low, and those with moderate intermolecular attractive forces will fall in between. For example, at 100° C, water is boiling into a gas, but the steel pot containing the water is nowhere near melting, let alone evaporating. At 0° C, water is transitioning into ice (a solid), but oxygen will still be a gas at 100 degrees below this.

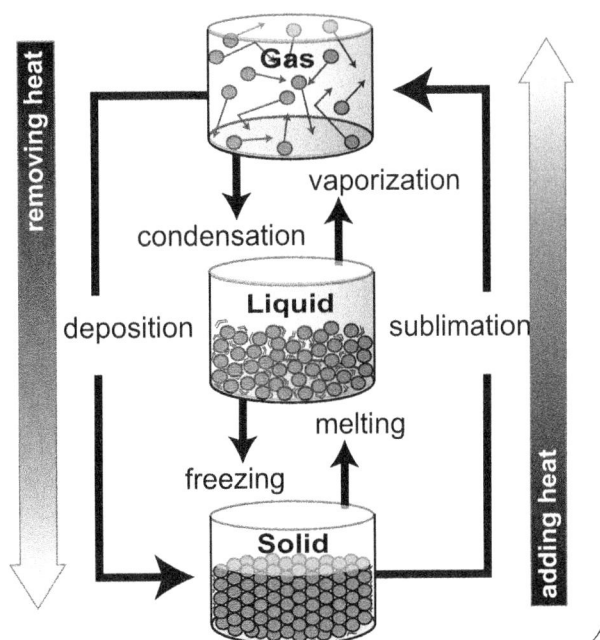

The transition from gas to liquid is called **condensation** and from liquid to gas is called **vaporization**. The transition from liquid to solid is called **freezing** and from solid to liquid is called **melting.** The transition from gas to solid is called **deposition** and from solid to gas is called **sublimation.**

Heat removed from a substance during condensation, freezing, or deposition permits new intermolecular bonds to form as molecules slow down and get close enough to each other to stick together, and heat added to a substance during vaporization, melting, or sublimation breaks intermolecular bonds. During these phase transitions, this latent heat is removed or added with **no change in the temperature** of the substance because the heat is not being used to alter the speed of the molecules or the kinetic energy when they strike each other or the container walls. Latent heat alters intermolecular bonds.

If a time graph was made for a pure substance being heated or cooled, it would look something like the graph below, for the heating of water. Different changes are taking place during each interval on the graph.

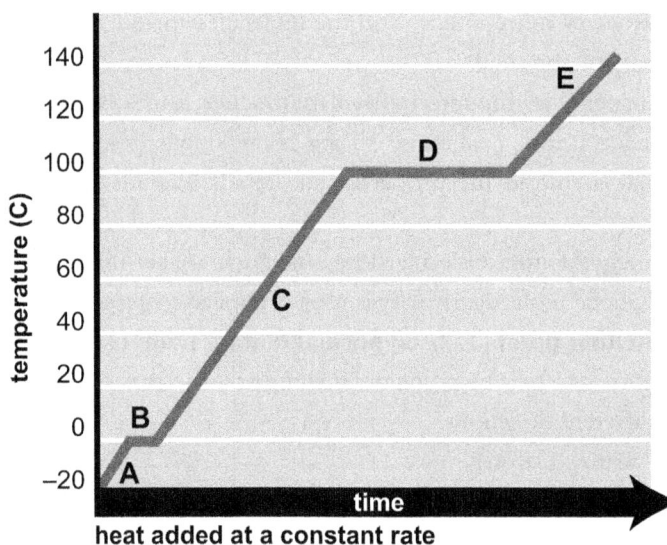

When the system is heated, energy is transferred into it. In response to the energy it receives, the system changes, either by increasing its temperature or changing phase.

During the interval marked A on the heating and cooling curve, below, energy is being absorbed by the water molecules to increase the temperature to water's melting point, 0°C. The slope of the line for this interval shows the increase in temperature and is related to the heat capacity of the substance.

During the interval marked B on the graph, energy is still being added to the water but the temperature remains the same, at 0° C or water's melting point temperature. The additional energy is being used to overcome the intermolecular forces holding the water molecules in their solid pattern. This energy is moving the particles apart, breaking or weakening the forces of attraction trying to keep the water molecules aligned. The solid water (ice) is being converted to liquid water; a phase change is occurring. The temperature will not increase until every solid particle has melted and the entire sample is liquid.

Temperature again increases during interval C on the graph. Energy is being absorbed by the liquid water molecules. Notice that the slope of the line during this interval is different than the slope of the line during interval A. Again, this is due to differences in the heat capacity of ice and liquid water.

The flat line during interval D indicates that a phase change is occurring. The additional energy is being used to overcome the attractive forces holding the liquid water molecules together. The water molecules increase their kinetic energies and move farther apart, changing to water vapor. This occurs at the boiling point temperature, or 100° C. The temperature stays at the boiling point temperature until all water molecules are converted to water vapor. Once this conversion occurs, the temperature increases as energy is added, at a rate determined by the heat capacity of the substance as a vapor.

A **phase diagram** is a graphical way to summarize the environmental conditions under which the different states of a substance are stable. The diagram is divided into three areas representing the three possible states of the substance (gas, liquid, or solid). Temperature and pressure determine the phase of a substance and are shown on the x-axis and y-axis of the phase diagram, respectively.

A typical phase diagram for a substance as a function of temperature and pressure in a closed system would look like this

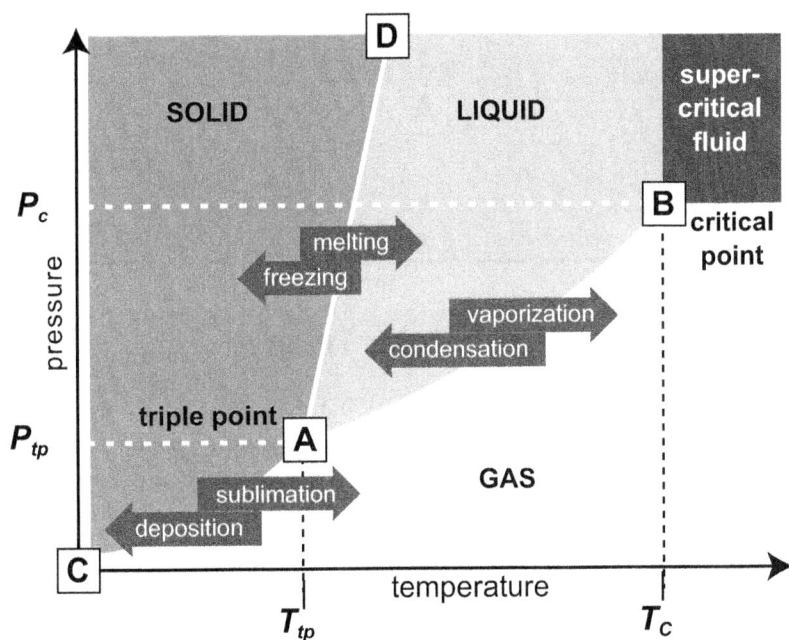

The line that connects points A and D separates the solid and liquid phases and shows how the melting point of a solid varies with pressure. The solid and liquid phases are in equilibrium all along this line; crossing the line horizontally corresponds to melting or freezing. The line that connects points A and B is the vapor pressure of the liquid. It ends at the **critical point**, beyond which the substance exists as a supercritical fluid. The line that connects points A and C is the vapor pressure curve of the *solid* phase. Along this line, the solid is in equilibrium with the vapor phase through sublimation and deposition. Finally, point A, where the solid/liquid, liquid/gas, and solid/gas lines intersect, is the

triple point; it is the *only* combination of temperature and pressure at which all three phases (solid, liquid, and gas) are in equilibrium and can therefore exist simultaneously. Because no more than three phases can ever coexist, a phase diagram can never have more than three lines intersecting at a single point.

Below is a phase diagram for water, H_2O. Due to the unique hydrogen bonding properties and lattice structure of water – the phase diagram is slightly different. The phase diagram for water is unusual in that the solid/liquid phase boundary slopes to the left with increasing pressure because the melting point of water decreases with increasing pressure. Note that the normal melting point of water is lower than its triple point. The diagram is not drawn to a uniform scale.

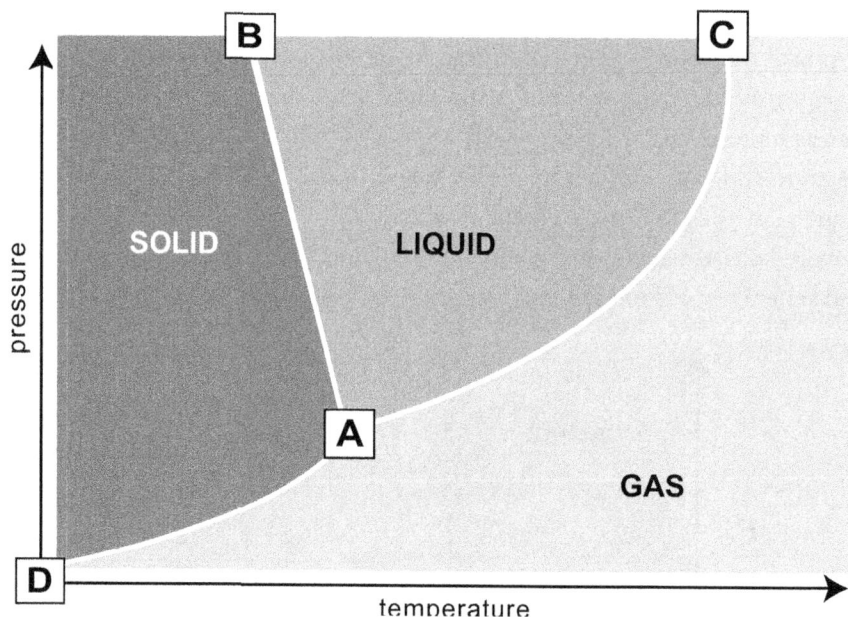

Chapter 2.4 Kinetic Molecular Theory

The relationship between **kinetic energy** and **intermolecular forces** determines whether a collection of molecules will be a gas, liquid, or solid. In a gas, the energy of intermolecular forces is much weaker than the kinetic energy of the molecules. Therefore, Kinetic Molecular Theory is usually applied to gases.

In a gas, the molecules (or single atoms, in the case of the noble gases) are in continuous random motion. Gas **pressure** results from molecular collisions with container walls. The number of molecules striking an **area** on the walls and the **average kinetic energy** per molecule are the only factors that contribute to pressure. A higher **temperature** means that the molecules have more kinetic energy; they are moving faster, as well as rotating, colliding, and vibrating more than molecules at low temperature. There are more collisions at higher temperatures, but the average distance between molecules does not change, and thus density does not change in a sealed container.

Kinetic molecular theory explains why the pressure and temperature of gases change and affect one another the way they do, using the following assumptions:

1   The energies of intermolecular attractive and repulsive forces are negligible.

2. Average kinetic energy of the molecules is proportional to absolute temperature.

3. Energy can be transferred between molecules during collisions and the collisions are elastic, so the average kinetic energy of the molecules doesn't change due to collisions.

4. The volume of the molecules in a gas is negligible compared to the total volume of the container. The movement of a molecule from one place to another is called translation. Translation and other types of movement (rotation and vibration) all are forms of kinetic energy. Translational kinetic energy is the form that is transferred by collisions, and the form that can result in heat transfer between substances. Kinetic molecular theory ignores other forms of kinetic energy because they are relatively small and can be considered negligible when it comes to temperature of gases. The average kinetic energy of a system is related to the mass and average velocity of the molecules by the following equation:

$$KE = \frac{1}{2}(mv^2)$$

You can use this to compare two gases. For example, if you have two gases with the same kinetic energy, the one with the smaller mass must have the larger velocity.

The following table summarizes the application of kinetic molecular theory to an increase in container volume, number of molecules, and temperature:

Effect of an increase in one variable holding the other two constant	Effect: – = decrease, 0 = no change, + = increase						
	Average distance between molecules	Density in a sealed container	Average speed of molecules	Average translational kinetic energy of molecules	Collisions with container walls per second	Collisions per unit area of |wall per second	Pressure (P)
Volume of container (V)	+	–	0	0	–	–	–
Number of molecules	–	+	0	0	+	+	+
Temperature (T)	0	0	+	+	+	+	+

Additional details on the kinetic molecular theory may be found at http://hyperphysics.phy-astr.gsu.edu/hbase/kinetic/ktcon.html. An animation of gas particles colliding is located at http://comp.uark.edu/~jgeabana/mol_dyn/.

The Maxwell-Boltzmann distribution

It is impossible to measure the velocity of each individual molecule of gas at every instance of time. Instead, the **Maxwell-Boltzmann distribution** can be used to characterize the distribution of velocities of the combined molecules of the gas. Using this distribution, velocities of gases can be compared under different conditions, and different gases can be compared.

This Maxwell-Boltzmann distribution shows how a gas at a higher temperature has both a higher average velocity than a gas at a lower temperature, and a broader distribution of velocities:

The size of the molecules in a gas affect the velocity of the particles as well, with larger molecules moving more slowly and with a narrower range of velocities. The following Maxwell-Boltzmann distribution shows the velocities for the noble gases at 25°C. Note

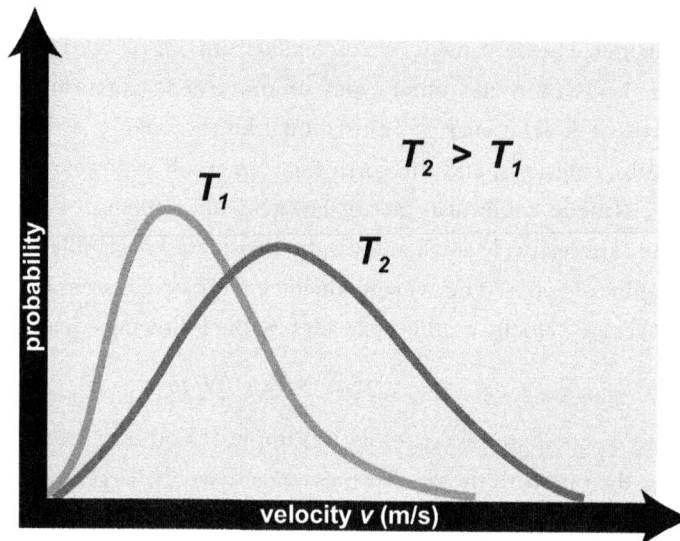

the extreme difference in shape between the very heavy noble gas xenon and the very light noble gas, He.

The behavior of a gas can be characterized qualitatively through the use of kinetic molecular theory and the Maxwell-Boltzmann distribution.

Chapter 2.5 The Gas Laws

Maxwell-Boltzmann Molecular Speed Distribution for Noble Gases

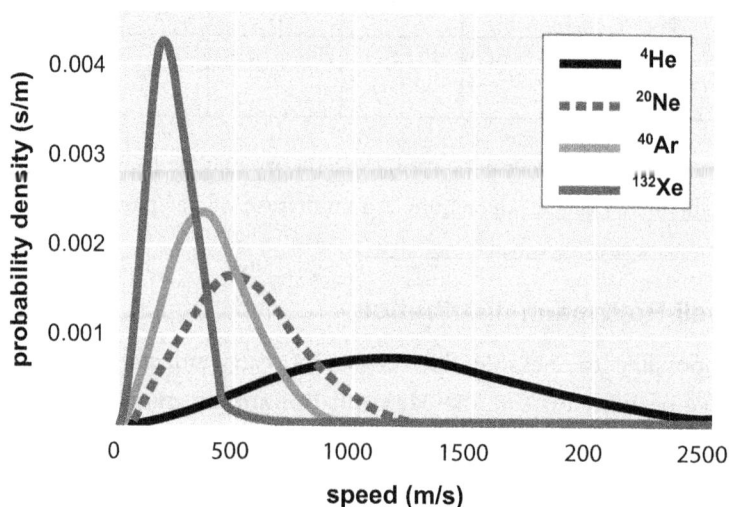

Amedeo Avogadro developed the hypothesis that **equal volumes of different gases contain equal numbers of molecules** if the gases are at the same temperature and pressure, even if the molecules themselves are different sizes. This is because of the very low density of gases. Whereas the volume of solids and liquids is mostly taken up by the matter itself, the volume of a gas mostly consists of the space between the molecules. The volume taken up by the molecules themselves is negligible; so tiny that it does not make a significant difference in the volume of any gas. What does make a difference is the *movement* of the molecules—their **kinetic energy**. At higher temperature, the molecules are moving around more, and as a result take up more space. At higher pressure, their movement is restricted, so the volume decreases.

The proportionality between volume and number of moles is called **Avogadro's Law**, and the number of molecules in a mole is called **Avogadro's Number**. Both were posthumously named in his honor. Avogadro's number is roughly equal to 6.022×10^{23}.

$$V \propto n$$

where V is the volume and n is the number of moles of gas.

Boyle's law states that the volume of a fixed amount of gas at constant temperature is inversely proportional to the gas pressure. In other words, increasing the pressure causes a gas to contract, in a mathematically proportional manner:

$$V \propto 1/P$$

When the pressure or volume of a gas is changed, but everything else is kept constant, you can calculate the change in the other parameter by using the equation:

$$(V_{original})(P_{original}) = (V_{final})(P_{final})$$

So, if you compress 2 liters of a gas into 1 liter, the gas in the 1 liter container will have twice as much pressure as the same gas when it was in a 2-liter container.

Charles' law states that the volume of a fixed amount of gas at constant pressure is directly proportional to absolute temperature. In other words, increasing the temperature causes a gas to expand, in a mathematically proportional manner:

$$V \propto T$$

Or $V = kT$ where k is a constant.

For any gas, changing the volume or temperature will result in a corresponding change in the other parameter, and this change can be calculated through the mathematical equation

$$(V_{original})(T_{original}) = (V_{final})(T_{final})$$

Changes in temperature or volume can be found using Charles' law.

Gay-Lussac's law states that the pressure of a fixed amount of gas in a fixed volume is proportional to absolute temperature, or:

$$P \propto T$$

Or $P = kT$ where k is a constant. This gives the mathematical equation.

$$\frac{P_1}{T_1} = \frac{P_2}{T_2}$$

Changes in temperature or pressure (with a constant volume) can be found using Gay-Lussac's law.

The **combined gas law** uses the above laws to determine a proportionality expression that is used for a constant quantity of gas.

Together, Avogadro's law and the combined gas law yield:

V α T/P

The relationships among temperature, pressure, volume and moles of gases were originally found by experimental observation. Now, they can be explained by the kinetic molecular theory.

If temperature, volume, or pressure of a gas is changed, there will be a corresponding change in one or both of the other parameters. For example, if you heat up a gas in a flexible balloon (so that the pressure is atmospheric pressure), to double the temperature, the volume of the balloon will also double. This combined gas law is often expressed as an equality between identical amounts of an ideal gas in two different states ($n_1 = n_2$):

$$\frac{(V_{original})(P_{original})}{(T_{original})} = \frac{(V_{final})(P_{final})}{(T_{final})}$$

or

$$(V_{original} * P_{original}) / T_{original} = (V_{final} * P_{final}) / T_{original}$$

The **ideal gas law** combines all of the information in Boyle's, Charles', Gay-Lussac's, and Avogadro's laws. It is expressed by the equation:

PV = nRT

R is the **universal gas constant** (also known as the ideal gas constant, the molar gas constant, or simply the gas constant). It is equal to 8.314 J/mol*K or 0.0821 L*atm/mol*K. P, V, and T are pressure, volume, and temperature, and n represents the number of moles of a gas, and is equal to the grams of the gas divided by the molar mass of the gas (g/mol), also known as the molecular weight. It may also be rearranged to determine gas molar density in moles per unit volume (molarity):

n/V = RT

Since it encompasses Boyle's, Charles', Gay-Lussac's, and Avogadro's laws, the ideal gas law can be used when considering any type of change in a gas. Simply cancel out the parameters that remain constant.

$$(P_{original}V_{original})/(P_{final}V_{final}) = (n_{original}RT_{original})/(n_{final}RT_{final})$$

which is equal to

$$(P_{original}V_{original})/(P_{final}Vfinal) = (n_{original}T_{original})/(n_{final}T_{final})$$

Airplanes and the ideal gas law

Have you ever wondered why shampoo bottles pop on an airplane, or why that water bottle you drank from during the flight ends up crumpled when you are once again on the ground? These phenomena can be explained by the ideal gas law, PV = nRT. When you screwed your shampoo bottle shut at atmospheric pressure, you trapped in a certain molar amount of gas that took up the volume of the shampoo bottle at atmospheric pressure. However, when you are high in the atmosphere in an airplane, the pressure is much lower. (Airplane cabins are pressurized, but the pressure is still substantially lower than on the ground.) When P decreases for a given molar amount of gas, V increases. Therefore, the volume of the gas in your shampoo bottle will increase, and, depending on the strength of your container, this can result in exploded shampoo! This aspect of the ideal gas law is Boyle's law, but it is reflected in the full ideal gas law as well, since n, R, and T are roughly constant. However, what would happen if T were not constant; would your shampoo be more or less likely to explode if the temperature were high or low? Looking at PV = nRT, if temperature goes up, volume goes up, so if the shampoo bottle were kept cold, the low temperature might counteract the decreased pressure and keep your shampoo contents intact. On the other hand, if your bottle was very cold when packed, and the cabin atmosphere was nicely heated, it would increase the volume, making a shampoo explosion even more likely! This is a factor to consider when packing sealed items from a refrigerator, such as a sealed yogurt parfait to snack on during the flight.

Gas density can also be calculated using the ideal gas law. Density, d, in grams per unit volume, is found through multiplication of n by the molar mass M (g/mol) and division by the volume:

$$d = (n * M) / V = P / RT$$

This principle underlies the phenomenon of warm air rising, which allows hot air balloons to float and is responsible for the formation of weather phenomena like storms when fronts of cold air and warm air meet. When temperature increases, density decreases, and the hotter, less dense, air will rise above the colder, more dense, air.

Molecular weight may also be determined from the density of an ideal gas by rearranging the equation above:

$$M = (d * V) / n$$

If pressure is given in atmospheres and volume is given in liters, a value for R of **0.08206 L-atm/(mol-K)** is used. If pressure is given in Pascals (newtons/m2) and volume in cubic meters, then the SI value for R of **8.314 J/(mol-K)** may be used. This is because a joule is defined as a Newton-meter. A value for R of **8.314 m^3-Pa/(mol-K)** is identical to the ideal gas constant using joules.

Many problems are given at "**standard temperature and pressure**" or "**STP.**" Standard conditions are *exactly* **1 atm** (101.325 kPa) and **0 °C (273.15 K)**. At STP, one mole of an ideal gas has a volume of:

$$V = nRT / P$$

The value of 22.4 L is known as the **standard molar volume of any gas at STP**.

For mixtures of gases in a container, each gas exerts a **partial pressure** that is the same as it would have if it were present in the container alone. **Dalton's law** of partial pressures states that the total pressure of a gas mixture is simply the sum of the partial pressures.

Dalton's law may be applied to the ideal gas law:

$$P_{total}V = (P_1 + P_2 + P_3 + ...)V = (n_1 + n_2 + n_3 ...)RT$$

Airplanes and Dalton's law of partial pressures

While the air pressure in airplane cabins is lower than that on the ground, they are actively pressurized relative to the air outside the airplane. This is necessary because human lungs require a certain minimal partial pressure of oxygen in order to absorb the oxygen adequately. When the pressure is decreased sufficiently, the partial pressure of oxygen can fall to levels too low to absorb, even though it is still 21% of the air, the same percent as on the ground. If there is a drop in cabin air pressure, oxygen masks will drop and passengers can breath pure oxygen. Dalton's law of partial pressures states that if the molar amount of one gas (n) goes up, the pressure will go up as well, allowing enough of a partial pressure of oxygen for adequate absorption.

Human lungs are also not optimized for too high a partial pressure of oxygen. A too-high partial pressure of oxygen can be a problem for divers. At depths below about 130 feet below the surface, the partial pressure of nitrogen at the normal air percentage of 78% increases enough to cause narcotic effects. To counteract this, divers can use a mixture of 40% oxygen, 60% nitrogen instead of normal compressed air, which is about 21% oxygen, 78% nitrogen. However, if a diver breathes this mix for extended periods of time, especially at high-pressure depths, the high partial pressure of oxygen can cause oxygen toxicity, leading to central nervous system damage. Safe diving therefore requires training in the principles of Dalton's laws and the ideal gas law in addition to other practical concerns.

Keeping these kinds of real-world examples in mind can help prevent errors on the AP exam. To protect against errors such as dividing instead of multiplying, take a step back after doing a gas law problem; is your answer logical, in terms of common sense and real world effects? Also, although solving gas law problems using these formulas is a straightforward process of algebraic manipulation, **errors commonly arise from using improper units**, particularly for the ideal gas constant R. An absolute temperature scale must be used (never °C) and is usually reported using the Kelvin scale, but volume and pressure units often vary from problem to problem. Temperature in Kelvin is found by adding 273.15 to the temperature in °C. Always include units in your equations, and make sure they cancel out (see math review chapter) properly.

Estimating absolute zero using the ideal gas law

As temperature decreases, the average kinetic energy in a gas decreases, and along with it, molecular velocities. Absolute zero is the temperature at which molecular velocities are zero and molecular motion has ceased. Since there is no molecular motion at absolute

"The strong electrostatic forces of attraction holding atoms together in a unit are called chemical bonds."

zero, pressure is equal to zero as well. To estimate absolute zero from data obtained under normal laboratory conditions, create a plot of pressure (Y axis) vs. temperature (X axis). Plot several datapoints for pressure under different temperatures and use the points to draw a line. Then, extend the line until it intersects with the X axis (temperature). (While it is well known that absolute zero is -273°C, it is important for you to understand how this can be determined experimentally.)

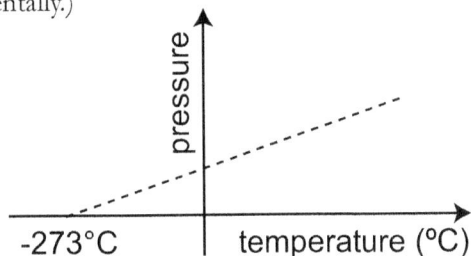

Real Gases

While the ideal gas law works under most temperatures and conditions, it breaks down when it is very cold or when pressure is very high. It also breaks down when gases are near the condensation point. Under these conditions, the gas becomes dense enough that the volume and intermolecular attractive forces of the molecules themselves are no longer negligible. Modifying the ideal gas equation $PV = nRT$, Johannes van der Waals postulated correction factors to adjust the value of V to reflect the effect of the volume of the gas molecules themselves, and, more importantly, P, to reflect the effect of attractive forces between the gas molecules. Van der Waals' final equation for real gases is:

$$(P + na/V)(V - bn) = nRT$$

where a is a measure of the attraction between the molecules and b is a measure of the volume excluded by a mole of particles.

Chapter 2.6 Intramolecular Forces

Chemical compounds form when two or more atoms join together. A stable compound occurs when the total energy of the combination of the atoms together is lower than the atoms separately. The combined state suggests an attractive force exists between the atoms. This attractive force is called a chemical bond. Intramolecular forces include ionic, covalent, and metallic bonds, and exist within molecules. These are the forces that hold molecules and solid structures together, and are generally much stronger than intermolecular forces. We will talk more about these in Chapter 3.

Single and multiple covalent bonds

A **covalent bond** forms when at least one pair of electrons is shared by two atoms. The shared electrons are found in the valence energy level and lead to a lower energy if they are shared in a way that creates a noble gas configuration (a full octet). Covalent, or molecular, bonds occur when a non-metal is bonding to a non-metal. This is due primarily

"In covalent bonding, electrons are shared between the nuclei of two atoms to form a molecule or polyatomic ion. Electronegativity differences between the two atoms account for the distribution of the shared electrons and the polarity of the bond."

"The type of bonding in the solid state can be deduced from the properties of the solid state."

to the fact that non-metals have high ionization energies and high electronegativities. Neither atom wants to "give up" electrons; both want to gain them. In order to satisfy both octets, the electrons can be shared between the two atoms.

Sharing of electrons can be equal or unequal, resulting in a separation of charge (polar) or an even distribution of charge (non-polar). The polarity of a bond can be determined through an examination of the **electronegativities** of the atoms involved in the bond. The more electronegative atom will have a stronger attraction to the electrons, thus possessing the electrons more of the time. This results in a partial negative charge ($\delta-$) on the more electronegative atom and a partial positive charge ($\delta+$) on the less electronegative atom.

The simplest covalent bond is between two hydrogen atoms. Covalent bonds may be represented by an electron pair (a pair of dots) or a line as shown below. The shared pair of electrons provides each H atom with two electrons in its valence shell (the $1s$ orbital), so both have the stable electron configuration of helium.

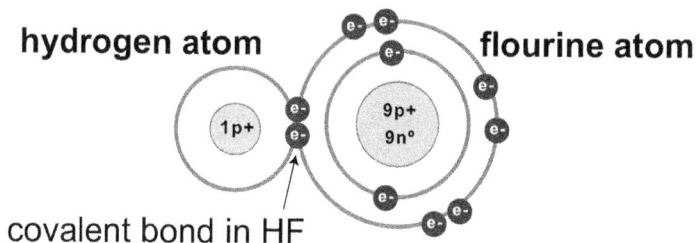

$$H\cdot + \cdot H \longrightarrow \begin{array}{c} H\!:\!H \\ H\!-\!H \end{array}$$

covalent bond in HF

Chlorine molecules have 7 electrons in their valence shell and share a pair of electrons so both have the stable electron configuration of argon.

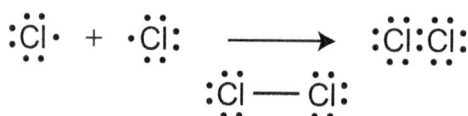

$$:\!\ddot{C}l\!\cdot + \cdot\ddot{C}l\!: \longrightarrow \begin{array}{c} :\!\ddot{C}l\!:\!\ddot{C}l\!: \\ :\!\ddot{C}l\!-\!\ddot{C}l\!: \end{array}$$

In the previous two examples, a single pair of electrons was shared, and the resulting bond is referred to as a **single bond**. When two electron pairs are shared, two lines are drawn, representing a **double bond**, and three shared pairs of electrons represents a **triple bond** as shown below for CO_2 and N_2. The remaining electrons are in **unshared pairs**.

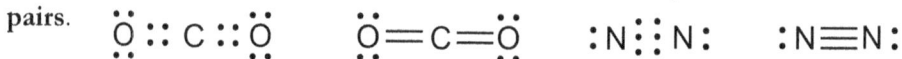

$$\ddot{O}\!::\!C\!::\!\ddot{O} \qquad \ddot{O}\!=\!C\!=\!\ddot{O} \qquad :\!N\!::\!N\!: \qquad :\!N\!\equiv\!N\!:$$

Polar/nonpolar covalent bonds

Electron pairs shared between **two atoms of the same element are shared equally**. At the other extreme, **for ionic bonding there is no electron sharing** because the electron is transferred completely from one atom to the other. Most bonds fall somewhere between these two extremes, and the electrons are **shared unequally (a polar bond)**.

The polarity of a bond can be determined through an examination of the electronegativities of the atoms involved in the bond. The more electronegative atom will have a stronger attraction to the electrons, thus possessing the electrons more of the time. This results in a partial negative charge ($\delta-$) on the more electronegative atom and a partial positive charge ($\delta+$) on the less electronegative atom as shown below for gaseous HCl. Such bonds are referred to as **polar bonds**. A particle with a positive and a negative region is called a **dipole**. A lower-case delta (δ) is used to indicate partial charge or an arrow is draw from the partial positive to the partial negative atom. (Remember, the sum of partial charges on any molecule or ion must add up to its overall charge.)

Molecular solids, nonmetals, diatomic elements, or compounds formed from two or more nonmetals, consist of distinct, individual units of covalently bonded molecules attracted to each other through relatively weak intermolecular forces. Because these intermolecular forces are weak, they tend to have low melting points. They can, however, form large polymers seen in biological systems. They are poor conductors of electricity. **Covalent network solids** are solids made up of a repeating network of covalently bonded atoms. They can only be made up of a single nonmetallic element or repeated patterns of two nonmetallic elements. Diamond and graphite are examples of elemental covalent network solids, while silicon dioxide and silicon carbide are examples of covalent network solids containing two different nonmetals. Three-dimensional covalent network solids tend to be rigid and hard because the covalent bond angles are fixed.

Elemental Carbon

Carbon exists stably in elemental form. Remember from the periodic table that carbon has four electrons that are free to react with other atoms. In **diamond**, each carbon is bonded to four other carbons in a tetrahedral crystal structure that is very hard.

A somewhat more complicated structure, **graphite**, consists of carbon atoms bonded together in six-member rings that form flat sheets. Within each ring, the atoms are bonded by sp^2 bonds, strong covalent bonds that include s and p orbitals of carbon. This leaves a fourth free electron, and this electron is used to bond the flat sheets of 6-carbon rings together in layers through bonding based on p orbitals, known as **pi bonding**. Pi bonds are covalent, but they are weaker than sigma bonds, allowing the layers to slide past each other. Graphite is the main substance in pencil "lead"; when you write with a pencil, the carbon sheets are actually sliding off of each other onto the paper. The pi bonds also allow the electrons to become dissociated, enabling graphite to conduct electricity.

A third form of elemental carbon is buckminsterfullerene, C60, which takes the form of large spheres made of fused pentagonal and hexagonal rings. Resembling soccer balls in appearance, this structure is also known as a "buckyball".

Another elemental covalent network solid is silicon. Silicon, which forms a geometry similar to diamond, is a semiconductor. Its conductivity increases with temperature. Adding a dopant to silicon changes its semiconducting properties. If the dopant is an element with

"Ionic bonding results from the net attraction between oppositely charged ions, closely packed together in a crystal lattice."

one extra valence electron, it will convert silicon into a negative charge-carrying (n-type semiconducting) material. If the dopant is an element with one fewer valence electron, it will convert silicon into a positive charge-carrying (p-type semiconducting) material. Because junctions between n-doped and p-doped materials can be used to control electron flow, these materials from the basis of modern electronics.

Ionic Bonds

An **ionic bond** describes the electrostatic forces that exist between particles of opposite charge. Elements that form an ionic bond with each other have a large difference in their electronegativity. Generally, bonds between a metal and a nonmetal tend to be ionic. When solid, ionically bonded substances form a crystal lattice composed of alternating positive and negative ions. This systematic structure maximizes attractive forces and minimizes repulsive forces. Coulomb's law determines the force of attraction between the bonded ions, with stronger bonding associated with larger charges as well as with smaller ions, since the force is inversely proportional to the square of the distance between the nuclei. In general, ionic solids are brittle, have low vapor pressure and high melting points, and only conduct electricity when molten or dissolved.

Metallic bonds occur when the bonding is between two metals. Metallic properties, such as low ionization energies, conductivity, and malleability, suggest that metals possess strong forces of attraction between atoms, but still have electrons that are able to move freely in all directions throughout the metal. This creates a "sea of electrons" model where electrons are quickly and easily transferred between metal atoms. In this model, the outer shell electrons are free to move. The transition metals, with their valence electrons in d-orbitals, exhibit the most highly metallic properties. The metallic bond is the force of attraction that results from the moving electrons and the positive nuclei left behind. The strength of metallic bonds usually results in regular structures and high melting and boiling points.

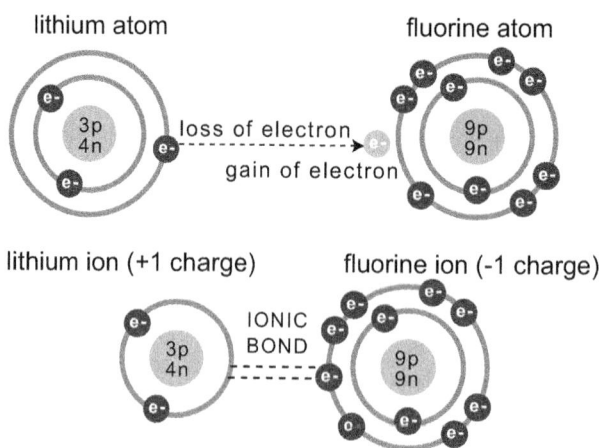

Physical properties relating to metallic character are summarized in the following table:

Element	Electrical/thermal conductivity	Malleable/ductile as solids?	Lustrous?	Melting point
Metals	High	Yes	Yes	High
Metalloids	Semiconductors. Intermediate thermal conductivity	No (brittle)	Varies	High
Nonmetals	Low (insulators)	No	No	Variable

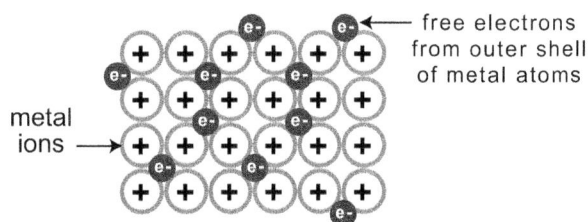

Mixtures of different metals are called **alloys.** Some properties of alloys can be understood in terms of the sizes of the component atoms. In **interstitial alloys**, smaller atoms fill the interstitial spaces between larger atoms. Steel is one example—carbon atoms occupy the spaces in between iron atoms, creating a very strong, dense material. **Substitutional alloys** form when the atoms are approximately the same size, and alternate in the lattice structure. Brass is an example of a substitutional alloy, with copper atoms substituted with zinc or another element. The density in substitutional alloys is generally in between those of the component metals, and they are less malleable and ductile than pure metals.

Bond energy graphs

It can be useful to plot the potential energy vs. distance to understand the nature and reason for a nonpolar covalent bond. The energy released when a bond is formed is the same amount as the energy absorbed when a bond breaks; the absolute value of this amount of potential energy is called the bond energy. Here is a graph of the bonding between two hydrogen atoms:

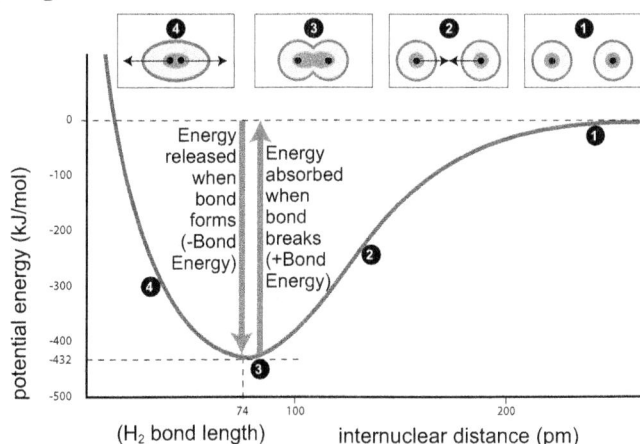

This plot clearly shows that the lowest energy state for two hydrogen atoms is when they are bonded together with a bond length of 74 pm. It also shows that while the energy level for separate, single hydrogen atoms is zero, the energy level for hydrogen atoms squeezed tightly together can go above zero, increasing exponentially. The repulsive forces of the atoms when they are very close together make this the least energetically favorable configuration.

Electronegativity and bond type

It is important to remember that ionic bonds are not fundamentally different from covalent bonds. Instead, bond type is a continuum, with some bonds having a highly ionic character (ionic bonds), some bonds having a partially ionic character (polar covalent bonds), and some bonds having very little ionic character (nonpolar covalent bonds). The best way to determine the type of bond is via the properties of the compound.

Electronegativity is a measure of the ability of an atom to attract electrons in a chemical bond. Metallic elements have low electronegativities and nonmetallic elements have high electronegativities. Relative values of electronegativity are shown in the table. Comparing the values allows us to determine the polarity of a bond, as described below.

H 2.2						
Li 1.0	Be 1.6	B 1.8	C 2.5	N 3.0	O 3.4	F 4.0
Na 0.9	Mg 1.3	Al 1.6	Si 1.9	P 2.2	S 2.6	Cl 3.2

Linus Pauling developed the concept of electronegativity and defined its relationship to different types of bonds in the 1930s.

A large electronegativity difference (greater than 1.7) results in an ionic bond. Any bond composed of two different atoms will be slightly polar, but for a small electronegativity difference (less than 0.4), the distribution of charge in the bond is so nearly equal that the result is called a nonpolar covalent bond. An intermediate electronegativity difference (from 0.4 to 1.7) results in a polar covalent bond. HCl is polar covalent because Cl has a very high electronegativity (it is near F in the periodic table) and H is a nonmetal (and so it will form a covalent bond with Cl), but H is near the dividing line between metals and nonmetals, so there is still a significant electronegativity difference between H and Cl. Using the numbers in the table above, the electronegativity for Cl is 3.2 and it is 2.2 for H. The difference of $3.2 - 2.2 = 1.0$ places this bond in the middle of the range for polar covalent bonds.

Chapter 2.7 Lewis Diagrams and the VSEPR Model _____

The valence-shell electron-pair repulsion (VSEPR) theory states that molecules will adopt a geometry that minimizes repulsion between valence electrons. This geometry can be determined through models called **Lewis structures.**

Lewis dot structures are used to keep track of each atom's valence electrons in a molecule. Drawing Lewis structures is a three-step process:

1. Determine the number of valence shell electrons for each atom. If the compound is an anion, add the charge of the ion to the total electron count because anions have "extra" electrons. If the compound is a cation, subtract the charge of the ion.

2. Write the symbols for each atom and show how the atoms within a molecule are bound to each other.

3. Draw a single bond (one pair of electron dots or a line) between each pair of connected atoms. Place the remaining electrons around the atoms as unshared pairs. If every atom has an octet of electrons except H atoms with two electrons, the Lewis structure is complete. Shared electrons count towards both atoms. If there are too few electron pairs to do this, draw multiple bonds (two or three pairs of electron dots between the atoms) until an octet is around each atom (except H atoms with two). If there are too many electron pairs to complete the octets with single bonds then the octet rule is broken for this compound.

ExampleSolution————Formal charge

Formal charge is the difference between the number of valence electrons of each atom in its ground state and the number of electrons the atom is associated with in a molecule. For maximum stability, formal charge should be zero, or minimized as much as possible. Mathematically, it can be expressed as:

$$\text{Formal charge} = V - N - (B/2)$$

where V = the number of valence electrons in the atom in its ground state; N = the number of non-bonding electrons associated with the atom; and B = the number of electrons in bonds associated with the atom.

In the Lewis structures of most molecules, when the octet rule has been fulfilled, the formal charge on each atom is zero. However, in larger molecules, there may be a negative or positive formal charge on certain atoms. This charge will need to be balanced by an opposite charge on an atom elsewhere in the molecule or the final molecule will be a polyatomic ion. When evaluating several different valid Lewis structures, minimizing formal charge is a way to determine which Lewis structure provides the best model for predicting structure and properties of the molecule.

Molecular orbital theory

Combining Lewis structures and the VSEPR theory can provide a powerful model for predicting properties of molecules. These properties include molecular geometry, bond

"The localized electron bonding model describes and predicts molecular geometry using Lewis diagrams and the VSEPR model."

angles, relative bond energies based on bond order, relative bond lengths (multiple bonds, effects of atomic radius), and presence of a dipole moment. As with any model, there are limitations to Lewis models, particularly when there are odd numbers of valence electrons.

Another way to describe bonds is through the interaction of their orbitals. Overlap between s orbitals leads to sigma bonds. Overlap between *p* orbitals leads to pi bonds. Overlap between multiple *p* orbitals can lead to delocalized bonds, as in benzene. The overlap is stronger in sigma than in pi bonds. Double and triple bonds consist of sigma and pi bonds.

In lines with the types of orbitals involved, the terms "hybridization" and "hybrid orbital" can be used to describe arrangements of electrons around a central atom. When a bond angle of 180° is present, the central atom is said to be sp hybridized; for a bond angle of 120°, the central atom is said to be *sp* hybridized; and for a bond angle of 109°, the central atom is said to be sp hybridized. For the AP test, you should know this terminology, and you should also be able to predict the shape of a molecule when the central atom has more than four pairs of electrons surrounding it.

Sample Test Questions (Big Idea #2) _____

Multiple Choice Questions

Instructions: This section consists of some practice multiple choice questions as well as one Free Response question related to this chapter. You may NOT use a calculator for the Multiple Choice. However calculators are permitted for the Free Response.

1. Which of the following graphs represents the plot of pressure versus volume at constant temperature?

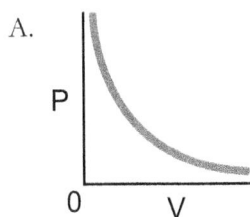

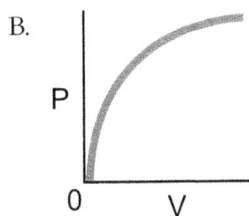

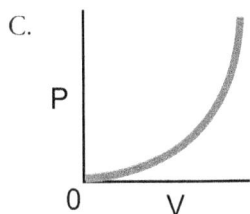

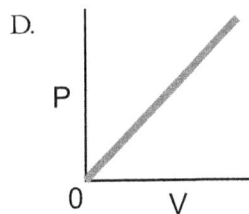

   A.

   P

   0   V

   B.

   P

   0   V

   C.

   P

   0   V

   D.

   P

   0   V

2. A sample of ozone gas is at 0°C. If both the volume and pressure double, what is the new Kelvin temperature?

   A.   135 K

   B.   273 K

   C.   546 K

   D.   1092 K

3. If the density of an unknown gas is 2.0 g/L at STP, what is the molar mass of the gas? R=0.0821 atm·L/mol·K)

   A.   11.4 g/mol

   B.   22.4 g/mol

   C.   23.8 g/mol

   D.   44.8 g/mol

4. Which of the following changes increases the pressure in a gaseous system?

   A.   increasing the volume

   B.   decreasing the temperature

   C.   increasing the number of gas molecules

   D.   all of the above

5. The fact that Br is a liquid and I is a solid is best explained by:

    A.   their ionic bonds which are very strong.

    B.   London dispersion forces which increase down the periodic table

    C.   The number of hydrogen bonds in Br is greater than for I

    D.   The number of valence electrons increases for I.

6. Which of the following Lewis Diagrams best represents CO

    A. $:\ddot{O} - C = \ddot{O}:$

    B. $:\ddot{O} - C - \ddot{O}:$

    C.   another "wrong" drawing like $:O = C = O:$

    D.   another "wrong" drawing like $::O - C - O::$

7. Which of the following illustrates the like dissolves like rule for two liquids?

    A.   a polar solute is soluble in a nonpolar solvent

    B.   a polar solute is insoluble in a polar solvent

    C.   a nonpolar solute is soluble in a nonpolar solvent

    D.   a nonpolar solute is soluble in a polar solvent

8. Coca-Cola is carbonated by injecting Coke with carbon dioxide gas. Under what conditions is the solubility of carbon dioxide gas the greatest?

    A.   low temperature, low pressure

    B.   low temperature, high pressure

    C.   low temperature, pressure is not a factor

    D.   high pressure, temperature is not a factor

9. Which separation method would be suitable to obtain pure water from a mixture of ink and water?

    A.   Chromatography

    B.   Simple distillation

    C.   Crystallization

    D.   evaporation

10. What is the molar concentration of water molecules in a beaker of water? (Recall that the density of water is 1.00 g/mL).

    A.   0.0555 M

    B.   1.00 M

    C.   18.0 M

    D.   55.5 M

11. Substances with greater intermolecular forces will have which of the following properties:

    I.   higher melting points

    II.   higher enthalpy of vaporization

    III. Lower viscosities

    IV. Lower vapor pressures

    A.   I only

    B.   I & II

    C.   II and III only

    D.   I, II and IV only

12. As heat is added to a system, which of the following will occur?

    I.   The energy of the system increases

    II.   The temperature of the system will always increase

    III. Intermolecular forces are broken

    IV. The vibrations of the molecules increase

    A.   All of the above are true

    B.   Only I, II, III are true

    C.   Only I, III and IV are true

    D.   Only II, III and IV are true

13. Maxwell-Boltzmann distribution is used to show the distribution of molecular velocities in a gas. According to this theory, which of the following are correct?

    A.   Gas as at a higher temperature has a higher average velocity than a gas at a lower temperature

    B.   The size of the molecule is not important

    C.   Equal volumes of different gases contain equal numbers of molecules

    D.   All of the above

14. How many moles of methane occupy a volume of 2.00 L at 50oC and 0.500 atm? (R= 0.0821 atm·L/mol·K)

    A.   0.0377 mol

    B.   0.151 mol

    C.   0.244 mol

    D.   4.11 mol

15. Whether a bond is polar or covalent is largely determined by

    A.   The number of valence electrons

    B.   the orbital of the valence electrons

    C.   the electronegativities of the atoms

    D.   All of the above

16. Given that Sodium(Na) has an atomic radius of 186 pm and a first ionization energy of 496kJ/mol, based on periodic trends, which of the following could be the atomic radius and first ionation energy of Magnesium?

    A.   160 pm, 737kJ/mol

    B.   186 pm, 898kJ/mol

    C.   135 pm, 423kJ/mol

    D.   197 pm, 423kJ/mol

17. Three 1.00 L flasks at 25.0 °C and 1atm pressure contain: CH (flask A), CO (flask B) and NH (flask C). Which flask (or none) contains 0.041 mol of gas? (assume all three flasks have the same volume and temperature).

    A.   Flask A

    B.   Flask B

    C.   Flask C

    D.   All of the flasks

18. A mixture of oxygen and helium is 96% by mass oxygen. It is collected at atmospheric pressure (745 torr). What is the partial pressure of oxygen in this mixture?

    A.   447 torr

    B.   600 torr

    C.   248 torr

    D.   149 torr

19. What is the volume of gas at 2.00 atm and 200.0 K if its original volume was 300.0 L at 0.250 atm and 400.0 K.

    A.   18.75 L

    B.   75 L

    C.   37.5 L

    D.   450 L

20. If the absolute temperature of a given quantity of gas is doubled and the pressure tripled, what happens to the volume of the gas?

    A.   the volume increases by 50%

    B.   the volume decreases by50%

    C.   the volume decreases to 66% of original volume

    D.   the volume increases to 1.5 times the original volume.

Free Response Question (Big Idea #2)

1. A.   A 1.5 L gas has a pressure of 0.56 atm. What will be the volume of the gas if the pressure doubles to 1.12 atm at constant temperature?

    B.   What is the new volume of a gas if 0.50 L of that gas at 25°C is heated to 35°C at constant pressure?

2. At STP, 0.250 L of an unknown gas has a mass of 0.429 g. Is the gas SO, NO, CH, or Ar? Support your answer.

3. A gas mixture composed of helium and argon has a density of 0.704 g/L at a 737 mmHg and 298 K. What is the percent composition of the mixture by (a) mass and by (b) volume.

Answers to Sample Questions (Big Idea #2) _____

1. **Answer A:**

   Using the Ideal Gas Equation

   $PV = nrT$ which rearranges to $P = (nrT) / V$

   At constant T a plot would show P inversely proportional to V. As V approaches zero, the equation blows up because the pressure would go to infinity. Therefore the correct answer is A

2. **Answer D:**

   Using the Gas Law $PV = nrT$

   If both P and V double ($P_2 = 2P_1$ and $V_2 = 2V_1$) then T must quadruple ($T=4T$). However, you must remember to use units of K. So the inital $T= 0oC$ becomes 273.15 K. Quadrupling the temperature results in a $t= 1092K$.

3. **Answer D:**

   STP represents 1 atm and 273.15 Kelvin temperature. The ideal gas law $PV = nRT$ can be rearranged to include density.

   First convert # of moles into mass/molecular weight then rearrange equation in terms of mass/volume

   $PV = m*R*T$ /molecular weight.

   $D = m/v =$ Pressure * molecular weight/ R * T

   Given Density = 1.95g/mol

   You can solve for molecular weight

   $MW = D *R * T$/Pressure = [2.0 * 0.0821 * 273] / 1 atm = 44.8 g/mol

4. **Answer C:**

   Increasing the volume will decrease the pressure as will decreasing the temperature. However increasing the number of molecules in the system (n) will increase the pressure.

5. **Answer: B:**

   London dispersion forces increase down the PT. tive forces (i.e. the stronger intermolecular bond) will require more energy to pull apart the molecules. Substances with greater intermolecular forces will have higher melting points.

6. **Answer A** is the correct Lewis structure as it shows the C=O double bonds.

7. **Answer B.**

   Like dissolves like refers to the fact that polar solutes prefer polar solvents and non-polar solutes

8. **Answer B**

9. **Answer B:**

   Explanation: Water changes into steam as it vaporizes and it rises and enters the condenser, where the steam condenses and changes back into pure water.

10. **Answer D:**

If the molar mass of water = 18g/mole (using H=1 and O=16), assume 1L, then there are 1000/18 = 55.556 moles in 1 liter, so concentration = 55.556 mole/L

11. **Answer D:**

Substances with greater intermolecular forces will have the following properties:

FOR SOLIDS:

> Higher melting point
>
> Higher enthalpy of fusion
>
> Greater hardness
>
> Lower vapor pressure

FOR LIQUIDS:

> Higher boiling point
>
> Higher critical temperature
>
> Higher critical pressure
>
> Higher enthalpy of vaporization
>
> Higher viscosity
>
> Higher surface tension
>
> Lower vapor pressure

12. **Answer C:**

As heat is added to the system, the energy of the system increases and molecular vibrations increase. This results in the breaking apart of intermolecular forces. Temperature increases until a phase change occurs. There is no increase in temperature during a phase change.

13. **Answer A:**

14. **Answer A:**

15. **Answer C:**

The polarity of a bond can be determined through an examination of the electronegativities of the atoms involved in the bond. The more electronegative atom will have a stronger attraction to the electrons, thus possessing the electrons more of the time.

16 **Answer A:**

Atomic raadii descrease left to right across the periodic table. Trends tell us that Mg should be smaller than Na In addition, FIE increase from left to right. Therefore, Mg will be smaller and have a larger FIE.

17. **Answer D:**

The above satisfies Avogaro's hypothesis: equal volumes of gases under the same conditions of pressure and temperature, contain equal number of molecules.

Therefore, all three flasks either all contain 0.041 mol or none does.

We will check flask A, using $PV = nRT$:

(1.00 atm) (1.00 L) = (n) (0.08206 L atm mol⁻1 K⁻1) (298 K)

n = 0.0409 mol

All three flasks contain 0.041 mol of the different gases.

18. **Answer D:**

Let's assume a big, big volume such that this volume holds 100 g of the mixture at 745 torr and whatever the temperature is. 96 g of the 100 g is O and 4 g is He.

Compute moles of each because pressure is proportional to the number of particles:

$O_2 \rightarrow$ 94 g / 32.0 g/mol = 2mol

$He \rightarrow$ 6 g / 4.0026 g/mol = 1.5 mol

We now will calculate what we need to solve this equation: partial pressure = mole fraction x total pressure. Get total moles:

2.0 + 1.5 = 3.5 mol

Get mole fraction of O:

2.0 / 3.5 = 0.57 = 0.600

Get partial pressure of O:

0.600 times 745 torr = 447 torr

19. **Answer A:**

x = [(0.25) (300) (200)] / [(400) (2)] = 18.75 L

20. **Answer C:**

$PV = nrT$ of $P_2 = 3P_1$ and $T_2 = 2T_2$

$PV = nrT$

$V = nRT / P$ $V_2 = nR(2T_1) / 3(P_1) =$ so volume is 2/3 original volume or 66%.

Answers to Free Response (Big Idea #2)

1. A. Solution:

    This is a pressure-volume relationship at constant temperature, so using Boyle's law:

    P_1 = 0.56 atm
    V_1 = 1.5 L
    P_2 = 1.12 atm
    V_2 = ?

    Use the equation $P_1V_1 = P_2V_2$, rearrange to solve for V_2 = $\dfrac{P_1V_1}{P_2}$

    Substitute and solve. V_2 = 0.75 L

    Pressure can be in atmospheres, Pascals, or mm Hg as long as it is the same units for P_1 and P_2.

    B. Solution:

    This is a volume-temperature change so use Charles' law. Temperature must be on the Kelvin scale. K = °C + 273.

    T_1 = 25°C + 273 = 298K
    V = 0.50 L
    T_2 = 308K
    V = ?

    Use the equation: $\dfrac{V_1}{T_1} = \dfrac{V_2}{T_2}$ and rearrange for V_2 = $\dfrac{T_2V_1}{T_1}$

    Substitute and solve
    V_2 = 0.52L.

2. Solution: Identify what is given and what is asked.

    Given: T_1 = 273K
    P_1 = 1.0 atm
    V_1 = 0.250 L
    Mass = 0.429 g

    Determine:

    Identity of the gas. In order to do this, you must find the molar mass (MM) of the gas.

    $$n = \frac{mass}{MM}$$

    Find the number of moles of gas present using PV = nRT and then determine the MM to compare to choices given in the problem.

    Solve for $n = \dfrac{PV}{RT}$ = (1.0 atm)

    (0.250 L) / (0.0821 atm L/mol KI)(273 K)
    n = 0.011 moles

    $MM = \dfrac{mass}{n}$ = 0.429 g/0.011 mol = 39.0 g/mol

Compare to:

MM of SO_2 (96 g/mol), NO_2 (46 g/mol), C_3H_8 (44 g/mol) and Ar (39.9 g/mol).

It is closest to Ar, so the gas is probably Argon.

3. A. **Solution 1:**

Calculate total moles of gases present:

Comment: assume that the volume of the gas mixture is 1.00 L

$PV = nRT \rightarrow n = PV / RT$

n = (737 torr / 760 torr/atm) (1.00 L) / (0.08206 L atm/mole K) (298 K)

n = 0.039656 mol

Solution 2:

Set up two simultaneous equations:

Comment: let x = mass He and y = mass Ar

Equation #1 → x + y = 0.704 g

Equation #2 → x/4.0026 g/mol + y/39.948 g/mol = 0.039656 mol

Solution 3:

Substitute x = 0.704 - y into the second equation and solve for y:

Comment: I left off the units until the end.

(0.704 - y)/4.0026 + y/39.948 = 0.039656

(39.948) (0.704 - y) + (4.0026) (y) = (0.039656) (39.948) (4.0026)

More algebra results in:

   y = 0.606 g Ar

   x = 0.098 g He

Solution 4:

Calculate mass percent of each gas:

Ar → (0.606 / 0.704) x 100 = 86.61%

He → 14.06%

B. **Solution 1:**

Let us determine the volume 0.606 g of Ar occupies at the stated T and P:

(737/760) (x) = (0.606/40) (0.08206) (298)

x = 0.382 L

Solution 2:

Since combined volume was 1.00 L, the volume percents are:

Ar → 38.2%

He → 61.8%

Chapter 3: Big Idea #3
Changes in Matter

Big Idea 3: Changes in matter involve the rearrangement and/or reorganization of atoms and/or the transfer of electrons.

What you should already know:

- the nature of atoms, elements, compounds, and bonds
- Lewis structures

What you will learn:

- stoichiometry
- 4 general types of chemical change: decomposition, synthesis, single replacement, and double replacement
- overview of specific reaction types, including combustion, acid-base neutralization, and redox
- redox reactions
- electrochemistry
- assessing chemical change by macroscopic observation, especially via temperature changes

---

Chapter 3.1 Stoichiometry

In a chemical reaction, reactants or reagents are transformed into different substances, called products. A chemical reaction is illustrated by writing the reactants on the left and the products on the right side of the equations. Both sides are separated by an arrow ($\rightarrow$) that indicates the direction of the reaction. A double arrow ($\leftrightarrow$) between two sides of a reaction or simply ($=$) is used for equilibrium reactions or to indicate that the reaction could be performed in both directions. This depiction of a reaction is called a **chemical equation.**

A properly written chemical equation must contain properly written formulas and must be **balanced**; that is, it must contain equal numbers of atoms of each element on both sides of the equation. Chemical equations are written to describe a certain number of moles of specific reactants becoming a certain number of moles of specific reaction products. Chemical equations obey the law of conservation of mass in that no atoms are created or destroyed; this is why each element has the same number of total atoms on the left of the arrow as it has on the right of the arrow when the equation is balanced.

Reaction stoichiometry

The number of moles of each compound is indicated by its stoichiometric coefficient. The number of moles of each element is determined by multiplying the coefficient by the subscript, with subscripts outside the parentheses being multiplied by subscripts inside the parentheses. This is done for each element in each compound. Then all the atoms of that element on that side of the arrow are added together.

For example, in the following reaction the moles of atoms are not conserved because the equation is not balanced.

$$H2 \ (g) + O \ (g) \rightarrow HO(l)$$

The balanced equation would be

$$2H_2 + O \rightarrow 2H_2O$$

Hydrogen has a stoichiometric coefficient of two, oxygen has a coefficient of one, and water has a coefficient of two because 2 moles of hydrogen react with 1 mole of oxygen to form two moles of water. The number of atoms of hydrogen on the left is 4, which is found by multiplying the coefficient of 2 by the subscript of 2. The number of atoms of hydrogen on the right is found in the same way to be 4. The number of atoms of oxygen on the left is 2 since the coefficient of 1 is multiplied by the subscript of 2. Using the same method on the right, the coefficient of 2 is multiplied by the subscript of 1 (no subscript means one) to obtain a total of 2 atoms of oxygen. Therefore, this reaction is balanced.

In a balanced equation, the stoichiometric coefficients are chosen so that the equation contains an equal number of each type of atom on each side. In our example, there are four H atoms and two O atoms on both sides. Therefore, the equation is balanced with respect to atoms of each element.

Balancing equations is a multi-step process. The following steps can help to achieve properly balanced chemical equations.

- Determine the **correct formulas** for all compounds.
- Write an **unbalanced equation**. This requires knowledge of the nature of the products that will be produced from the reactants. Reactants are written on the left and products are written on the right.
- Determine the number of each type of atom on each side of the equation to determine whether the equation is already balanced. Under the reactants, list all the elements in the reactants starting with metals, then nonmetals, listing oxygen last and hydrogen next to last. Under the products, list all the elements in the same order as those under the reactants – preferably, straight across from them.
- Count the atoms of each element on the left side and list the numbers next to the elements. Repeat for products. Don't forget that subscripts outside a set of parentheses are multiplied by subscripts inside the parentheses. If each element has the same number of atoms on the right that it has on the left, the equation is balanced. If not, proceed to Step 5.

- For the first element in the list that has unequal numbers of atoms, use a coefficient (whole number to the left of the compound or element) on either the left of the arrow or the right of the arrow to give an equal number of atoms. NEVER change the subscripts to balance an equation.
- Go to the next unbalanced element and balance it, moving down the list until all are balanced.
- Start back at the beginning of the list and actually count the atoms of each element on each side of the arrow to make sure the number listed is the actual number. Re-balance and re-check as needed.

Example:

1. The structural formula of methanol is CH_3OH, so its molecular formula is CH_4O. The formula for carbon dioxide is CO_2. Therefore the unbalanced equation is:

$$CH_4O + O_2 \rightarrow CO_2 + H_2O$$

2. On the left there are: one C, four H, and three O atoms. On the right, there are one C, two H, and three O atoms. The equation is close to being balanced but there is still work to do.

3. Assuming that CH_4O has a stoichiometric coefficient of one means that the left side has one C and four H that are currently missing on the right. Therefore the stoichiometric coefficient of CO will be 1 to balance C and the stoichiometric coefficient of HO will be 2 to balance H. Now we have:

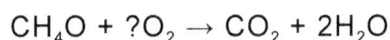

$$CH_4O + ?O_2 \rightarrow CO_2 + 2H_2O$$

Only oxygen remains unbalanced. There are 4 O on the right and three on the left. However, one of these is accounted for by methanol, which we would rather not change, since we have already balanced the carbon and hydrogen. This leaves 3 O to be accounted for by O2. This gives a stoichiometric coefficient of 3/2 and a balanced equation.

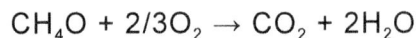

$$CH_4O + 2/3O_2 \rightarrow CO_2 + 2H_2O$$

4. Whole-number coefficients are achieved by multiplying the entire reaction equation by the denominator; in this case, 2:

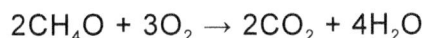

$$2CH_4O + 3O_2 \rightarrow 2CO_2 + 4H_2O$$

Chapter 3.2 Limiting reagents, percent composition, yield, and practical stoichiometry

The **limiting reagent** of a reaction is the reactant that runs out first. This reactant **determines the amount of products formed**, and any **other reactants remain unconverted** to product and are called excess reagents.

Example.

Consider the reaction below and suppose that 3 mol H_2 and 2 mol N_2 are available for this reaction. What is the limiting agent?

$$3H + N \rightarrow 2NH$$

Solution.

The equation tells us that 3 mol H_2 will react with on mol N_2 to produce 2 mol NH_3. This means that H_2 is the limiting reagent because when it is completely used up, 2 mol N_2 will still remain.

The limiting reagent may be determined by **dividing the number of moles of each reactant by its stoichiometric coefficient**. This determines the moles of reactant if each reactant were limiting. The **lowest result** will indicate the actual limiting reagent. Remember to use moles and not grams for these calculations.

The percent composition of a substance is the percentage by mass of each element. Chemical composition is used to verfiy the purtiy of a compound in the lab. An impurity will make the actual composition vary from the expected one.

To determine percent composition from a formula, follow these steps:

1. Write down the **number of atoms each element contributes** to the formula.

2. Multiply these values by the molecular weight of the corresponding element to determine the **grams of each element** in one mole of the formula.

3. Add the values from step 2 to obtain the **formula mass**.

4. Divide each value from step 2 by the formula mass from step 3 and multiply by 100% to obtain the **percent composition of each element**.

The yield of a chemical reaction is the amount of the obtained products. This value is usually less than what would be predicted from the stoichiometric calculation. The reasons for this are numerous, including: i) secondary reactions may produce different products; ii) the reverse reaction may occur; and iii) some material may be lost during the procedure—because of evaporation, for example. The yield calculated from the stoichiometry based on the limiting reagent is called the **theoretical yield**:

Percent yield is defined as the actual yield divided by the theoretical yield times 100%:

Example 1

In the reaction of formation of CO2 from solid carbon and oxygen gas, $C(s) + O2(g) \rightarrow CO2(g)$. it is theoretically expected that 1 mol of carbon burned will yield one mol of carbon dioxide. However, this does not always happen. If 12 grams of carbon (1 mol) are burned to make CO2, then the amount of carbon dioxide is expected to be one mol of CO2 or 44 grams of CO2. However, the true amount produced will be less than 44 grams and more like 34 grams of CO2. The reason lies in competing or secondary reactions such as formation of CO.

$$\rightarrow \qquad 2C(s) + O_2(g) \rightarrow 2\ CO(g) .$$

Hence, the reaction will not yield 100% of the expected CO; 34 grams of CO is only 77% and not 100% of the expected 44 grams. The percent yield for this reaction is 77%.

Stoichiometry in the "real world"

Stoichiometry is used throughout chemistry. One area that makes heavy use of stoichiometry is solution chemistry, including titrations. Solution chemistry, titrations, and acid-base chemistry in general are reviewed in detail in Chapter 6. Another is oxidation-reduction reactions, reviewed later in this chapter.

In addition, stoichiometry is not just used in the chemistry lab. Numerous applications require the use of stoichiometry, including other types of labs, such as biology and medical labs, and manufacturing, such as pharmaceutical production. In addition, stoichiometry crops up in some very unexpected places. One is in your car.

Stoichiometry and air bags

Have you ever thought about where the "air" in airbags comes from? Airbags inflate at a rate of 150-250 miles per hour! How can a bag inflate with gas that fast? The answer lies in chemistry. The "air" in airbags is actually nitrogen gas, produced by igniting sodium azide, NaN. The unbalanced reaction for this is:

$$NaN_3(s) \rightarrow Na(s) +\ N_2(g)$$

However, to be safe and effective, the airbag must generate a precise quantity of nitrogen gas (too much, and the bag could cause injury; too little, and the bag will fail to prevent injury). To calculate the precise amount of sodium azide that must be loaded into the airbag, a balanced equation is needed:

$$2\ NaN_3(s) \rightarrow 2\ Na(s) + 3\ N_2(g)$$

Now, we can use stoichiometry to calculate the amount of sodium azide needed to fill the airbag. If a given airbag is designed to inflate with 60 L of N gas, how much solid sodium azide needs to be ignited?

The density of nitrogen will vary with temperature, but let us use an approximate value of 1 g/L. This means we will need to generate 60 grams of N gas. From the periodic table, we know that the molar mass of Na is approximately 23 g/mol; that of N is approximately $2(14) = 28$ g/mol; and that of NaN is $23 + 3(14) = 65$ g/mol.

"Quantitative information can be derived from stoichiometric calculations that utilize the mole ratios from the balanced chemical equations. The role of stoichiometry in real-world applications is important to note, so that it does not seem to be simply an exercise done only by chemists."

Now, we solve for the grams of NaN needed:

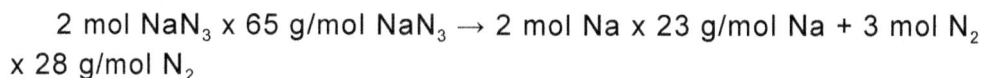

$$2 \text{ mol NaN}_3 \times 65 \text{ g/mol NaN}_3 \rightarrow 2 \text{ mol Na} \times 23 \text{ g/mol Na} + 3 \text{ mol N}_2 \times 28 \text{ g/mol N}_2$$

$$130 \text{ g NaN}_3 \rightarrow 46 \text{ g Na} + 84 \text{ g N}_2$$

solving for the amount of NaN needed to generate 1 g of N, we divide by 84. We will need 1.55 g of NaN to generate one gram of N.

Since we want our airbag to inflate to 60 L, however, and the density of N is approximately 1 g/L, we need to generate about 60 grams of N. Multiplying 60 by 1.55, we will need to ignite about 93 grams of NaN to inflate the airbag.

But that's not the end of the air bag stoichiometry story. Air bag designers will have to take into account that the yield of the reaction will be less than 100%; they will have to determine the percent yield experimentally and add that into the equation when calculating the number of grams of sodium azide to add to the air bag.

Also, remember the other product in the reaction, Na? This is sodium metal. Sodium metal is highly reactive, and definitely not something you want to be left with after an auto accident. Therefore, the airbag must also be loaded with chemicals that will react with the sodium metal to produce a safe product. KNO is used to react with the sodium metal, and SiO is used to react with the product of that reaction, NaO, which is also highly reactive and needs to be removed. In the process, more nitrogen gas is generated that also fills the airbag, and this needs to also be taken into account when designing the airbag. As you can imagine, the stoichiometry involved in airbags is quite complex!

Stoichiometry in space and on Earth

Stoichiometry is also used in space, where the CO breathed out by astronauts needs to be "scrubbed" by reacting with LiOH in special canisters known as LiOH scrubbers. The amount of LiOH must be calculated before a mission to make sure there is enough LiOH to scrub all of the CO that will be generated by the astronauts on that mission. Of course, in order to get into space in the first place, the amount of solid rocket fuel must also be calculated using stoichiometry.

In daily life on Earth, stoichiometry must be used when adding chlorine to the city water supply, in farming when adding phosphate as fertilizer to crops, and in the production of silicon for computer chips. Candymakers and other cooks who need to produce exact food products (such as companies that make brand-name processed food) also need to use stoichiometry. Even firefighters use stoichiometry in understanding the fuel-air dynamics that can produce hazardous fire events.

Chapter 3.3 Reaction Types _____

Reactions can be classified in a number of different ways: according to products,

"Chemical reactions can be classified by considering what the reactants are, what the products are, or how they change from one into the other. Classes of chemical reactions include synthesis, decomposition, acid-base, and oxidation-reduction reactions."

according to reactants, or according to how the products change into the reactants. Classes of chemical reactions include synthesis, decomposition, acid-base (neutralization), and oxidation-reduction reactions. At the most fundamental level, however, there are four general types of reactions.

Synthesis or Combination: formation of a compound from two elements ($A + B \rightarrow AB$), two compounds ($AB + CD \rightarrow ABCD$), or a compound and element ($A + BC \rightarrow ABC$). This is a single product from two reactants.

Examples:
$$N_2 + 3 H_2 \rightarrow 2 NH_3 \quad \underline{OR} \quad ZnS + 2 O_2 \rightarrow ZnSO_4$$

Decomposition:

The opposite of syntheis - this is the coming apart of a compound into two elements or an element and a compound of two compouns. This is a single reactant with two products:

$$AB \rightarrow A + B$$

Example 1.
$$2 H_2O \rightarrow 2 H_2 + O_2 \quad \underline{OR} \quad NH_4NO_3 \rightarrow N_2O + 2 H_2O$$

Example: $Na2CO3 \rightarrow Na2O + CO2\uparrow \quad \underline{OR} \quad MgSO_4 \rightarrow MgO + SO_3$

Decomposition of –*ate* polyatomic ions results in a metal oxide and water or carbon dioxide or sulfur trioxide.

Single Replacement: An element and a compound becoming a different element and different compound. The reactant element replaces one of the elements in the reactant compound so that the products are a different compound that contains the original elemental reactant and an element that was originally part of the reactant compound:

$$A + BC \rightarrow AC + B$$

Example: $Zn(s) + H_2SO_4 (aq) \rightarrow ZnSO_4 (aq) + H_2\uparrow$

A single replacement reaction involving a metal and water yields a metal hydroxide and hydrogen gas: $Ca + 2 H_2O \rightarrow Ca(OH)_2 + H_2\uparrow$

Double Displacement or Double Replacement. This is two compounds switching ions or polyatomic ions and becoming two different compounds.

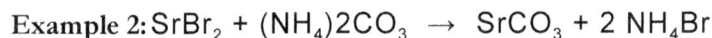

$$AB + CD \rightarrow AD + CB$$

Example 2: $SrBr_2 + (NH_4)2CO_3 \rightarrow SrCO_3 + 2 NH_4Br$

Note: Double displacement reactions are not oxidation-reduction reactions.

Once we have an idea of the **reaction type**, we can make a good prediction about the products of chemical equations, and also balance the reactions. **General reaction types** are listed in the following table. Some reaction types have multiple names.

"Synthesis reactions are those in which atoms and/or molecules combine to form a new compound. Decomposition is the reverse of synthesis, a process whereby molecules are decomposed, often by the use of heat."

Reaction type	General equation	Example
Combination/ Synthesis	$A + B \rightarrow AB$	$N_2 + H_2 \rightarrow 2NH_3$
Decomposition	$AB \rightarrow A + B$	$2 H2O \rightarrow 2 H2 + O2$
Single substitution/ Single displacement/ Single replacement	$A + BC \rightarrow AC + B$	$Zn(s) + H_2SO_4(aq) \rightarrow ZnSO_4(aq) + H2\uparrow$
Double substitution/ Double displacement/ Double replacement/ Ion exchange/ Metathesis/	$AB+CD \rightarrow AD+CB$	$SrBr_2 + (NH_4)_2CO_3 \rightarrow SrCO_3 + 2 NH_4Br$

"In a neutralization reaction, protons are transferred from an acid to a base."

Many **specific reaction types** also exist. One of these is **precipitation**, in which reactants that are in solution form an insoluble product. One example of a precipitation reaction is the formation of kidney stones in the body; kidney stones form when soluble ions combine into an insoluble solid, the stone. One of the most common solids that make up kidney stones is calcium phosphate, Ca(PO). Calcium phosphate forms from dissolved calcium and phosphate ions:

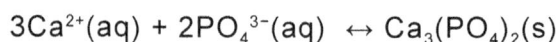

$$3Ca^{2+}(aq) + 2PO_4^{3-}(aq) \leftrightarrow Ca_3(PO_4)_2(s)$$

Another is **combustion** the use of oxygen to burn a reactant (usually organic) to produce carbon dioxide and water. For example, the burning of methane:

$$CH_4 + 2 O_2 \rightarrow CO_2 + 2 H_2O$$

Acid-base reactions involve the donation of a proton by the acid and acceptance of a proton by the base. This type of reaction is called a **neutralization**, and in the case of strong acids and bases, the final product is a salt and water. In the case of weak acids and bases, there will be an equilibrium between the acid and base forms of the molecule. Chapter 6 includes a detailed discussion of acid-base reactions.

One of the most important and common types of reactions is **oxidation-reduction**, or redox, reactions. (Note: combustion reactions are a subtype of redox reaction.)

Chapter 3.4. Redox Reactions

Oxidation and reduction

"In oxidation-reduction (redox) reactions, there is a net transfer of electrons. The species that loses electrons is oxidized, and the species that gains electrons is reduced."

Redox is a term for reduction and oxidation reactions. **Reduction** is a process whereby a molecule, atom, or ion gains one or more electrons. **Oxidation** is the opposite process; a molecule, atom, or ion loses one or more electrons. Oxidation and reduction processes always occur together, with electrons lost by one substance gained by another, resulting in movement of electrons (so called electron transfer). In a redox process, the **oxidation numbers** of atoms are altered. Reduction decreases, or reduces, the oxidation number of an atom; the atom gains one or more electrons, and since electrons are negative, the ion becomes "more negative". Oxidation increases the oxidation number; the atom loses one or more electrons and becomes "more positive".

In redox chemistry, oxidation numbers, also called states, are somewhat hypothetical, but very useful representations of the number of electrons involved in redox reactions. They represent, essentially, the electric charge the atom would possess if it were to dissociate itself from the formula unit. Hence, a redox reaction could be defined as a chemical reaction in which the oxidation state of at least two atoms change.

Oxidation states can be defined as formal charge attributed to an atom (either alone, in ionic form, or in a compound) on the basis of its electronegativity. In other words, knowledge of the oxidation state of an atom helps to determine its positive or negative character. Oxidation states are frequently utilized to construct chemical formulas and help balance equations where transfer of electrons is involved. In themselves, oxidation numbers have no physical meaning; they are simply utilized to make redox processes more understandable.

Rules for determining oxidation state

1. Free elements are assigned an oxidation state of 0.

   Example: Al, Na, Fe, H, O, N, Cl etc. have zero oxidation states

2. The oxidation state for any simple one-atom ion is equal to its charge. **Example:** The oxidation state of Na is +1, Be+, +2, and of F-, -1

3. The alkali metals (Li, Na, K, Rb, Cs and Fr) in compounds are always assigned an oxidation state of +1.

   Example: In LiOH, Li = +1; in Na2SO4, Na = +1

4. Fluorine in compounds is always assigned an oxidation state of -1.

   Example: In HF- and BF-, F=-1

5. The alkaline earth metals (Be, Mg, Ca, Sr, Ba, and Ra) and also Zn and Cd in compounds are always assigned an oxidation state of +2. Similarly, Al & Ga are always +3.

   Example: In CaSO, Ca = +2; in AlCl, Al = +3

6. Hydrogen in compounds is assigned an oxidation state of +1, except i. hydrides, where it is assigned -1.

   Example: In HSO, H = +1; in LiH, H = -1

7. Oxygen in compounds is assigned an oxidation state of -2. An exception is peroxide, where in HO, O = -1.

   Example: In HPO, O = -2

8. The oxidation states of many other atoms may vary from compound to compound. However, their oxidation states can frequently be determined based on the following rule: the sum of the oxidation states of all the atoms in a species

must be equal to the net charge on the species.

Example: The net charge of HClO = 0. Applying the above rules, H = +1 and $4O = -2 \times 4 = -8$. Therefore, the oxidation state of Cl must be +7 so that the overall charge = 1 + -8 + 7 = 0.

Example. The net charge of CrO- = -2. Applying the above rules, $4O = -2 \times 4 = -8$. Therefore, the oxidation state of Cr must be +6 so that the overall charge = -8 + 6 = -2.

The easiest redox processes to identify are those involving monatomic ions with altered charges. For example, the reaction

$$Zn(s) + Cu\ (aq) \rightarrow Zn\ (aq) + Cu(s)$$

is a redox process because electrons are transferred from Zn to Cu.

In another example, the reaction:

$$H_2 + F_2 \rightarrow 2HF$$

is a redox process because the oxidation numbers of atoms are altered. The oxidation numbers of elements are always zero, and oxidation numbers in a compound are never zero. Fluorine is the more electronegative element, so in HF it has an oxidation number of −1 and hydrogen has an oxidation number of +1. This is a redox process where electrons are transferred from H to F to create HF.

On the other hand, in the reaction:

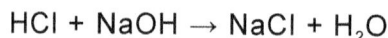

$$HCl + NaOH \rightarrow NaCl + H_2O$$

the H-atoms on both sides of the reaction have an oxidation number of +1, the Cl atom has an oxidation number of −1, the Na atom has an oxidation number of +1, and the O atom has an oxidation number of −2. **This is not a redox process because oxidation numbers remain unchanged** by the reaction.

Redox reactions may always be written as **two half-reactions**, a **reduction half-reaction** with **electrons as a reactant** and an **oxidation half-reaction** with **electrons as a product**.

Example: The redox reactions considered in the previous section:

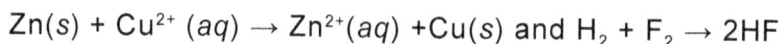

$$Zn(s) + Cu^{2+}\ (aq) \rightarrow Zn^{2+}(aq) + Cu(s) \text{ and } H_2 + F_2 \rightarrow 2HF$$

may be written in terms of half reactions:

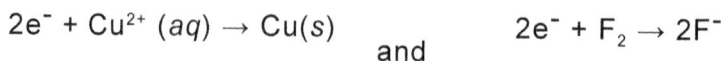

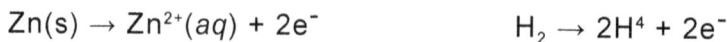

$$2e^- + Cu^{2+}\ (aq) \rightarrow Cu(s) \qquad 2e^- + F_2 \rightarrow 2F^-$$
$$\text{and}$$
$$Zn(s) \rightarrow Zn^{2+}(aq) + 2e^- \qquad H_2 \rightarrow 2H^4 + 2e^-$$

An additional (non-redox) reaction, $2F^- + 2H^4 \rightarrow 2HF$, achieves the final products for the second reaction.

Redox reactions must be balanced to observe the Law of Conservation of Mass. This process is a little more complicated than balancing other reactions because the number of electrons lost must equal the number of electrons gained. Balancing redox reactions, then, conserves not only mass but also charge or electrons. It can be accomplished by slightly varying our balancing process.

For example, the equation:

$$Sn^{2+} + 2Fe^{3+} \rightarrow Sn^{4+} + 2Fe^{2+}$$

contains one Sn and one Fe on each side, but it is not balanced because the sum of charges on the left side of the equation is +5 and the sum on the right side is +6. A charge balance is obtained by considering each half-reaction separately before multiplying by an appropriate factor so that the number of electrons gained by one half-reaction is the same as the number lost in the other. The two half-reactions may then be combined again into one reaction. From the example above, one electron is gained in the reduction half-reaction ($Fe^{3+} + e^- \rightarrow Fe^{2+}$) but two are lost in the oxidation half-reaction

$$(Sn^{2+} \rightarrow Sn^{4+} + 2e^-)$$

A charge balance is obtained by multiplying the reduction half-reaction by two to obtain the half-reactions:

$$2Fe^{3+} + 2e^- \rightarrow 2Fe^{2+}$$

$$Sn^{2+} \rightarrow Sn^{4+} + 2e^-$$

the equation $Sn^{2+} + 2Fe^{3+} \rightarrow Sn^{4+} + 2Fe^{2+}$ is properly balanced because both sides contain the same sum of charges (8+) and electrons cancel from the half-reactions.

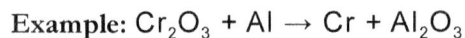

Example: $Cr_2O_3 + Al \rightarrow Cr + Al_2O_3$

Solution: Assign oxidation numbers to each atom in order to identify which atoms are losing and gaining electrons:

$$Cr_2O_3 + Al \rightarrow Cr + Al_2O_3$$
$$3+ \ 2- \quad 0 \quad \ \ 0 \quad \ 3+ \ 2-$$

Identify those atoms gaining and losing electrons:

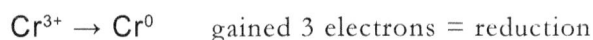

$Cr^{3+} \rightarrow Cr^0$ gained 3 electrons = reduction

$Al^0 \rightarrow Al^{3+}$.   lost 3 electrons = oxidation

Balance the atoms and electrons:

$$Cr_2O_3 \rightarrow 2Cr + 6e^-$$

$$2Al + 6e^- \rightarrow Al_2O_3$$

Balance the half reactions by adding missing elements. Ignore elements whose oxidation number does not change. Add HO to add oxygen and H to add hydrogen:

$$Cr_2O_3 + 6H^+ \rightarrow 2Cr + 6e^- + 3H_2O$$

$$2Al + 6e^- + 3H_2O \rightarrow Al_2O_3 + 6H^+$$

There is a need for 3 oxygen atoms on the reactant side and 6 H^+ on the product side. Put the two half reactions together and add the species. Then, cancel out the species that occur in both the reactant and product sides of the equation.

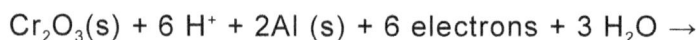

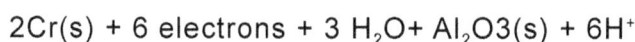

$$Cr_2O_3(s) + 6\ H^+ + 2Al\ (s) + 6\ \text{electrons} + 3\ H_2O \rightarrow$$

$$2Cr(s) + 6\ \text{electrons} + 3\ H_2O + Al_2O3(s) + 6H^+$$

Therefore, the balanced equation is:

$$Cr_2O_3(s) + 2Al(s) \rightarrow 2Cr(s) + Al_2O_3\ (s)$$

Redox reactions in the "real world"

Redox reactions do not just take place in the chemistry lab; in fact, they are taking place all around us. Rust on an old car and the blue-green patina on the Statue of Liberty are examples of metals that have been oxidized, as is the blackish tarnish on silver. The equations for these reactions are:

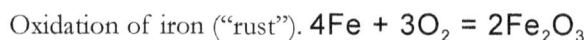

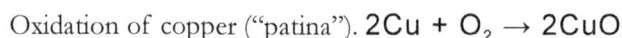

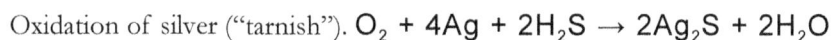

Oxidation of iron ("rust"). $4Fe + 3O_2 = 2Fe_2O_3$

Oxidation of copper ("patina"). $2Cu + O_2 \rightarrow 2CuO$

Oxidation of silver ("tarnish"). $O_2 + 4Ag + 2H_2S \rightarrow 2Ag_2S + 2H_2O$

While a copper patina is considered beautiful and desirable, rust and tarnish are not. In the case of silver tarnish, while you can polish it off with an abrasive, in the process, you will also be removing a tiny layer of silver as well. Since silver is a valuable metal, why not use chemistry to remove the tarnish without harming the silver? To reduce the silver sulfide (tarnish) to silver metal, use boiling water, baking soda, and aluminum foil to accomplish a "reverse redox" reaction. The aluminum will donate electrons to the silver, converting it back into its metal form, while the aluminum becomes oxidized to aluminum sulfide; the black silver sulfide tarnish disappears and off-white/eggshell-colored flakes of aluminum sulfide will appear. The essential redox reaction is:

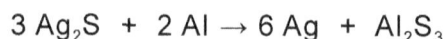

$$3\ Ag_2S\ +\ 2\ Al \rightarrow 6\ Ag\ +\ Al_2S_3$$

(For detailed instructions on how to do this at home or in the lab, an excellent description can be found at: http://scifun.chem.wisc.edu/homeexpts/tarnish.html)

In addition to reactions involving metals, redox reactions are responsible for transportation and keeping us warm; the burning of gasoline, wood, and other organic fuels are redox reactions. In these exothermic reactions, oxygen oxidizes large organic molecules to produce carbon dioxide and water. Because these reactions are exothermic, they can produce heat from a campfire or the energy can be harnessed in an engine and used to move vehicles.

Redox reactions are also extremely important in biology. In fact, life would not be possible without them. Our bodies burn organic molecules (mainly carbohydrates and fats, but also others, such as proteins and ethanol) to yield energy just as a campfire does, except that the body has set up a fine-tuned system of enzymes that work together to break the combustion reaction into tiny steps that yield usable energy rather than destructive heat. The actual chemical reactions involve many steps and are quite complex, but the general, overall reaction is:

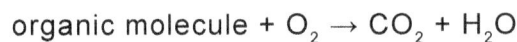

$$\text{organic molecule} + O_2 \rightarrow CO_2 + H_2O$$

Of course, to obtain a balanced equation, you would need to know which specific organic molecule is being burned. In the case of proteins burned as fuel, nitrogen-containing compounds will be an additional product.

Another practical, real-world use of redox reactions is batteries—whenever you pop a couple of AA batteries in a flashlight, use your cell phone, or start your car, you are harnessing the power of redox chemistry. This branch of redox chemistry is called electrochemistry.

Chapter 3.5 Electrochemistry

Electrochemistry deals with the flow of electricity generated by chemical reactions. The reactions either generate electrical current in galvanic (voltaic) cells, or are driven by an externally applied electrical potential in electrolytic cells. In electrochemical systems, chemical and electrical energy can be converted back and forth.

Both systems contain two electrodes. An electrode is a piece of conducting metal that is used to make contact with a nonmetallic material. One electrode is an **anode.** An **oxidation reaction** occurs at the anode, so electrons are removed from a substance there. The other electrode is a **cathode**. A **reduction reaction** occurs at the cathode, so electrons are added to a substance there. Electrons flow from anode to cathode. Visual representations of galvanic and electrolytic cells are tools of analysis to identify where half-reactions occur and the direction of current flow.

The emf or *electromotive force* of a cell measured under standard conditions is given the symbol E°cell. One can measure the cell potential, Ecell, in volts, of any galvanic cell with the aid of a potentiometer. A standard cell potential, E^O_{cell} , is the voltage generated by an galvanic cell at 100 kPa and 25° C when all components of the reaction are pure materials or solutes at a concentration of 1 M. Older textbooks may use 1 atm instead of 100 kPa. Standard solute concentrations may differ from 1 M for solutions that behave in a non-ideal way, but this difference is beyond the scope of AP chemistry.

Electrochemical systems

An **electrochemical cell** separates the half-reactions of a redox process into two compartments or half-cells. The whole system (both half-cells together) is referred to as a cell; the terms *electrochemical cell*, *galvanic cell*, and *voltaic cell* all refer to this type of system. Galvanic cells are set up so that electron transfer from the oxidation to the reduction reaction may only take place through an external circuit. The movement of electrons can thus be "harvested" as usable electricity.

In electrochemical cells, the two half reactions occur in different beakers (each beaker is a half-cell). Electrical neutrality is maintained in the half-cells by **ions migrating** through a **salt bridge**. A salt bridge in the simplest cells is an inverted U-tube filled with a non-reacting electrolyte and plugged at both ends with a material like cotton or glass wool that permits ion migration but prevents the electrolyte from falling out.

The **electromotive force** produced in this way is also called **voltage** or **cell potential** and is measured in volts. Electrons are allowed to leave the chemical process at the anode and permitted to enter at the cathode. The result is a **negatively charged anode and a positively charged cathode**.

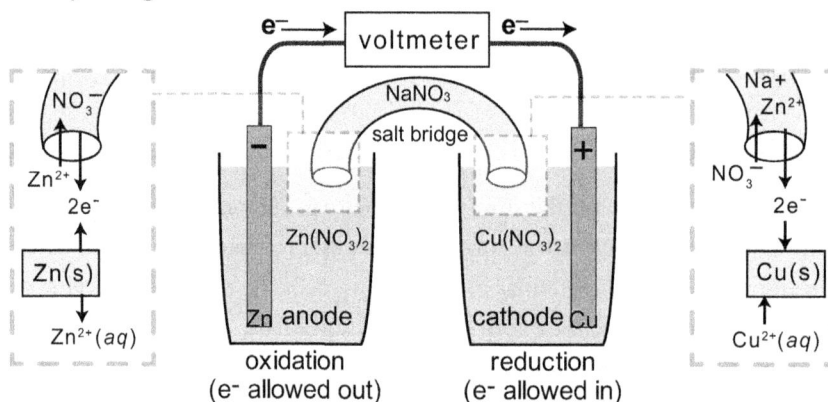

In the cell above, the spontaneous redox reaction:

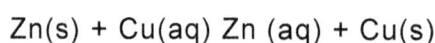

$$Zn(s) + Cu(aq) \rightleftharpoons Zn(aq) + Cu(s)$$

generates a voltage in the cell pictured above. The oxidation half-reaction,

$$Zn(s) \rightarrow Zn^{2+}(aq) + 2e^-,$$ occurs at the anode. Electrons are allowed to flow through a voltmeter before they are consumed by the reduction half-reaction at the cathode. Zinc dissolves away from the anode into solution, and copper from the solution builds up onto the cathode.

To maintain electrical neutrality in both compartments, positive ions (Zn and Na) migrate through the salt bridge from the anode half-cell to the cathode half-cell and negative ions (NO^-) migrate in the opposite direction.

A **battery** consists of one or more galvanic cells connected together.

An animation of an electrochemical cell and its relation to redox reactions can be found at: http://www.wiley.com/college/pratt/0471393878/student/review/redox/3_half_reactions.html

Electrolytic systems

Electrolysis, essentially the opposite of the process above, is a chemical process driven by a **battery** or another source of electromotive force. This source pulls electrons out of the chemical process at the anode and forces electrons into the cathode. The result is a **negatively charged cathode and a positively charged anode**.

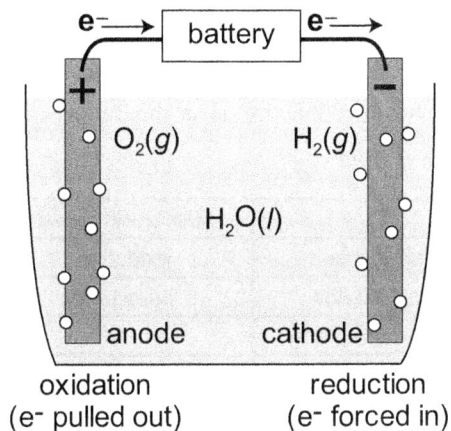

oxidation
(e⁻ pulled out)

reduction
(e⁻ forced in)

Electrolysis of pure water forms O bubbles at the anode by the oxidation half-reaction:

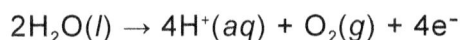

$$2H_2O(l) \rightarrow 4H^+(aq) + O_2(g) + 4e^-$$

and forms H bubbles at the cathode by the reduction halfreaction:

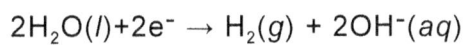

$$2H_2O(l)+2e^- \rightarrow H_2(g) + 2OH^-(aq)$$

The net redox reaction is:

$$2H_2O(l) \rightarrow 2H_2(g) + O_2(g)$$

Neither electrode took part in the reaction described above. An electrode that is only used to contact the reaction and deliver or remove electrons is called an **inert electrode.** An electrode that takes part in the reaction is called an **active electrode**.

Here is a summary of some key differences between galvanic and electrolytic cells:

Galvanic Cell	Electrolytic Cell
Converts chemical energy into electrical energy.	Converts electrical energy into chemical energy.
Based on spontaneous redox reactions ("generates energy")	Based on non-spontaneous redox reactions that take place only when energy is supplied ("uses up energy")
Different chemical changes occur in two beakers	Only one chemical compound undergoes decomposition.
Oxidation takes place at the anode; the anode is negative	Oxidation takes place at the anode; the anode is positive
Reduction takes place at the cathode; the cathode is positive	Reduction takes place at the cathode; the cathode is negative

One of the most confusing aspects of these two types of cells is the polarity of the anode and cathode. A summary of anode and cathode properties for both cell types is therefore contained in the table below. Importantly, while the electrode polarity is different in electrolytic vs. galvanic cells, **oxidation always occurs at the anode and reduction always occurs at the cathode**, in both electrolytic and galvanic cells. This conceptualization in terms of oxidation-reduction, rather than positive or negative electrode poles, is the essential concept the creators of the AP Chemistry exam want you to understand. One helpful way to remember this is to imagine a shrinking cat—cat for cathode--the cat is being reduced in size; this image can remind you that **reduction always occurs at the cathode**.

		Electrolytic cell	Electrochemical cell
Anode	half-reaction	oxidation	oxidation
	electron flow	pulled-out	allowed out
	electrode polarity	+	-
Cathode	half-reaction	reduction	reduction
	electron flow	forced in	allowed in
	electrode polearity	-	+

Electroplating is the process of depositing dissolved metal cations in a smooth even coat onto an object used as an active electrode. Electroplating is used to protect metal surfaces or for decoration. For example, to electroplate a metal surface with copper, a copper rod is used for the anode and the metal object is used for the cathode. CuSO (aq) or another substance with free Cu^{2+} ions is used in the electrolytic cell. Oxidation occurs at the anode and reduction occurs at the cathode.

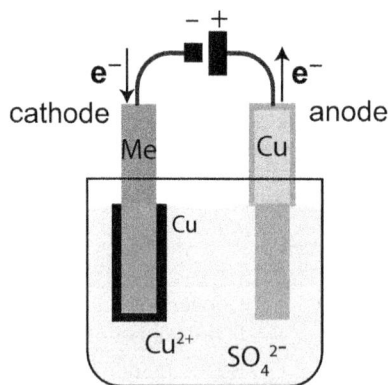

The reducing and oxidizing agents in a standard galvanic cell are depleted with time. This will result in "running down the battery". In a **rechargeable battery** (e.g., lead storage batteries in cars or plug-in chargers for AA batteries at home) the direction of the spontaneous redox reaction is reversed and **reactants are regenerated** when electrical energy is added into the system. A **fuel cell** has the same components as a standard galvanic cell except that **reactants are continuously supplied**.

Reduction potentials (cell potentials)

It is impossible to directly measure the potential of each individual half-cell. Therefore, chemists have compiled tables of half-cell potentials relative to the potential of the standard hydrogen electrode.

That is, all half-reaction potentials are relative to the reduction of H+ to the form H. This potential is assigned a value of zero:

For 2H$^+$(*aq* at 1M) + 2e   H$_2$(*g* at 199kPa), E^o_{red} + 0V

The emf measured when a metal / metal ion electrode is coupled to a hydrogen electrode under standard conditions is known as the standard electrode potential of that metal / metal ion combination.

Standard cell potentials are calculated from the **sum of the two half-reaction potentials** for the reduction and oxidation reactions occurring in the cell:

$$E^{\circ}_{cell} = E^{\circ}_{red}(\text{cathode}) + E^{\circ}_{ox}(\text{anode})$$

All half-reaction potentials are relative to the reduction of H$^+$ to form H. This potential is assigned a value of zero:

For 2H$^+$(*aq* at 1M) + 2e $\rightarrow$ H$_2$(*g* at 199kPa), E^o_{red} + 0V

The standard potential of an oxidation half-reaction $\boldsymbol{E^o_{ox}}$ **is equal in magnitude to the potential of the reduction half-reaction but has the opposite sign**.

All tables of standard reduction potentials list all half-reactions as reductions. The standard potential of an oxidation half-reaction $\boldsymbol{E^o_{ox}}$ **is equal in magnitude but has the opposite sign to the potential of the reverse reduction reaction.** These are sometimes referred to as **standard electrode potentials E°**. Therefore,

$$E^{\circ}_{cell} = E^{\circ}(\text{cathode}) \rightarrow E^{\circ}(\text{anode})$$

Looking at a table of standard cell reduction potentials one notices that the more electronegative species (those with the greatest attraction for electrons) are easily reduced, i.e. given an electron. The most electronegative element, F, has the largest reduction potential whereas one of the least electronegative elements, Li, has the smallest reduction potential.

On the AP exam you will be given any necessary values for the standard reduction potential of the half-reactions for which they are required or a table from which you can look up the values you need. These questions show up very often in the Free Response section of the exam.

It is now possible to calculate the cell potential, Eo cell, of any arbitrary redox reaction.

Example:

E° = 0.34 V   Cu^{2+} (*aq*) + 2e$^-$ $\rightarrow$ Cu (s)

E° = −0.76 V   Zn^{2+} (*aq*) + 2e$^-$ $\rightarrow$ Zn (s) find the standard cell potential of the system:

$$Zn(s) + Cu^{2+}(aq) \rightarrow Zn^{2+}(aq) + Cu(s)$$

Solution:

$$E^0_{cell} = E^0(\text{cathode}) - E^0(\text{anode})$$

$$= E^0\left(Cu^{2+}(aq) + 2e^- \rightarrow Cu(s) - E^0(Zn2+ +2e- \rightarrow Zn(s))\right)$$

$$= 0.34 \text{ V} - (-0.76 \text{ V}) = 1.10 \text{ V}.$$

Spontaneity

By knowing the sign of Eo cell, we can predict whether a reaction is spontaneous at standard conditions. If Eo cell is positive, then the reaction is spontaneous. Conversely, if Eo cell is negative, then the reaction is non-spontaneous as written but spontaneous in the reverse direction. If the E° value is negative, an outside energy source is necessary for the reaction to occur. In the above example, the E° is a positive 1.10 V, therefore this reaction is spontaneous.

Electrolytic cells use electricity to force non-spontaneous redox reactions to occur. **Galvanic cells generate electricity** by permitting spontaneous redox reactions to occur.

ΔG° (standard Gibbs free energy) is proportional to the negative of the cell potential for the redox reaction from which it is constructed. When ΔG° is negative, a process is spontaneous. When E° is positive, ΔG° is negative, and the movement of electrons will proceed spontaneously (electricity will be generated). ΔG° is discussed in detail in Chapter 5.

Real systems and concentration

Two major issues need to be taken into account when considering electrical potential in real systems. First, most real systems do not operate at standard conditions. Second, the concentration of the reactants has a large effect on potential. Potential can be thought of as a driving force towards equilibrium; the farther from equilibrium it is, the greater the potential will be. If this is not immediately obvious, think about a U-shaped pipe with water poured into one end. The water will flow upwards in the opposite pipe, but only until the water level is equal on both sides; at equilibrium, the difference in potential energy provided by gravity is no longer there. When reactants in a cell are near equilibrium, they will likewise show lowered potential.

Concentration cells are a special type of cell that contains the same type of ions in both electrodes, but in different concentrations. As the concentrations move towards equilibrium, a potential results. Note: The makers of the AP Chemistry exam consider the Nernst equation to be beyond the scope of the exam because they would rather you qualitatively understand the effect of concentration on potential; they will not ask you to determine this effect quantitatively through the Nernst equation.

Faraday's Laws

The stoichiometry of the redox reactions that form the basis of electrochemical cells can be determined using **Faraday's Laws**.

Faraday's First Law of Electrolysis states that the mass of a substance altered at an electrode during electrolysis is directly proportional to the quantity of electricity transferred at that electrode. Quantity of electricity refers to the quantity of electrical charge.

Faraday's Second Law of Electrolysis states that, for a given quantity of electric charge, the mass of an elemental material altered at an electrode is directly proportional to the element's molar mass divided by the change in oxidation state it undergoes upon electrolysis (often equal to its charge or valence).

Faraday's laws can be used to determine.

i.   The number of electrons transferred
ii. The mass of material deposited or removed from an electrode
iii. The current
iv. The time elapsed
v.   The charge of ionic species

The laws can be summarized in the following equation:

$$m = (Q/F)\,(M/z)$$

where m = mass of the substance altered at the electrode (g); Q = total electric charge passed through the substance (C); F = Faraday's constant (96,485 C mol^{-1}); M = molar mass of the substance (g/mol); z = number of electrons transferred per ion

OR $It = nF$

where I is the current, t is the time, n is the moles of electrons, and F is Faradays's constant.

The current is determined by:

$$I = q/t$$

where I is the current (in amperes, A), q is the charge in coulombs © and t is the time in seconds. (In other words, current = charge per second.)

It is particularly important to do a careful unit analysis when solving these types of problems.

Example 1:

Calculate the quantity of electricity necessary to generate 100.00 g of copper from a $CuSO_4$ solution.

Solution:

1.   Determine moles of copper plated out:

100.00 g divided by 63.546 g/mole = 1.573663 mol

"Production of heat or light, formation of a gas, and formation of a precipitate and/ or a color change are possible evidences that a chemical change has occurred."

2. Determine moles of copper plated out:

$$Cu^{2+} + 2e^- \rightarrow Cu$$

therefore, every mole of Cu plated out requires two moles of electrons

1.573663 mol x 2 = 3.147326 mol e⁻ required

3. Convert moles of electrons to Coulombs of charge:
3.147326 mol e⁻ x 96,485.309 C/mol = 3.0367 x 105 C

Example 2:

If a 2.00A current is run through a solution of iron (III) Chloride for 10 minutes, how much iron would plate out? (remember to convert time to seconds)

Solution:

Since 10 minutes is equal to 60 seconds,
Mass of iron =
600s * 2.5C/s * 1 mol e-/96500 C * 1 mol Fe/3 mol e- * 55.85 g/mol Fe

Mass of iron = 0.289 g Fe

Employing the Faraday equation one can also calculate the quantity of electricity (Coulombs) necessary to deposit or electroplate a metal onto another surface. Note – due to the complex calculations involved these questions routinely appear in the Free Response section of the exam.

Example 3:

A constant electric current deposits 0.360 g of silver metal in 12,960 seconds from a solution of silver nitrate. What is the current? What is the half reaction for the deposition of silver?

Solution:

1. Determing the moles of silver deposited:

0.3650 g divided by 107.8682 g/mol = 0.00338376 mol

2. Determin the moles of electrons required:
$$Ag+ + e^- \rightarrow Ag$$
therefore, every mole of Ag plated out requires one mole of electrons.
0.00338376 mol Ag x 1 mol e- / mol Ag = 0.00338376 mol e⁻ required

3. Determing the moles of silver deposited:
0.00338376 mol e⁻ times 96,485.309 C/mol = 326.483 C

4. current (remember that 1 A = 1 C/sec):
326.483 C / 12960 sec = 0.0252 amp

Chapter 3.6 Energy and the observation of chemical changes

Chemical changes can be observed in the laboratory on the macroscopic level; some examples of evidence that a chemical change has occurred are production of heat or light, production of a gas, formation of a precipitate, or color change. Physical changes, however, also can be observed macroscopically, and distinguishing between the two can be challenging. This is one reason for the importance of understanding different types of chemical reactions; in particular, precipitation, redox, and acid-base reactions. (Acid-base reactions are considered in detail in Chapter 6. In order to be able to distinguish between physical, chemical, and ambiguous changes, knowledge of both the chemical rearrangements (the nature of chemical reactions) and macroscopic patterns associated with certain types of reactions is necessary.

One method of macroscopic observation is measurement of changes in temperature, which represents the heat energy in a system. Chemical energy is the **energy stored in substances due to the arrangement of atoms** within the substance. When atoms are rearranged during chemical reactions, energy is either released or consumed. It is the energy released from chemical reactions that fuels our cars and powers our bodies. Most of the electricity produced on the planet comes from chemical energy released by the burning of petroleum, coal, and natural gas. ATP is the molecule used by our bodies to carry chemical energy from cell to cell.

transformations may be observed in several ways and typically involve a change in energy."

Bond energies

Temperature is among the most useful ways to detect chemical change because of its relation to energy. The energy in molecules is located in the **bonds between the atoms,** and breaking these bonds requires energy. Once broken apart, the atoms, ions, or molecules rearrange themselves to form new substances, making new bonds. Making these new bonds releases energy.

If during a chemical reaction, **more energy is needed to break the reactant bonds than is released when the products form new bonds,** the reaction is **endothermic** and heat is absorbed. The environment becomes colder.

On the other hand, if **more energy is released when the products form new bonds than is needed to break the reactant bonds,** the reaction is **exothermic** and the excess energy is released to the environment as heat. The temperature of the environment goes up.

"Net changes in energy for a chemical reaction can be endothermic or exothermic."

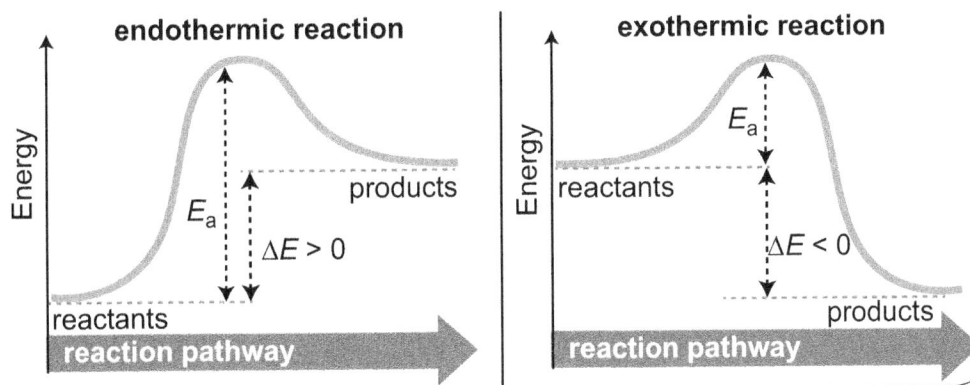

endothermic reaction — Energy vs reaction pathway: E_a, $\Delta E > 0$, reactants, products

exothermic reaction — Energy vs reaction pathway: E_a, $\Delta E < 0$, reactants, products

The total energy absorbed or released in the reaction can be determined by using **heats of formation** or **bond energies**. The total energy change of the reaction is equal to the total energy of all of the bonds of the products minus the total energy of all of the bonds of the reactants.

Propane (CH) is a common fuel used in heating homes and backyard grills. When burned, the combustion reaction shown below takes place and excess energy is released and used for heating or cooking:

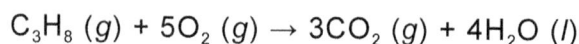

$$C_3H_8 \ (g) + 5O_2 \ (g) \rightarrow 3CO_2 \ (g) + 4H_2O \ (l)$$

The total energy of the products is found from the bonds in the carbon dioxide molecules and the water molecules:

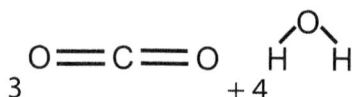

$$3 \quad O = C = O \quad + 4 \quad \overset{O}{\underset{H \quad H}{\diagup\diagdown}}$$

or 6 C=O bonds and 8 H–O bonds.

A table of bond energies gives the following information:

C=O   743 kJ/mol

H–O   463 kJ/mol

For these molecules there would be.

$$(6 \times -743 \text{ kJ/mol}) + (8 \times -463 \text{ kJ/mol}) = -8162 \text{ kJ}$$

energy released when these molecules form. Negative values are used to indicate energy released for this exothermic process of bond formation.

While a negative **enthalpy** value (meaning heat energy is released) favors a chemical reaction, there is another quality that can drive a reaction forward: **entropy**. Entropy is the degree of randomness in a system. Any system will naturally tend toward maximum entropy, so if a given reaction increases the entropy in the system, this will also tend to drive the reaction forward.

A detailed discussion of **thermodynamics**, the branch of chemistry concerned with changes in energy, can be found in Chapter 5.

Big Idea 3: Practice Problems

1. **Balance the following redox reaction:**

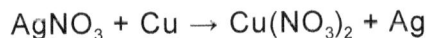

$$AgNO_3 + Cu \rightarrow Cu(NO_3)_2 + Ag$$

Solution: Assign oxidation numbers to identify which atoms are losing and gaining electrons:

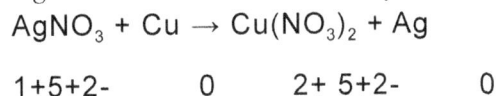

$$AgNO_3 + Cu \rightarrow Cu(NO_3)_2 + Ag$$

$$1+5+2- \quad\quad 0 \quad\quad 2+ 5+2- \quad\quad 0$$

Identify those atoms gaining and losing electrons:

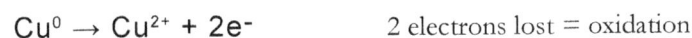

$$Ag^+ + e^- \rightarrow Ag° \quad\quad\quad \text{1 electron gained = reduction}$$

$$Cu^0 \rightarrow Cu^{2+} + 2e^- \quad\quad\quad \text{2 electrons lost = oxidation}$$

Balance the atoms and electrons:

$$2AgNO_3 + 2e^- \rightarrow 2Ag°$$

$$Cu^0 \rightarrow Cu(NO_3)_2 + 2e^-$$

Balance the electrons:

$$AgNO_3 + 1e^- \rightarrow Ag°$$

$$Cu^00 \rightarrow Cu(NO_3)_2 + 2e^-$$

1 electron gained and 2 electrons lost. Needs to be equal so 2 electrons need to be gained.

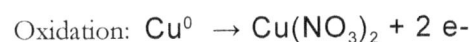

$$2 [Ag + 1 e^- \rightarrow Ag^0] = 2AgNO_3 + 2 e^- \rightarrow 2 Ag^0+$$

Reduction: $2AgNO_3 + 2e^- \rightarrow 2Ag^0$

Oxidation: $Cu^0 \rightarrow Cu(NO_3)_2 + 2 e^-$

Balance the half reactions by adding missing elements. Ignore elements whose oxidation number does not change. Add H2O for oxygen and H+ for hydrogen.

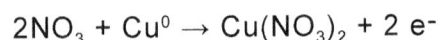

$$2 AgNO_3 + 2e^- \rightarrow 2 Ag° + 2 NO_3$$

$$2NO_3 + Cu^0 \rightarrow Cu(NO_3)_2 + 2 e^-$$

Put the two half reactions together and add the species. Cancel out the species that occur in both the reactants and products. The balanced reaction is:

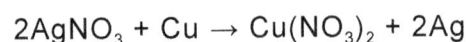

$$2AgNO_3 + Cu \rightarrow Cu(NO_3)_2 + 2Ag$$

2. **Balance the reaction:**

$$Ag_2S + HNO_3 \rightarrow AgNO_3 + NO + S + H_2O$$

Solution: Assign oxidation numbers:

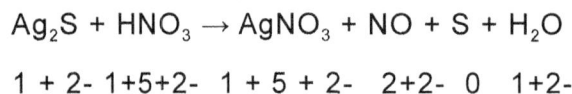

$$Ag_2S + HNO_3 \rightarrow AgNO_3 + NO + S + H_2O$$

1 + 2- 1+5+2- 1 + 5 + 2-   2+2- 0   1+2-

Identify those atoms gaining and losing electrons:

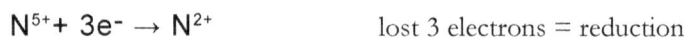

$$S^{2-} \rightarrow S^0 + 2e^-$$ gained 2 electrons = oxidation

$$N^{5+} + 3e^- \rightarrow N^{2+}$$ lost 3 electrons = reduction

Balance the atoms:

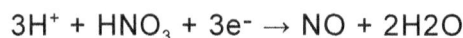

$$2NO_3^- + Ag_2S \rightarrow S + 2e^- + 2AgNO_3$$

$$3H^+ + HNO_3 + 3e^- \rightarrow NO + 2H2O$$

Balance electrons lost and gained:

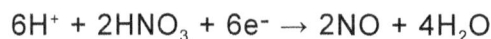

$$6NO_3^- + 3Ag_2S \rightarrow 3S + 6e^-- + 6AgNO_3$$

$$6H^+ + 2HNO_3 + 6e^- \rightarrow 2NO + 4H_2O$$

Put the two half reactions together and add the species. Cancel out the species that occur in both the reactants and products. The balanced reaction is:

$$3Ag_2S + 8HNO_3 \rightarrow 6AgNO_3 + 2NO + 3S + 4H_2O$$

To balance a redox reactyion which occurs in basic solution is very similar to balancing a redox reaction which occurs in acidic conditions. First, balance the reaction as you would for an acidic solution and then adjust for the basic solution.

3. **Solid chromium(III) hydroxide, $Cr(OH)_3$, reacts with aqueous chlorate ion, ClO_3-, in basic conditions to form chromate ions, CrO_4^{2-}, and chloride ions,**

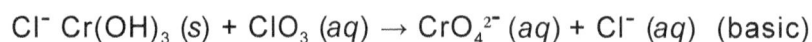

Cl^- $Cr(OH)_3$ (s) + ClO_3 (aq) $\rightarrow CrO_4^{2-}$ (aq) + Cl^- (aq) (basic)

Balance this reaction:

Solution: Write the half-reactions:

$$Cr(OH)_3 \text{ (s)} \rightarrow CrO_4^{2-} \text{ (aq)}$$

$$ClO_3^- \text{ (aq)} \rightarrow Cl^- \text{ (aq)}$$

Balance the atoms in each half-reaction. Use H2O to add oxygen a toms and H+ to add hydrogen atoms:

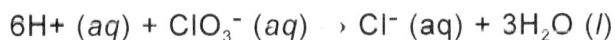

$$H_2O\ (l) + Cr(OH)_3\ (s) \rightarrow CrO_4^{2-}\ (aq) + 5H^+\ (aq)$$

$$6H+\ (aq) + ClO_3^-\ (aq)\ \rangle\ Cl^-\ (aq) + 3H_2O\ (l)$$

Balance the charges of both half-reactions by adding electrons:

$$H_2O\ (l) + Cr(OH)_3\ (s) \rightarrow CrO_4^{2-}\ (aq) + 5H^+\ (aq) + 3e^-$$

$$6e^- + 6H\ (aq) + ClO_3^-\ (aq)) \rightarrow Cl^-\ (aq) + 3H_2O\ (l)$$

The number of electrons lost must equal the number of electrons gained so multiply each half- section by a number that will give equal numbers of elextrons lost and gained:

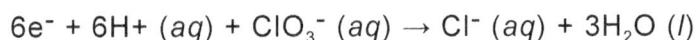

$$2H_2O\ (l) + 2Cr(OH)_3\ (s) \rightarrow 2CrO_4^{2-}\ (aq) + 10H^+\ (aq) + 6e^-$$

$$6e^- + 6H+\ (aq) + ClO_3^-\ (aq) \rightarrow Cl^-\ (aq) + 3H_2O\ (l)$$

Add the two half-reactions together, canceling out species that appear on both sides of the reaction. Since the reaction occurs in basic solution and there are 4 H^+ ions on the right side, 4 OH need to be added to both sides. Combine the H+ and OH- where appropriate to make water molecules to write the final balanced equation:

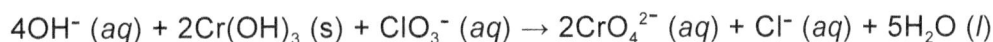

$$4OH^-\ (aq) + 2Cr(OH)_3\ (s) + ClO_3^-\ (aq) \rightarrow 2CrO_4^{2-}\ (aq) + Cl^-\ (aq) + 5H_2O\ (l)$$

4. **Balance the following redox reaction:**

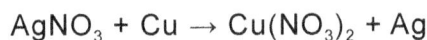

$$AgNO_3 + Cu \rightarrow Cu(NO_3)_2 + Ag$$

Solution: First, assign oxidation numbers to each atom in order to identify which are losing or gaining electrons.

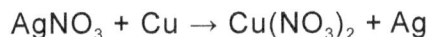

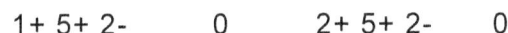

$$AgNO_3 + Cu \rightarrow Cu(NO_3)_2 + Ag$$

1+ 5+ 2-   0   2+ 5+ 2-   0

Then identify the atoms that are gaining and losing electrons:

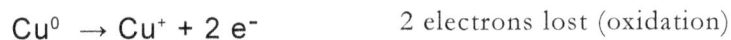

$$Ag + 1\ e^- \rightarrow Ag^0 \qquad \text{1 electron gained (reduction)}$$

$$Cu^0 \rightarrow Cu^+ + 2\ e^- \qquad \text{2 electrons lost (oxidation)}$$

Balance each of the atoms and electrons:

$$AgNO3 + 1\ e\text{-} \rightarrow Ag^0$$

$$Cu^0 \rightarrow Cu(NO_3)_2 + 2\ e^-$$

$$Cu^0 \rightarrow Cu^+ + 2\ e^-$$

There is one electron gained and two electrons lost. This must be balanced, so 2 electrons must be gained.

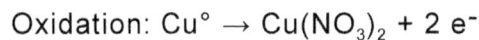

$$2 [Ag + 1 e^- \rightarrow Ag^0] = \textbf{2 AgNO}_3 \textbf{+ 2 e}^- \rightarrow \textbf{2 Ag}^{0+}$$

Reduction: $2 AgNO_3 + 2 e^- \rightarrow 2 Ag^+$

Oxidation: $Cu^\circ \rightarrow Cu(NO_3)_2 + 2 e^-$

Balance the half reactions by adding missing elements. Ignore elements whose oxidation number does not change. Add H2O for oxygen and H+ for hydrogen.

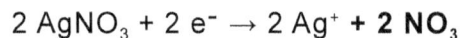

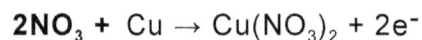

$$2 AgNO_3 + 2 e^- \rightarrow 2 Ag^+ \textbf{+ 2 NO}_3$$

$$\textbf{2NO}_3 \textbf{+ } Cu \rightarrow Cu(NO_3)_2 + 2e^-$$

Put the two half reactions together and add the species. Cancel out the species that occur in both the reactants and products.

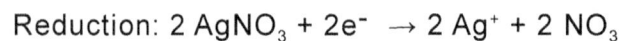

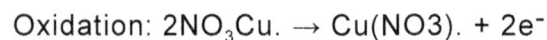

Reduction: $2 AgNO_3 + 2e^- \rightarrow 2 Ag^+ + 2 NO_3$

Oxidation: $2NO_3Cu. \rightarrow Cu(NO3). + 2e^-$

The balanced reaction is then:

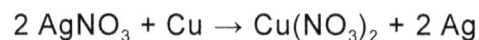

$$2 AgNO_3 + Cu \rightarrow Cu(NO_3)_2 + 2 Ag$$

Multiple Choice Questions

Instructions: This section consists of some practice multiple choice questions as well as one Free Response question related to this chapter. You may NOT use a calculator for the Multiple Choice. However calculators are permitted for the Free Response.

1. Write a balanced net ionic equation for the following reaction:

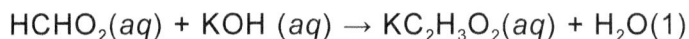

$$HCHO_2(aq) + KOH\ (aq) \rightarrow KC_2H_3O_2(aq) + H_2O(1)$$

   A. $H+(aq) + OH-(aq) \rightarrow H_2O(1)$

   B. $C_2H_3O_2^-(aq) + K+(aq)^- \rightarrow KC_2H_3O_2(aq)$

   C. $H+(aq) + KOH\ (aq) \rightarrow K+\ (aq) + H_2\ O(1)$

   D. $HC_2H_3O_2(aq) + OH-(aq) \rightarrow C_2H_3O-2(aq) + H_2O(1)$

2. In order for an ionic redox reaction to be balanced, what quantities must be equal on the reactants and products side of the equation?

   A. electron gain must equal electron loss

   B. atoms of each reactant element must equal atoms of products

   C. total ionic charge for the reactants must equal total for the products

   D. all of the above

3. The percent composition of O in KNO_3 is approximately

   A. 33%

   B. 20%

   C. 48%

   D. 38%

4. All three substances (CHO, CHO, CHCOOH) have the same percentage composition because

   A. They were weighed out precisely

   B. They have the same Empirical Formula

   C. They are alternate ways of writing the same compound

   D. They are isomers of the same compound

5. An oxide of an element with a valance of 6 contains 48% oxygen. Considering the atomic weight of this element, identify the element..

   A. Cr

   B. O

   C. Te

   D. Si

6. What is the chemical composition of ammonium carbonate $(NH_4)2CO_3$?

    A. 60%N, 16%H, 6%C, 25%O

    B. 50% N, 10% H, 20% C, 20% O

    C. 30%N, 8%H, 12%C, 50%O

    D. 30% N, 16% H, 4% C, 50% O

7. What is the term for a process characterized by the gain of electrons?

    A. electrochemistry

    B. oxidation

    C. redox

    D. reduction

8. After balancing the following redox reaction, what is the coefficient of H?

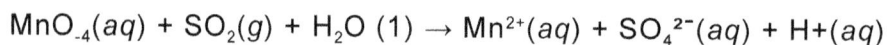

    $$MnO_{-4}(aq) + SO_2(g) + H_2O\ (1) \rightarrow Mn^{2+}(aq) + SO_4^{2-}(aq) + H+(aq)$$

    A. 1

    B. 2

    C. 4

    D. 6

9. After balancing the following redox reaction in basic solution, what is the coefficient of OH-? (add HOH and OH-1ions as needed)

    $$MnO_{-4}(aq) + SO_3 \rightarrow MnO_2(aq) + SO_4^{2-}(aq)$$

    A. 1

    B. 2

    C. 3

    D. 6

10. What is the molar concentration of water molecules in a beaker of water? (Recall that the density of water is 1.00 g/mL).

    A. 0.0555 M

    B. 1.00 M

    C. 18.0 M

    D. 55.5 M

11. In order to determine the quantity of electricity (Coulombs) necessary to deposit 100.00 g of copper from a $CuSO_4$ solution, which of the following calculations should be used ?

A. 100.00 / 63.5 * 2 * 96500

B. 100 * 2 * 96500

C. 100/ 63.5 * 96500

D. 100. /63.5 * 2

12. How many hours will it take to plate out 5790 C using a current of 0.2 amps

A. 28900

B. 8

C. 1158

D. Cannot be determine from the given information

13. For the reaction $2\ Al + 3\ I_2 \rightarrow 2\ AlI_3$ what is the limiting reagent if 1.2 mol of Al react with 2.4 mol of I?

A. Al

B. I

C. none of the above

D. cannot be determined

14. Based on the balanced equation:

$$CH + 6O \rightarrow 4CO + 4HO$$

calculate the number of excess reagent units remaining when 28 CH molecules and 228 O molecules react?

A. There will be 60 units of exces. oxygen

B. There will be 38 units of excess oxygen

C. there will be 28 units of excess butane

D. there will be 10 units of butane

15. Consider the overall electrochemical reaction of a discharged lead battery:

$$Pb + PbO_2 + H_2SO_4 \rightarrow 2PbSO_4 + 2H_2O$$

Which species is oxidized during the process of battery discharge?

A. Pb

B. H_2O

C. PbO_2

D. $PbSO_4$

16. In $KHCO_3$, what is the oxidation number of carbon?

    A.   +8
    B.   +1
    C.   +4
    D.   -2

17. Consider the following reaction: Hg_2+ (aq) + 2I- (aq) → Hg(l) + I_2(s)

    The potential for the overall reaction is +0.32 V and the potential of redox couple Hg/ Hg is 0.855 V. What is the reduction potential of the redox couple I/I⁻?

    A.   -1.120 V
    B.   - 0.535 V
    C.   +0.400 V
    D.   +1.120 V

18. In the reaction of formation of CO_2 :

    $C(s) + O_2(g) → CO_2(g)$. If 34 grams of CO_2-
    is predicted what is the % yield ?

    A.   44%
    B.   77%
    C.   100%
    D.   38%

19. In the reaction:

    $HF (aq) + H_2O (l) ↔ F^- (aq) + H_3O+ (aq)$

    A.   HF gains a proton and is oxidized
    B.   HF is the Bronsted acid
    C.   HF gains an electron and is reduced
    D.   H0 is the he conjugate base of HF

20. In order for an ionic redox reaction to be balanced, what quantities must be equal on the reactants and products side of the equation?

    A.   electron gain must equal electron loss
    B.   atoms of each reactant element must equal atoms of products
    C.   total ionic charge for the reactants must equal total for the products
    D.   all of the above

Free Response Question (Big Idea #3)

Problem 1: A certain metal hydroxide, $M(OH)_2$, contains 32.80% oxygen by mass. What is the density of the metal M?

Answers to Sample Questions (Big Idea #3)

1. **Answer D.**

2. **Answer D.**

3. **Answer C.**
   Molar mass = 101.1 g/mol
   Potassium: (39.10 / 101.1) x 100 = 38.67%
   Nitrogen: (14.01 / 101.1) x 100 = 13.86%
   Oxygen: (48.00 / 101.1) x 100 = 47.48%

4. **Answer B.**

5. **Answer A.**
   An element, M, with a valence of +6 will form an oxide with the formula MO.
   Let us assume 100 g of the compound is present. Therefore, 48 g is oxygen and 52 g is M.
   The moles of oxygen → 48 g / 16 g/mol = 3 mol
   From the 1:3 molar ratio of M to O, we have this:
   1 is to 3 as x is to 3 moles
   x = 1 mole (This is how many moles of M are in the 100 g of MO.)
   52 g / 1 mol = 52 g/mol ← the atomic weight of M
   The element is chromium.

6. **Answer C.**
   $(NH4)2CO3$ contains 2 N, 8 H, 1 C, and 3 O atoms.

   $$\frac{2 \text{ mol N}}{\text{mol } (NH_4)CO_3} \times \frac{14.0 \text{ g N}}{\text{mol N}} = 28.0 \text{ g N/mol } (NH_4)CO_3$$

   $$8(1.0) = 8.0 \text{ g H/mol } (NH_4)CO_3$$

   $$1(12.0) = 12.0 \text{ g C/mol } (NH_4)CO_3$$

   $$3(16.0) = 48.0 \text{ g O/mol } (NH_4)CO_3$$

   Sum is $\overline{96.0 \text{ g } (NH_4)CO_3/\text{mol } (NH_4)CO_3}$

   $$\%N = \frac{28.0 \text{ g N/mol } (NH_4)_2CO_3}{96.0 \text{ g } (NH_4)_2CO_3/\text{mol } (NH_4)_2CO_3} = 0.292 \text{ g N/g } (NH_4)_2CO_3 \times 100\% = 29.2\%$$

   $$\%H = \frac{8.0}{96.0} \times 100\% = 8.3\% \quad \%C = \frac{12.0}{96.0} \times 100\% = 12.5\% \quad \%O = \frac{48.0}{96.0} \times 100\% = 50.0\%$$

7. **Answer D.**

8. **Answer C.**

9. **Answer B.**

10 **Answer D.**

11. **Answer A.**
    100.00 g divided by 63.546 g/mole = 1.573663 mol
    Determine moles of electrons required:

    $$Cu + 2e^- \rightarrow Cu$$

    therefore, every mole of Cu plated out requires two moles of electrons.
    1.573663 mol x 2 = 3.147326 mol e⁻ required
    Convert moles of electrons to Coulombs of charge:
    3.147326 mol e⁻ x 96,485.309 C/mol

12. Convert to seconds required to deliver the Coulomb. (remember, 1 A = 1 C/sec):
    5789.12 C divided by 0.200 C/sec = 28945.6 sec
    Convert seconds to hours:
    28945.6 sec divided by 3600 sec/hr = 8.04 hours

13. **Answer A.**
    For aluminum 1.2 / 2 mo. = 0.6
    For . 2.4/. = 0.6
    Therefore you will run out of Aluminum before running out of Iodine.

14. **Answer A.**
    1. Determine the limiting reagent:
       butane: 28/1 = 28
       oxygen: 228 / 6 = 38
       Butane is the limiting reagent.
    2. Determine how much oxygen
       reacts with 28 C4H8 molecules:
       the butane:oxygen molar ratio is 1:6
       28 x 6 = 168 oxygen molecules react
    3. Determine excess oxygen:
       228 - 168 = 60

15. **Answer A.** Explanation: During the discharge process, at the anode Pb is oxidized to
    $PbSO_4$ ($Pb(s) + SO_4^{2-}(aq) \rightarrow PbSO_4(s) + 2e^-$) and the oxidation number of Pb changes from
    0 to +2. At the cathode $PbO_2(s)$ is reduced ($PbO_2(s) + SO_4^{2-}(aq) + 4H+(aq) + 2e^- \rightarrow$
    $PbSO_4(s) + 2H^2O(\ell)$.

16. **Answer D.**
    Explanation: The sum of the oxidation numbers is a neutral compound equals zero. The oxidation
    number of oxygen is always -2, except in peroxide where the oxidation number of -1 or when oxygen
    combines with fluorine. In O, O (-2) × 3 oxygen atoms equals a charge of -6. K is always +1 and when

H acts as a metal, the charge is +1.

consequently, +2 plus -6 results in a charge of -4. Therefore each carbon atom must have a charge of +4 to result in 0 charge.

17. **Answer B.**
    E°cell = E° ox + E°red = 0.855V + (-0.535V) = 0.32

18. **Answer B.**
    Normally we expect a 1 mol yield of carbon dioxide for every mol of carbon burned. If you burn 12 grams of carbon to make CO, then amount of carbon dioxide expected is one mol of CO or 44 grams of CO. If you only get 34g then %-yield

    Percent yield = 100 x (34 grams CO2 actual / 44 grams CO predicted) = 77 %

19. **Answer B.**

20. **Answer D.**

Answer to Free Response Question (Big Idea #3) _____

Problem #1:

   A certain metal hydroxide, M(OH)2, contains 32.80% oxygen by mass.
   What is the identity of the metal M?

Solution #1:

1. Write an expression for the molar mass of the compound:
    Let M = the molar mass of the metal.
    M + (2 x 16.00) + (2 x 1.01)
    M + 34.02

2. Write an expression for the given mass percent of oxygen:
    [32.00 / (M + 34.02)] x 100% = 32.8%

3. Algebra!
    32.00 = (0.3280)(M + 34.02)
    32.00 = 0.3280M + 11.16
    20.84 = 0.3280M
    M = 63.54
    The metal is copper.

Chapter 4: Big Idea #4
Rates of chemical reactions

Big Idea 4: Rates of chemical reactions are determined by details of the molecular collisions.

What you should already know:
- moles, mass, volume, and pressure
- math as reviewed in this book

What you will learn:
- reaction rate
- factors affecting the rate constant
- spectroscopy and Beer's Law
- reaction mechanisms
- activation energy
- catalysis

The rate of a reaction is influenced by the concentration or pressure of reactants, the phase of the reactants and products, and environmental factors such as temperature and solvent."

Chapter 4.1 Obtaining reaction rates from concentration data

The reaction rate is the rate of change in the concentration of reactants and products over time. Reaction rates are determined experimentally, but can also be predicted on the basis of previous experimental data, the concentration (or pressure) of reactants, and environmental factors such as temperature. Reaction rates are also influenced by the phase of the reactants and products.

Just as the speed of a car is measured by its change in position with time using units of miles per hour, the speed of a chemical reaction is usually measured by a change in the concentration of a reactant or product over time using units of **molarity per second** (M/s). The molarity of a chemical is represented in mathematical equations using brackets.

The **average reaction rate** is the change in concentration of either reactant or product per unit time during a time interval:

"Reaction rates that depend on temperature and other environmental factors are determined by measuring changes in concentrations of reactants or products over time."

$$\text{Average reaction rate} = \frac{\text{Change in concentration}}{\text{Change in time}}$$

Reaction rates are positive quantities. Product concentrations increase and reactant concentrations decrease with time, so a different formula is required depending on the identity of the component of interest:

$$\text{Average reaction rate} = \frac{\text{product}_{final} - \text{product}_{initial}}{\text{time}_{final} - \text{time}_{initial}}$$

$$= \frac{\text{reactant}_{initial} - \text{reactant}_{final}}{\text{time}_{final} - \text{time}_{initial}}$$

The **reaction rate** at a given time refers to the **instantaneous reaction rate**. This is found from the absolute value of the **slope of a curve of concentration vs. time**. The **initial reaction rate** is the instantaneous reaction rate at the beginning of the reaction. An estimate of the reaction rate at time t may be found from the average reaction rate over a small time interval surrounding t. For those familiar with calculus notation, the following equations define reaction rate, but calculus is not needed:

$$\text{Reaction rate at time } t = \frac{d[\text{product}]}{dt} = -\frac{d[\text{reactant}]}{dt}$$

Example:

The following concentration data describes the decomposition of N_2O_5 according to the reaction $2N_2O_5 \rightarrow 4NO_2 + O_2$.

Determine the average reaction rate from 1000 to 5000 seconds and the instantaneous reaction rate at 0 and at 4000 seconds.

Time (sec)	$[N_2O_5]$ (M)
0	0.0200
1000	0.0120
2000	0.0074
3000	0.0046
4000	0.0029
5000	0.0018
7500	0.0006
10000	0.0002

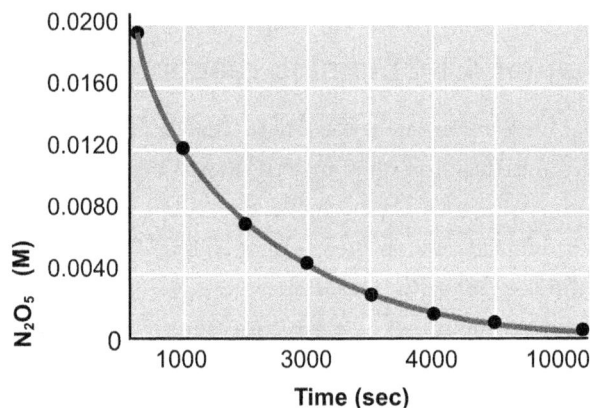

Solution: Average reaction rate from 0 to 7500 seconds is found from:

$$\frac{\text{reactant}_{initial} - \text{reactant}_{final}}{\text{time}_{final} - \text{time}_{initial}} = \frac{0.0120\ M - 0.0018\ M}{5000\ \sec - 1000\ \sec} = 2.55 \times 10^{-6}\ \frac{M}{s}$$

Instantaneous reaction rates are found by drawing lines tangent to the curve, finding the slopes of these lines, and forcing these slopes to be positive values.

At 0 seconds

$$\text{rate} = \text{slope} = \frac{0.0200M}{2000\ s}$$

$$= 1.00 \times 10^{-5}\ \frac{M}{s}$$

At 4000 seconds

$$\text{rate} = \text{slope} = \frac{0.0090M}{6000\ s}$$

$$= 1.5 \times 10^{-6}\ \frac{M}{s}$$

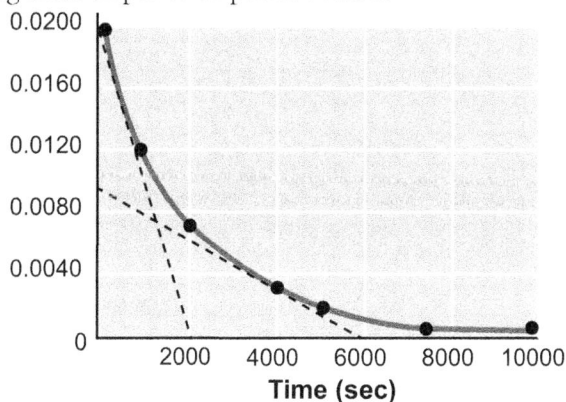

Spectroscopy and Beer's Law.

A number of experimental techniques are used to determine reaction rates. Either the loss of reactants or the gain of products must be measured. One way to do this is through spectroscopy and calculation using Beer's Law (also known as the Beer–Lambert Law and the Beer–Lambert–Bouguer Law). Since most compounds absorb light (within and outside the visible range), and specific compounds absorb light at specific wavelengths, measuring the amount of light transmitted through a substance or absorbed by the substance provides a way to measure the concentration of the substance. For example, DNA and RNA absorb UV light with a wavelength of 260 nm. If light at this wavelength is shined through a sample, the amount of light that passes through can be measured as the optical density of the sample. Through Beer's Law, this absorbance value can be converted into a measurement of concentration through the equation:

$$A = ebc$$

where

A = the absorbance of the light [the spectrophotometer reading (unitless)]

e = the molar absorptivity [units: L/(mol*cm)]

b = the path length of the sample - that is, the width or diameter of the cuvette in which the sample is contained [units: cm]

c = the concentration of the compound in solution [units: mol/L]

Spectrophotometry can also clarify the order of a reaction. A plot of absorbance vs. time will give a straight line for a zero order reaction and curved lines for higher order reactions. Here are the shapes of absorbance plots of zero, first, and second order reactions:

Zero order reaction:

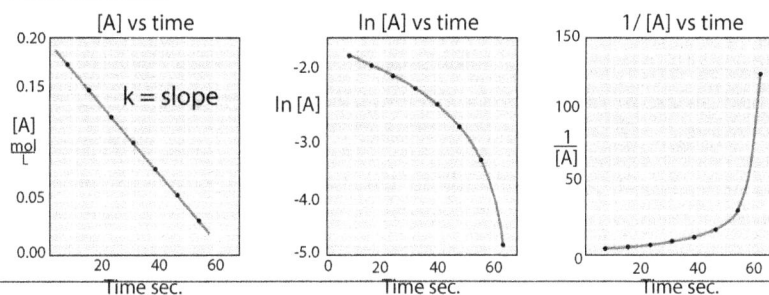

First order reaction:

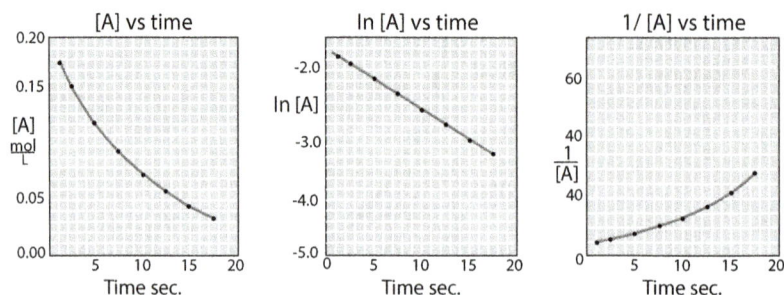

[A] vs time	ln [A] vs time	1/ [A] vs time

Second order reaction:

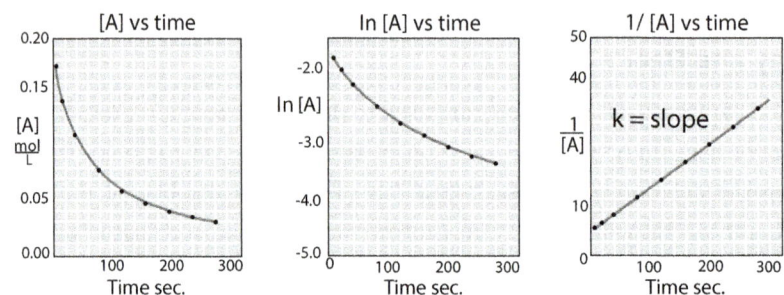

[A] vs time	ln [A] vs time	1/ [A] vs time

k = slope

Chapter 4.2. Deriving rate laws from reaction rates _____

A **rate law** is an **equation relating a reaction rate to concentration**. The rate laws for most reactions discussed in first-year college chemistry are of the form:

$$\text{Rate} = k\left[\text{reactant 1}\right]^{a}\left[\text{reactant 2}\right]^{b}\ldots$$

In the above general equation, k is called the **rate constant**, and a and b are called **reaction orders**. The rate constant is an important, measurable quantity, and is determined experimentally. Rate constants are highly variable—differing by orders of magnitude—because reaction rates vary widely. Factors affecting the rate constant include temperature, surface area, the presence of a catalyst (described in detail later in this chapter), and the nature of the solvent. Most reactions considered in introductory chemistry have a reaction order of zero, one, or two. The sum of all reaction orders for a reaction is called the **overall reaction order**. Rate laws cannot be predicted from the stoichiometry of a reaction. They must be determined by experiment or derived from knowledge of reaction mechanism.

If a reaction is **zero order** for a reactant, the concentration of that reactant has no impact on the rate as long as some reactant is present. If a reaction is first order for a reactant, the reaction rate is proportional to the reactant's concentration. For a reaction that is second order with respect to a reactant, **doubling that reactant's concentration increases reaction rate by a factor of four**. Rate laws are determined by finding the appropriate reaction order describing **the impact of reactant concentration on reaction rate**.

"The rate law shows how the rate depends on reactant concentrations."

Reaction rates typically have units of M/s (moles/liter-sec) and concentrations have units of M (moles/liter). For units to cancel properly in the expression above, the units of the rate constant k must vary with overall reaction order as shown in the following table. The value of k may be determined by finding the slope of a plot charting a function of concentration against time. These functions may be memorized or computed using calculus.

Overall reaction order	Units of rate constant k	Method to determine k for rate laws with one reactant
0	M/sec	–(slope) of a chart of [reactant] vs. t
1	sec^{-1}	–(slope) of a chart of ln [reactant] vs. t
2	$M-1sec^{-1}$	slope of a chart of 1/[reactant] vs. t

As an alternative to using the rate constant k, the course of first order reactions may be expressed in terms of a **half-life**, $t_{halflife}$. The half-life of a reaction is the time required for reactant concentration to reach half of its initial value. It is inversely proportional to the rate constant k, and is independent of reactant concentration. The rate of reaction of radioactive decay processes is often expressed in terms of half-life.

Example: Derive a rate law for the reaction

$$2N_2O_5 \rightarrow 4NO_2 + O_2$$

using data from the previous example.

Solution: Three methods will be used to solve this problem.

1. In the previous example, we found the following two instantaneous reaction rates

Time (sec)	$[N_2O_5]$ (M)	Reaction rate (M/sec)
0	0.0200	1.00×10^{-5}
4000	0.0029	1.5×10^{-6}

A decrease in reactant concentration to 0.0029/0.0200=14.5% of its initial value led to a nearly proportional decrease in reaction rate to 15% of its initial value. In other words, reaction rate remains proportional to reactant concentration. The reaction is first order:

$$Rate = k\left[N_2O_5\right].$$

We may estimate a value for the rate constant by dividing reaction rates by the concentration:

$$k_{first\ order} = Rate\ /\ [N_2O_5]s$$

Time (sec)	$[N_2O_5]$ (M)	Reaction rate (M/sec)	$k(sec^{-1})$
0	0.0200	1.00×10^{-5}	5.00×10^{-4}
4000	0.0029	1.5×10^{-6}	5.2×10^{-4}

2. We could estimate this rate constant by finding **average reaction rates** in each small time interval and assuming this rate occurs halfway between the two concentrations:

Time (sec)	$[N_2O_5]$ (M)	Average rate (M/sec)	Halfway $[N_2O_5]$ (M)	$k(sec^{-1})$
0	0.0200	8.00×10^{-6}	0.0160	5.00×10^{-4}
1000	0.0120	4.6×10^{-6}	0.0097	4.7×10^{-4}
2000	0.0074	2.8×10^{-6}	0.0060	4.7×10^{-4}
3000	0.0046	1.7×10^{-6}	0.0038	4.5×10^{-4}
4000	0.0029	1.1×10^{-6}	0.0024	4.7×10^{-4}
5000	0.0018	4.8×10^{-7}	0.0012	4.0×10^{-4}
7500	0.0006	2×10^{-7}	0.0004	4×10^{-4}
10000	0.0002			

3. If concentration data are given then no rate data need to be found in order to determine a rate constant. For a first order reaction, chart the natural logarithm of concentration against time and find the slope.

Time (sec)	$[N_2O_5]$ (M)	$\ln[N_2O_5]$
0	0.0200	-3.91
1000	0.0120	-4.41
2000	0.0074	-4.90
3000	0.0046	-5.37
4000	0.0029	-5.83
5000	0.0018	-6.30
7500	0.0006	-7.39
10000	0.0002	-8.46

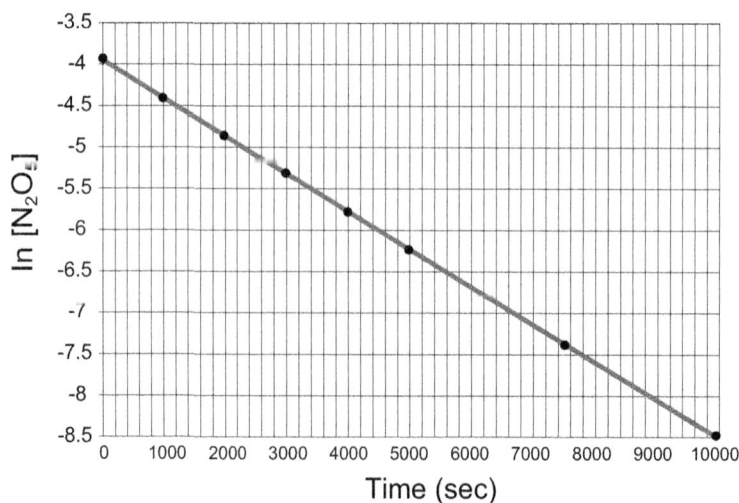

The slope may be determined from a best-fit method or it may be estimated from

$$\frac{-8.46 - (-3.91)}{10000} = -5 \times 10^{-4}$$

The rate law describing this reaction is:

$$\text{Rate} = \left(5 \times 10^{-4} \, \frac{M}{\text{sec}}\right)\left[N_2O_5\right]$$

Chapter 4.3 Reaction mechanisms

"Elementary reactions can be unimolecular or involve collisions between two or more molecules."

When a reaction is represented stoichiometrically, in a balanced chemical equation, reactants and products are shown in their initial and final states. In reality, though, reactions consist of steps that take place in between these two states, so that the overall reaction is a summary of a series of small steps that leads from the initial state (reactants) to the final state (products). The series of steps that lead from the reactants to the products in a reaction is called the **reaction mechanism.** The steps that make up the reaction mechanism are called **elementary reactions, elementary steps,** or **elementary processes.**

Derive rate laws from simple reaction mechanisms

Elementary steps represent a **single event**. This might be a collision between two molecules or a single rearrangement of electrons within a molecule. The simplest reaction mechanisms consist of a single elementary reaction. The number of molecules required determines the rate laws for these processes. A **unimolecular process** can be depicted as:

A → product(s)

Unimolecular processes include **decomposition** (including radioactive decay) and molecular rearrangement, such as cis-trans isomerism. The number of molecules of A that decompose in a given time will be proportional to the number of molecules of A present. Therefore, unimolecular processes are first order; that is, solely based on the rate constant and the concentration of the reactant:

Rate = k [A]

Bimolecular processes involve the collision of two particles. They can be depicted as:

A + B → product(s)

The rate law for such reactions will be second order:

Rate = k [A][B]

If the reaction is the collision of two of the same type of molecules to make a product, the rate will be second order because there are still two molecules involved, even if they are the same type:

Rate = k [A]^2

This same pattern continues for reactions involving any number of reactants, although elementary reactions involving the simultaneous reaction of 3 or more reactants are rare. For example, the rate for a **termolecular process** ($A+B+C \rightarrow$ products) is third order:

$$\text{Rate} = k\,[A][B][C]$$

The reason for this is **collision theory**; simply put, in order to be eligible to react, molecules need to encounter each other. Only a certain percentage of molecules that collide with each other will have enough energy to react, known as the **activation energy**. The more concentrated the reactants are, the more likely they will be to collide and therefore the more likely it is that some of them will have sufficient energy to react.

Most reaction mechanisms are multi-step processes involving **reaction intermediates**. Intermediates are chemicals that are formed during one elementary step and consumed during another, but they are not overall reactants or products. In many cases one elementary reaction in particular is the slowest and determines the overall reaction rate. This slowest reaction in the series is called the **rate-limiting step** or rate-determining step.

Example: The overall reaction:

$$NO_2\ (g) + CO\ (g) \rightarrow NO\ (g) + CO_2\ (g)$$

is composed of the following elementary reactions in the gas phase:

$$NO_2 + NO_2 \rightarrow NO + NO_3$$
$$NO_3 + CO \rightarrow NO_2 + CO_2.$$

The first elementary reaction is very slow compared to the second. Determine the rate law for the overall reaction if NO_2 and CO are both present in sufficient quantity for the reaction to occur. Also name all reaction intermediates.

Solution: The first step will be rate limiting because it is slower. In other words, almost as soon as NO_3 is available, it reacts with CO, so the rate-limiting step is the formation of NO_3. The first step is bimolecular. Therefore, the rate law for the entire reaction is:

$$\text{Rate} = k\left[NO_2\right]^2.$$

NO_3 is formed during the first step and consumed during the second. NO_3 is the only reaction intermediate because it is neither a reactant nor a product of the overall reaction.

A number of possible reaction mechanisms may be proposed for a given balanced reaction, and determining the dominant mechanism is a central activity in chemistry. Reaction intermediates are present only during the reaction, so they are neither detected as final products nor initial reactants, but they can be detected experimentally. The specific techniques used to do this are beyond the scope of the AP chemistry exam, but it will be expected that you understand that detection of intermediates is an important tool for figuring out the mechanism of a given reaction.

Chapter 4.4. Kinetic energy and reaction pathways _____

Kinetic molecular theory also predicts that **reaction rate constants (values for k) increase with temperature** for two possible reasons:
1. More reactant molecules will collide with each other per second.
2. These collisions will each occur at a higher energy that is more likely to overcome the activation energy of the reaction.

Interpreting energy diagrams for chemical reactions

The collisions between the reactants determine how fast the reaction takes place. However, during a chemical reaction, only a fraction of the collisions between the appropriate reactant molecules convert them into product molecules.

This occurs for two reasons:
1. Not all collisions occur with a **sufficiently high energy** for the reaction to occur.
2. Not all collisions **orient the molecules properly** for the reaction to occur.

The activation energy, E_a, of a reaction is the minimum energy to overcome the barrier to the formation of products (to break bonds and/or form bonds) and allow the reaction to occur.

At the scale of individual molecules, a reaction typically involves a very small period of time when old bonds are broken and new bonds are formed. During this time, the molecules involved are in a **transition state** between reactants and products. A threshold of maximum energy is crossed when the arrangement of molecules is in an unfavorable intermediate state between reactants and products known as the **activated complex**. Formulas and diagrams of activated complexes are often written within brackets to indicate they are transition states that are present for extremely small periods of time.

The activation energy, E_a, is the difference between the energy of reactants and the energy of the activated complex. The energy change during the reaction, ΔE, is the difference between the energy of the products and the energy of the reactants. The activation energy of the reverse reaction is $E_a \rightarrow \Delta E$. These energy levels are represented in an **energy diagram** such as the one shown below. This is an exothermic reaction because products are lower in energy than reactants.

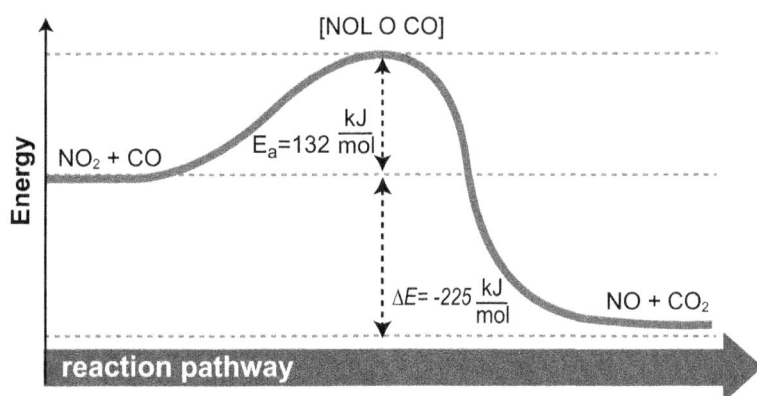

> "Elementary reactions are mediated by collisions between molecules. Only collisions having sufficient energy and proper relative orientation of reactants lead to products."

> "Not all collisions are successful. To get over the activation energy barrier, the colliding species need sufficient energy. Also, the orientations of the reactant molecules during the collision must allow for the rearrangement of reactant bonds to form product bonds."

An energy diagram is a conceptual tool, so there is some variability in how its axes are labeled. The y-axis of the diagram is usually labeled energy (E), but it is sometimes labeled "enthalpy (H)" or (rarely) "free energy (G)." There is an even greater variability in how the x-axis is labeled. The terms "reaction pathway," "reaction coordinate," "course of reaction," or "reaction progress" may be used on the x-axis, or the x-axis may remain without a label.

The energy diagrams of an endothermic and exothermic reaction are compared below.

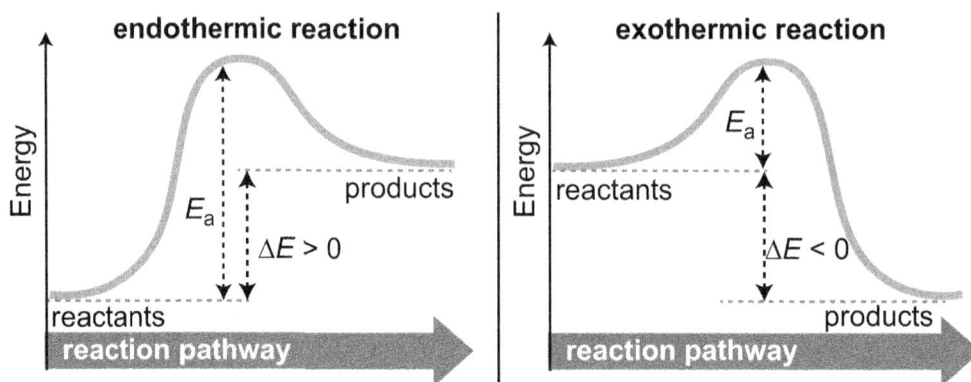

The rate of most simple reactions **increases with temperature** because a **greater fraction of molecules have the kinetic energy** required to overcome the reaction's activation energy.

Recall the **Maxwell-Boltzmann distributions** in Chapter 2.4. The effect of temperature on the distribution of kinetic energies in a sample of molecules can be particularly useful in understanding the influence of environmental factors such as temperature on reaction kinetics. Only a fraction of molecules contain sufficient kinetic energy for a reaction to occur. This fraction is larger at a higher temperature; so more molecules are above the activation energy and more molecules react per second.

On the AP chemistry exam, you will be expected to be able to use a Maxwell-Boltzmann distribution to qualitatively estimate the fraction of collisions between molecules that can be expected to lead to a reaction, as well as how this fraction will be expected to change with changes in temperature.

The role of temperature in reaction kinetics can be expressed by the Arrhenius equation:

$$k = Ae^{-E_a/RT}$$

where E_a is the activation energy and RT is the average kinetic energy of the reactants. The pre-exponential factor A is the rate at which the reaction would proceed if the activation energy were zero, or if all of the molecules had sufficient kinetic energy to exceed the activation energy.

You will NOT be expected to do calculations involving the Arrhenius equation on the AP chemistry exam, but you will be expected to understand it conceptually and be able to interpret graphs on the basis of this understanding.

"A successful collision can be viewed as following a reaction path with an associated energy profile."

"Reaction rates may be increased by the presence of a catalyst."

Chapter 4.5. Catalysis

A **catalyst** is a material that increases the rate of a chemical reaction without changing itself permanently in the process. **Catalysts reduce the activation energy** of a reaction. This is the amount of energy needed for the reaction to proceed. One way they can do this is by lowering the activation energy of an early step in the reaction by stabilizing a transition state, without otherwise changing the reaction mechanism. Catalysts can also change the reaction pathway by providing an alternate reaction mechanism, such as through the formation of a new reaction intermediate that allows the reaction to proceed faster. Either way, the reaction can still proceed in the forward and in the reverse direction; the thermodynamic profile remains unchanged. Therefore, catalysts **have no impact on the chemical equilibrium** of a reaction. They will not make a less favorable reaction more favorable; they will only increase the rate at which the reactants reach that equilibrium.

The impact of a catalyst may also be represented on an energy diagram (as shown below). **A catalyst increases the rate of both the forward and reverse reactions by lowering the activation energy for the reaction.** Catalysts provide a different activated complex for the reaction at a lower energy state.

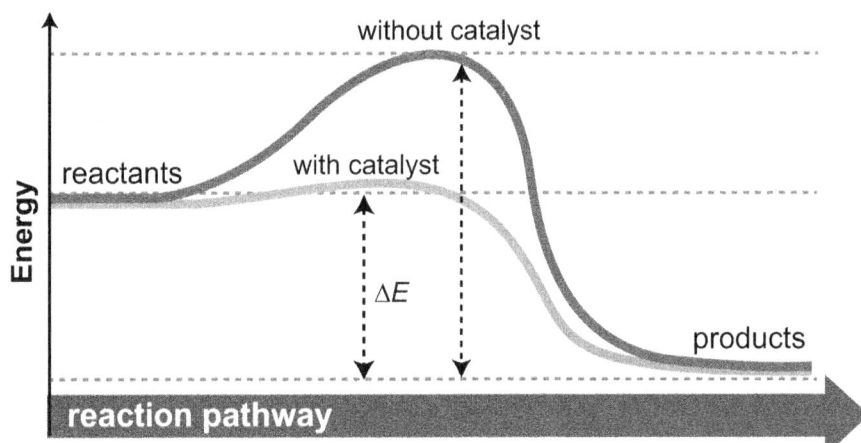

There are two types of catalysts: **Homogeneous catalysts** are in the same physical phase as the reactants. Biological catalysts are called **enzymes**, and most are homogeneous catalysts. A typical homogenous catalytic reaction mechanism involves an initial reaction with one reactant followed by a reaction with a second reactant and release of the catalyst.

Heterogeneous catalysts are present in a different physical state from the reactants. A typical heterogeneous catalytic reaction involves a solid surface onto which molecules in a fluid phase temporarily attach themselves in a way that favors a rapid reaction. Catalytic converters in cars utilize heterogeneous catalysis to break down harmful chemicals in exhaust.

Acid-base catalysis

Acid-base catalysts are often used in organic chemistry reactions. An acid catalyst will donate a hydrogen to a reactant, and a base catalyst will accept a proton from a reactant, in order to form a more stable intermediate that can then allow the final step to proceed faster. In many cases, both acids and bases will catalyze a reaction. Acid/base catalysts include soluble catalysts, such as phosphoric acid, or solid acids or bases, such as metal oxides, which are used in industrial scale chemistry. In the following example, a base catalyst takes a proton from an amine, converting it into a highly reactive intermediate that is able to overcome the activation energy needed to bond with another organic molecule. An acid also helps, by donating a proton that stabilizes the structure of this intermediate:

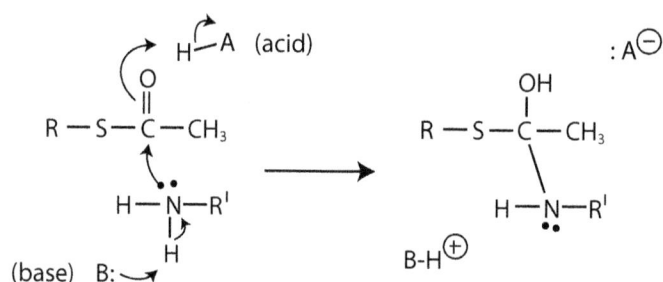

Then, the conjugate base of the acid acts as a base catalyst, taking back the proton. The conjugate acid of the original base catalyst takes back its proton, causing the intermediate to break apart, but not into the original reactants—it will break apart into two new products that are thermodynamically more stable (lower free energy) than the original two reactants:

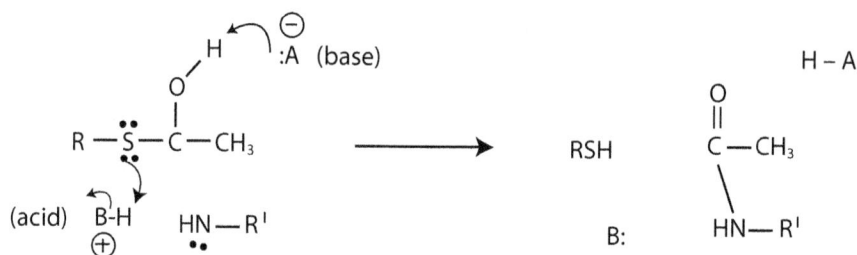

Surface catalysis

Surface catalysis takes place when reactants are adsorbed to the surface of a solid. For example, it is energetically favorable to produce ethane by adding hydrogen gas to ethene but the activation energy is too high for this to proceed spontaneously because hydrogen gas is very stable. However, if a metal such as platinum is added to the reaction, the hydrogen can adsorb to the metal as two separate hydrogen atoms. Now, the atoms are free to bond with the carbons in the ethene, leading to the product, ethane.

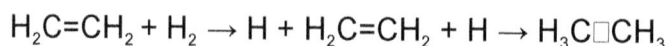

$$H_2C=CH_2 + H_2 \rightarrow H + H_2C=CH_2 + H \rightarrow H_3C\square CH_3$$

Enzyme catalysis

Enzymes are very important catalysts in living systems. Mainly, enzymes are proteins, although they can also be made of RNA (ribozymes). These large molecules have a specific shape, and can catalyze biochemical reactions by binding to reactants and holding them together in a particular configuration until the reaction takes place. They can also react with the reactants themselves, forming an intermediate that in turn reacts again, releasing the enzyme. As with all catalysts, at the end of the reaction, the enzyme is in the same state as at the beginning of the reaction, and is ready to catalyze again.

Sample Test Questions (Big Idea #4) _____

Multiple Choice Questions

Instructions: This section consists of some practice multiple choice questions as well as one Free Response question related to this chapter. You may NOT use a calculator for the Multiple Choice. However calculators are permitted for the Free Response.

1. There are a number of ways to determine the rate of a reaction. The most common is
    A. Spectropscopy and Beer's Law
    B. Spectroscopy and Hess's Law
    C. Titrations
    D. transmission electron microscopy

2. A second order reaction will have a straight line plot for
    A. a plot of $1/[A]$ vs time
    B. a plot of $[A]$ vs time
    C. a plot of $\log [A]$ vs time
    D. cannot determine without more information.

3. If the concentration of the reaction has no impact on the rate of the reaction:
    A. The reaction must involve a catalyst.
    B. The reaction must be zero-order
    C. Doubling the amount of products will increase the amount of reactants by a factor of four.
    D. The reaction must be uni-molecular.

4. For a bi-molecular process $A + B \rightarrow$ product(s)
    A. The higher the temperature the more likely the reaction
    B. The higher the concentration of $[B]$ the more likely the reaction
    C. The reaction is second order
    D. All of the above

5. Most reactions are multi-step processes. It is important to remember that :
    A. The fastest elementary reaction determines the over-all rate.
    B. Inter-mediate steps have no effect on the reaction.
    C. The intermediate steps in the reaction is the rate-limiting step
    D. Intermediates which may not be involved in the overall reaction play an important role.

6. Which of the following statements regarding Collision Theory is FALSE?
    A. a small percentage of molecules that collide with each other have enough energy to react
    B. increasing the Temperature of the system increases the number of collisions
    C. the E_a of a reaction is determined by the number of collisions
    D. the proper orientation of the molecules is necessary for the reaction to occur.

7. For the reaction A + B → C

    a plot of [A] vs time yields a graph with a straight line with a slope of -2.

    If the initial [A] = 100 moles, how much is left after 30m ?

    A. 50 moles

    B. 40 moles

    C. 60 moles

    D. cannot be determined

8. Reactant A decomposes according to the table below. What is the rate law for this reaction?

t(min)	[ ] moles
0	0.5
1	0.25
2	0.125
3	0.625

    A. Rate = k[A]
    B. Rate = k[A]^2
    C. Rate = ½ k[A]
    D. Rate = 2 k[A]

9. Which of the following statements regarding catalysts are TRUE

    I. They reduce the activation energy of a reaction

    II. Many enzymes work as catalysts

    III. They are not consumed during the course of the reaction.

    IV. Are never intermediates in a reaction.

    A. I, and II are TRUE

    B. I, II, and III are TRUE

    C. I and IV are TRUE

    D. All of the statements are TRUE

Questions 10- 12 refer to the following diagram

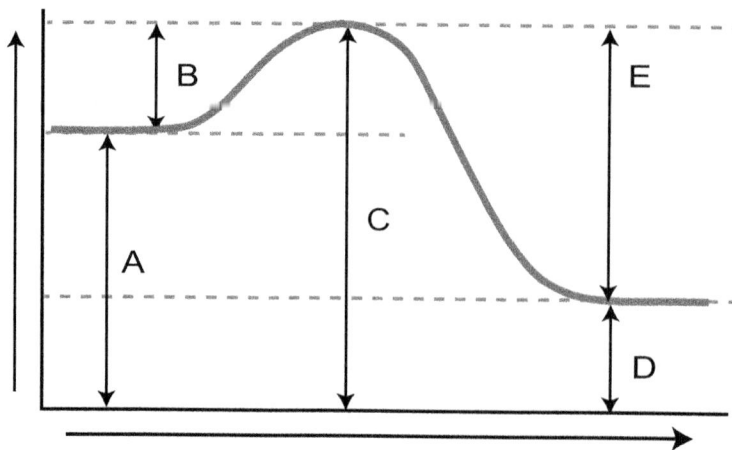

10. In the reaction profile on page 156, A and D refer to
    A.   the potential energy of the reactants and the potential energy of the products
    B.   the activation energy of the reactants and the potential energy of the products
    C.   the kinetic energy of the reactants and the kinetic energy of the products
    D.   the ΔH of the reactants and the ΔH of the reactants

11. Addition of a catalyst will
    A.   Lower the initial value of A
    B.   Lower the height of C.
    C.   raise the height of D
    D.   all of the above

12. For this reaction:
    A.   The reverse E_a < forward E_a so ΔH is exothermic
    B.   The reverse E_a is > forward E_a so ΔH is endothermic
    C.   The forward E_a is > reverse E_a so ΔH is endothermic
    D.   The forward E_a is < reverse E_a so ΔH is exothermic

13. An Arrhenius plot is a graph of
    A.   ln (k) vs 1/T
    B.   ln (k) vs [A]
    C.   ln(k) vs T
    D.   k vs ln(T)

14. Which of the following factors can affect reaction rates?
    A.   Concentration of the reactants
    B.   Temperature
    C.   number of particle collisions
    D.   All of the above.

15. Use the image below to determine which of the following statement is FALSE.

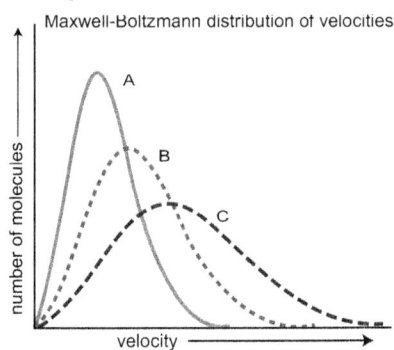
Maxwell-Boltzmann distribution of velocities

    A.   Particle C will moving faster
    B.   Particle C is at a higher temperature
    C.   Particle A will have the most collisions
    D.   Curve B represents the distribution of molecules at T_B

Free Response Question (Big Idea #4)

For the reaction A + B→ products, the following initial rates were found. What is the rate law for this reaction?

Trial 1: [A] = 0.50 M; [B] = 1.50 M; Initial rate = 4.2 x 10-3 M/min

Trial 2: [A] = 1.50 M; [B] = 1.50 M; Initial rate = 1.3 x 10-2 M/min

Trial 3: [A] = 3.00 M; [B] = 3.00 M; Initial rate = 5.2 x 10-2 M/min

Answers to Sample Questions (Big Idea #4)

1. **Answer D.**

   The most common way to determine the rate of a reaction includes the use of a spectrophotometer to determine the concentration using Beer's Law.

2. **Answer A.**

   a second order plot will be a straight line for a plot of 1/[A] vs time

   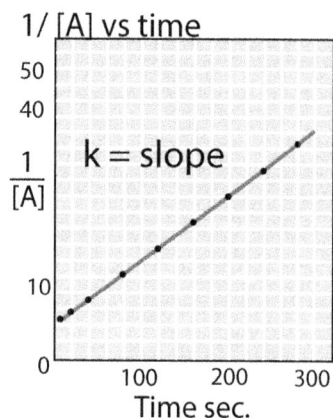

3. **Answer B.**

   The reaction must be zero-order. If a **reaction is zero order for a reactant, the concentration of that reactant has no impa**ct on the rate as long as some reactant is present.

4. **Answer D**.

   All of the above are true for a bimolecular process. An increase in temperature and concentration will increase the likelihood of collisions and therefore the reaction. The reaction is second order.

5. **Answer D**.

   The overall rate is the sum of the individual rates and is determined mostly by the slowest or rate-limiting step. Intermediates which are formed during the reaction but not present in the overall reaction, play an important role in multistep reactions.

6. **Answer C.**

   The E_a of a reaction is not determined by the number of collisions. To get over the activation energy barrier, the colliding species need sufficient energy and orientation

7. **Answer C.**

The plot is a straight line with a slope of -2moles/min. Then we can use $y = mx + b$

b is the interecept – or the concentration at t = 0 = 100moles

therefore $y = mx + b$

$y = (-2 \text{ moles/min}) \ (30 \text{ min}) + 100 \text{ moles}$

$= (-60 \text{ moles}) + 100 \text{ moles} = 40 \text{ moles}$

8. **Answer A.**

First order reaction, dependent on the concentration of the reactant.

9. **Answer B.**

These three statements are all true statements about catalysts: I. They reduce the activation energy of a reaction II. Many enzymes work as catalysts III. They are not consumed during the course of the reaction.

10. **Answer A.**

In the reaction profile lines A and D refer to the potential energy of the reactants and the products respectively.

11. **Answer B.**

The addition of a catalyst lowers the activation energy shown by line C.

12. **Answer D.**

The forward reaction has a smaller Ea (shown by line B) than the reverse reaction (line E)
AH is exothermic.

13. **Answer A.**

An Arrhenius plot is a graph of $\ln (k)$ vs $1/T$

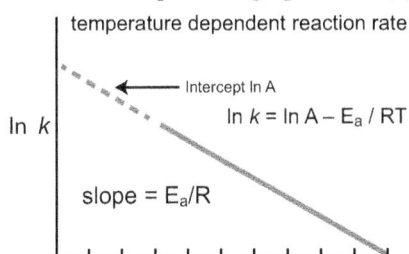

14. **Answer D.**

All of the above. Concentration of the reactants, temperature, and number of particle collisions all affect reaction rates

15. **Answer C.**

This graph is a Maxwell-Boltzmann distribution of velocities. As the temperature increases, the distribution becomes flatter and more particles have higher velocities. At lower temperatures there is a small distribution of velocities and most ofthem are at a lower velocity. Therefore all the statements are TRUE except C. Particles A will have the smallest distributions of velocity and fewer collisions.

Answer to Free Response (Big Idea #4)

1. Order with respect to A:

   Look at trials 2 and 3. The key to this is that we already know that the order for A is first order. Both concentrations were doubled from 2 to 3 and the rate goes up by a factor of 4. Since A is first order, we know that doubling of the rate is due to consentration of A being doubled.

   So, we look at the concentration change for B (a doubling) and the consequent rate change (another doubling – remember the overall increase was a factor of 4 – think of 4 as being a doubled doubling

   Conclusion: the order for B is the first order.

   The rate law is rate= k [A] [B]

Chapter 5: Big Idea #5
Thermodynamics

Big Idea 5: The laws of thermodynamics describe the essential role of energy and explain and predict the direction of changes in matter.

What you should already know:
- the temperature scales and conversion factors
- units of energy and unit conversion

What you will learn in this chapter:
- how to calculate various energy terms
- heats of formation /heat of reaction / heat of combustion
- predicting if a reaction is spontaneous
- using experimental data to calculate energies
- calorimetry

Chapter 5.1 Heat and Temperature

The temperature of an object is proportional to the average kinetic energy of the particles in the substance. Increase the temperature of a substance and its particles move faster, so their average kinetic energies increase as well. When the average kinetic energy of the particles in the sample doubles, the Kelvin temperature is doubled. As the temperature approaches 0 K (zero Kelvin), the average kinetic energy of a system approaches a minimum near zero.

As discussed in Chapter 2, the **Maxwell-Boltzmann** equation, which forms the basis of the Kinetic Molecular Theory of gases, can be used to describe the distribution of speeds (energies) for molecules of a gas at a certain temperature. From this distribution function, the most probable speed, the average speed, and the root-mean-square speed can be derived. At a higher temperature a gas has both a higher average velocity and a broader distribution of velocities.

However, temperature is NOT energy; therefore, it is not conserved. **Heat**, on the other hand, is a form of energy and is transferred between objects that differ in their temperatures. Heat passes spontaneously from an object of higher temperature to one of lower temperature. This transfer continues until both objects reach the same temperature. Eventually, thermal equilibrium is reached and both systems are at the same temperature.

"Temperature is the measure of the average kinetic energy of the atoms and molecules."

Technically, heat is the internal energy gained or lost when two objects come in contact with one another. The temperature difference between two such objects is used as an indirect measurement of heat because heat itself can not be directly measured. Instead, the heat lost or gained is calculated from the mass of the objects, the heat capacity of the objects, and the change in temperature. The heat capacity represents how much heat a substance can absorb in order to raise the temperature a given amount. It is specific to the type of matter; some elements or molecules can absorb a large amount of heat with little change in temperature, whereas in others, a small change in heat will change the temperature dramatically. This has important real-life effects, such as the effect of water on environments. Liquid water has a high heat capacity, so it tends to moderate temperatures. In environments without large bodies of water, such as deserts, temperatures vacillate between extremes, with very hot temperatures during the day and very cold temperatures at night.

A substance's **molar heat capacity** is the heat required to **change the temperature of one mole of the substance by one degree Celsius**. Heat capacity has units of joules per mol-kelvin or joules per mol-°C. The two units are interchangeable because we are only concerned with differences between one temperature and another. A Kelvin degree and a Celsius degree are the same size.

The **specific heat** of a substance (also called specific heat capacity) is the heat required to **change the temperature of one gram or kilogram by one degree**. Specific heat has units of joules per gram or joules per kilogram.

Heat Lost = Heat Gained

$$m_{lost} * C_{lost} * \Delta T_{lost} = m_{gained} * C_{gained} * \Delta T_{gained}$$

where m= mass of substance undergoing temperature change,

c= heat capacity (specific heat) of the substance undergoing temperature change

and ΔT = change in temperature.

These terms are used to solve thermochemistry problems involving a change in temperature by applying the formula:

$q = n \times C \times \Delta T$ where $q \rightarrow$ heat added (positive) or evolved (negative)

$\quad n \rightarrow$ amount of material

$\quad C \rightarrow$ molar heat capacity if n is in moles, specific heat if n is in mass

$\quad \Delta T \rightarrow$ change in temperature $T_{final} - T_{initial}$

Every substance has its own specific heat and each phase has its own distinct value. The specific heat values of water are 2.02 J/g (gas phase), 4.184 J/g (liquid phase) and 2.06 J/g (solid phase). The transfer of energy will not produce the same temperature change in objects with different **specific heat capacities**. Specific heat values can be looked up in reference books. They are determined by experiment. Typically, in the classroom, you will not be asked to memorize any specific heat values. However, you may be asked to recall that the values are different for the three phases of water.

Chapter 5.2 Energy

Technically, **energy is the ability to do work or supply heat**. Work is the transfer of energy to move an object a certain distance. It is motion against an opposing force. Lifting a chair into the air is work; the opposing force is gravity. Pushing a chair across the floor is work; the opposing force is friction.

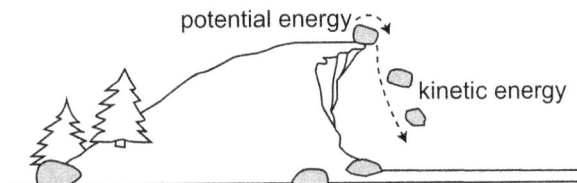
potential energy / kinetic energy

"Energy Is transferred between systems either through heat transfer or through one system doing work on the other system."

Energy is the **driving force for change**. Energy has units of joules (J). Energy exists in two basic forms, **potential and kinetic.** Kinetic energy is the energy of a moving object. Potential energy is the energy stored in matter due to position relative to other objects.

In any object, solid, liquid or gas, the atoms and molecules that make up the object are constantly moving (vibrational, translation and rotational motion) and colliding with each other. They are not stationary.

The **internal energy** of a material is the **sum of the total kinetic energy** of its molecules and the **potential energy** of interactions between those molecules. Potential energy due to position is equal to the mass of the object times the gravitational pull on the object times the height of the object, or:

$$PE = mgh$$

Where PE = potential energy; m = mass of object;

g = gravity; and h = height.

Potential energy includes **energy stored in the form of resisting intermolecular attractions** between molecules. Total kinetic energy includes contributions from translational motion and other types of motion such as rotation.

Both kinetic energy and potential energy can be transformed into heat energy. When you step on the brakes in your car, the kinetic energy of the car is changed to heat energy by friction between the brake and the wheels. Kinetic energy can also be transformed into potential energy. Since most of the energy in our world is in a form that is not easily used, humans have developed some clever ways of changing one form of energy into another:

Chemical Potential Energy

The energy generated from chemical reactions in which the chemical bonds of a substance are broken and rearranged to form new substances is called chemical potential energy. One example of this is combustion of the organic molecules that make up gasoline. Another example is the burning of calories; all living organisms harness chemical potential energy to fuel the processes of life.

Light or Radiant Kinetic Energy

Radiant energy comes from a light source, such as the sun. Energy released from the sun is in the form of photons. These tiny particles, invisible to the human eye, move in a way similar to a wave. This energy has long been harnessed by plants through photosynthesis, and more recently by humans through solar cells to produce electricity.

Energy transformations make it possible for us to use energy to do work. Here are some examples of how energy is transformed to do work:

1. Different types of stoves are used to transform the chemical energy of organic fuel (gas, coal, wood, etc.) into heat.

2. Solar collectors can be used to transform solar energy into electrical energy, and passive solar heating transforms light energy into heat to heat homes.

3. Wind mills make use of the kinetic energy of moving air molecules, transforming it into mechanical or electrical energy.

4. Hydroelectric plants transform the kinetic energy of falling water into electrical energy.

5. A flashlight converts chemical energy stored in batteries to light energy and heat.

A common example of the transfer of energy between systems is the expansion of a gas in response to a piston; this example also illustrates the gas laws found in **Chapter 2.**

Chapter 5.3 Thermodynamics

Kinetic molecular theory says that the particles of a substance are in constant random motion. What causes this motion? Energy.

While temperature is not the same thing as energy, it is a representative measure of energy in the form of heat. A liter of water at 50° C has more energy than a liter of water at 25° C. As heat energy is added to a system, the matter absorbs the heat. This increased heat causes the atoms or molecules to move more (increasing kinetic energy). In the solid state, the motion is mainly vibration. In melting, this vibration separates the atoms from each other, moving them apart, decreasing the force of attraction between the atoms or molecules, causing a phase change to liquid. As a liquid, the molecules still have attractive forces acting between them. The addition of more heat forces the atoms or molecules farther apart, weakening the forces to the point where they can no longer keep the atoms or molecules together. The substance has entered the gas phase. The amount of energy needed to vaporize one mole of a pure substance is the molar enthalpy of vaporization, and the energy released in condensation has an equal magnitude.

The kinetic energy of molecules is unaltered during phase changes, but the freedom of molecules to move relative to one another increases dramatically. The following table summarizes the application of kinetic molecular theory to the addition of heat to ice, first changing it to liquid water and then to water vapor.

"When two systems are in contact with each other and are otherwise isolated, the energy that comes out of one system is equal to the energy that goes into the other system. The combined energy of the two systems remains fixed. Energy transfer can occur through either heat exchange or work."

"Chemical systems undergo three main processes that change their energy: heating/cooling, phase transitions, and chemical-reactions."

Effect at 1 atm of the addition of heat to:	0 = no change, + = increase, ++ = strong increase			
	Temperature	Average speed of molecules	Average translational kinetic energy of molecules	Intermolecular freedom of motion
Ice at less than 0 °C	+	+	+	+
Ice at 0 °C	0	0	0	++ (melting)
Liquid water at 0 °C	+	+	+	+
Liquid water at 100 °C	0	0	0	++ (boiling)
Water vapor at 100 °C	+	+	+	0 (complete freedom for an ideal gas)

The third way a chemical system can change its energy is through chemical reaction. If the potential energy decreases as a result of a chemical reaction, it is called an exothermic reaction. If it increases, it is called an endothermic reaction. For exothermic reactions, the energy lost by the reacting molecules (system) is transferred to the surroundings, in the form of either heat or work. Likewise, for endothermic reactions, the system gains energy from the surroundings by heat transfer or work done on the system.

Work done on a system is mainly in the domain of physics. For the AP chemistry exam, calculations involving work will be limited to the work associated with changes in the volume of a gas. For example, you may be asked about the transfer of energy between systems through work during the expansion of gas in a steam engine or car piston. Chapter 2 reviews molecular collision theory, which explains the behavior of gasses with respect to heat and volume as a function of energy exchange.

Laws of thermodynamics

Thermodynamics is the study of energy flow in natural systems. Findings in this area have been codified into three important physical laws that describe how energy behaves throughout the universe.

In addition to the three traditional laws, a "zeroth law" is often included:

Zeroth Law: If two thermodynamic systems are in thermal equilibrium with a third, they are also in thermal equilibrium with each other. This law simply establishes equivalence in thermodynamic systems.

First Law: The increase in the energy of a closed system is equal to the amount of energy added to the system by heating, minus the amount lost in the form of work done by the system on its surroundings. This law means that **energy is conserved**: energy can be transferred from one system to another, but not created or destroyed. Thus, the total amount of energy available in the Universe is constant. Einstein's famous equation ($E=mc^2$) describes the relationship between energy and matter.

Second Law: The total **entropy** (disorder) of any isolated thermodynamic system **tends to increase over time**, approaching a maximum value. This law indicates that disorder increases with every reaction and some energy is always lost to the increase in that disorder. As a result of this law, it is also true that energy transfer always occurs in one direction (heat can pass spontaneously only from a hotter to a colder body).

Third Law: As a system asymptotically approaches absolute zero of temperature all processes virtually cease and the entropy of the system asymptotically approaches a minimum value. This law **defines absolute zero (0 K or −273° C), at which all thermal motion stops**. In nature, absolute zero cannot be achieved and even empty outer space has a temperature around 3 K. As reported in Nature in 2013, physicists have accomplished instantaneous, quantum sub-zero K temperatures in the laboratory, but at such temperatures, atoms behave strangely, resembling dark energy, the force that is believed to underlie the constant expansion of the universe against the force of gravity.

Chapter 5.4 Enthalpy

The **enthalpy** (H) of a material is the **sum of its internal energy and the mechanical work** it can do. A change in the **enthalpy** of a substance is the total **energy** change caused by **adding or removing heat** at constant pressure.

When a material is heated and experiences a phase change, **thermal energy is used to break the intermolecular bonds** holding the material together. Similarly, bonds are formed and thermal energy is released when a material changes its phase during cooling. Therefore, **the energy of a material increases during a phase change that requires heat and decreases during a phase change that releases heat**. For example, the energy of H_2O increases when ice melts and decreases when water freezes.

Standard thermodynamic state is defined as the state when all components are at 25° C and 100 kPa. This thermodynamic standard state is slightly different from *standard temperature and pressure* (STP), often used for gas law problems (0° C and 1 atm = 101.325 kPa). Standard thermodynamic values of common chemicals are listed in tables.

Just as specific heat is the heat required to change the temperature of one gram or kilogram by one degree, one can determine a substance's enthalpies of fusion, vaporization and sublimation as well enthalpies of specific reactions. A substance's **enthalpy of fusion** (ΔH_{fusion}) is the heat required to **change one mole of a substance from a solid to a liquid** by melting. This is also the heat released from the substance when it changes from a liquid to a solid (freezes). The amount of energy needed to vaporize one mole of a pure substance is the molar enthalpy of vaporization, and the energy released in condensation has an equal magnitude.

A substance's **enthalpy of sumblimation** ($\Delta H_{sublimation}$) is the heat required to change the substance directly from a solid to a gas by sublimation or the heat released by deposition (transition from gas to solid).

These three values are also called "heats" or "latent heats" of fusion, vaporization and sublimation. They have units of joules per mole, and are negative values when heat is released.

$$\text{Solid} \xrightarrow[\text{melting}]{\Delta H_{fusion}} \text{Liquid} \xrightarrow[\text{vaporization}]{\Delta H_{vaporization}} \text{Gas} \qquad \text{Solid} \xrightarrow[\text{sublimation}]{\Delta H_{sublimation}} \text{Gas}$$

$$\text{Gas} \xrightarrow[\text{condensation}]{^-\Delta H_{vaporization}} \text{Liquid} \xrightarrow[\text{freezing}]{^-\Delta H_{fusion}} \text{Solid} \qquad \text{Gas} \xrightarrow[\text{deposition}]{^-\Delta H_{sublimation}} \text{Solid}$$

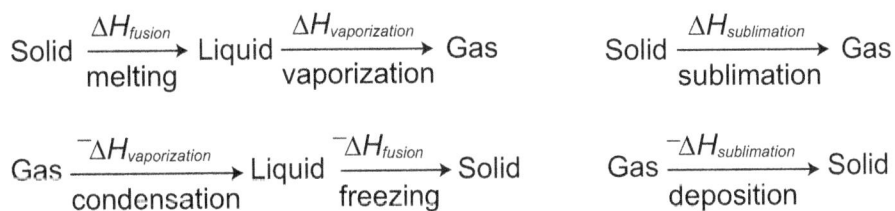

These terms are used to solve thermochemistry problems involving a change of phase by applying the formula:

Example: What is the change in energy of 10 g of gold at 25° C when it is heated beyond its melting point to 1300° C. You will need the following data for gold:

Solid heat capacity: 28 J/mol-K

Molten heat capacity: 20 J/mol-K

Enthalpy of fusion: 12.6 kJ/mol

Melting point: 1064 **°C**

Solution: First determine the number of moles used: .

$$10 \text{ g} \times \frac{1 \text{ mol}}{197 \text{ g}} = 0.051 \text{ mol}$$

There are then three steps. 1) Heat the solid. 2) Melt the solid. 3) Heat the liquid. All three require energy so they will be positive numbers.

Heat the solid:

$$q_1 = n \times C \times \Delta T = 0.051 \text{ mol} \times 28 \frac{J}{\text{mol-K}} \times (1064 \text{ °C} - 25 \text{ °C})$$

$$= 1.48 \times 10^3 \text{ J} = 1.48 \text{ kJ}$$

Melt the solid:

$$q_2 = n \times \Delta H_{fusion} = 0.051 \text{ mol} \times 12.6 \frac{kJ}{mol}$$

$$= 0.64 \text{ kJ}$$

Heat the liquid:

$$q_3 = n \times C \times \Delta T = 0.051 \text{ mol} \times 20 \frac{J}{\text{mol-K}} \times (1300 \text{ °C} - 1064 \text{ °C})$$

$$= 2.4 \times 10^2 \text{ J} = 0.24 \text{ kJ}$$

The sum of the three processes is the total change in energy of the gold:

$$q = q_1 + q_2 + q_3 = 1.48 \text{ kJ} + 0.64 \text{ kJ} + 0.24 \text{ kJ} = 2.36 \text{ kJ}$$

$$= 2.4 \text{ kJ}$$

Chapter 5.4.1 Bond Energies _____

Atoms bond together to form compounds because in doing so they attain lower energies than they possess as individual atoms. Chemical bonds arise from attractive forces between the atoms. As atoms approach each other the attraction increases. Sharing electrons results in a lower energy state and energy is released. Thus **energy is released when bonds form**. This distance at which the energy of interaction in optimal is called the **bond length**. If the atoms become too close there is a repulsion between nuclei as shown in the figure below.

"Breaking Bonds requires energy, and making bonds releases energy."

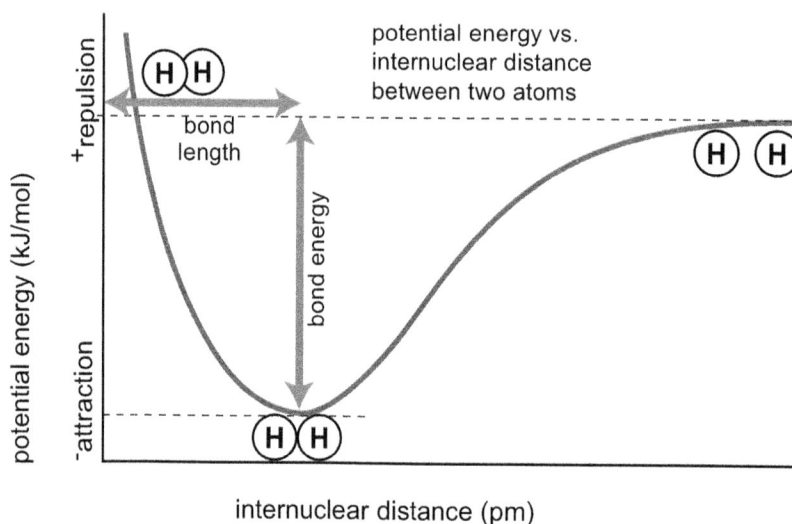

potential energy vs. internuclear distance between two atoms

Bond lengths vary depending on many factors, but in general, they are very consistent. Tables of **interatomic distances** or bond lengths can be found in standard handbooks.

The amount of energy required to break a bond is called **bond dissociation energy** or simply bond energy. Since bond lengths are consistent, bond energies of similar bonds are also consistent. Bond lengths and energies are variable and dependent on the other atoms in the molecule. (The C—H bond length can vary by as much as 4%, and C—O and C—C bonds can vary by almost 20%!) Therefore, average bond energies are approximate. Some typical bond lengths, in picometers (1 pm = 10^{-12}), and bond energies, in kJ/mol, are given here to illustrate a general trend:

H — H	74pm	436 kJ/mol	H — C	109pm	413 kJ/mol
C — C	154pm	348 kJ/mol	O — O	148pm	145 kJ/mol
C = C	134pm	614 kJ/mol	O = O	121pm	498 kJ/mol
C ≡ C	120pm	839 kJ/mol			

Notice that single bonds are longer and have lower energies, double bonds are shorter with higher energies and triple bonds are the shortest and have the highest energy. This makes sense as triple bonds are stronger because they share more electrons. Bond energy is a measure of the strength of a chemical bond. The larger the bond energy, the stronger the bond.

The bond energy is essentially the average enthalpy change for a gas reaction to break all of that type of bond. For the methane molecule, $C(-H)_4$, 435 kJ is required to break a single C-H bond for a mole of methane, but breaking all four C-H bonds for a mole requires 1662 kJ. Thus the average bond energy is (1662/4) 416 (not 436) kJ/mol.

Chapter 5.4.2 Heat of Formation

The **enthalpy of formation** or **heat of formation, ΔH_f,** of a chemical is the heat required for formation (positive) or emitted during formation (negative), when elements react to form the chemical. The **standard heat of formation ΔH_f°** is the heat of formation with all reactants and products at 25 °C and 100 kPa. The energy released when a bond is formed is the same amount as the energy absorbed when a bond breaks; the absolute value of this amount of potential energy is called the bond energy.

Elements in their **most stable form** are assigned a value of $\Delta H_f^\circ = 0$ kJ/mol. Different forms of an element in the same phase of matter are known as **allotropes**. The different forms of carbon provide a good example of allotropes.

Example: The heat of formation for carbon as a gas is:

$$\Delta H_f^\circ \text{ for } C(g) = 718.4 \ \frac{kJ}{mol}$$

Carbon in the solid phase exists in three allotropes. A C_{60} *buckyball* (one face is shown on the left), contains C atoms linked with aromatic bonds and arranged in the shape of a soccer ball. C_{60} was discovered in 1985. *Diamonds* (below center) contains single C–C bonds in a three dimensional network. The most stable form at 25 °C is *graphite* (below right). Graphite is composed of C atoms with aromatic bonds in sheets.

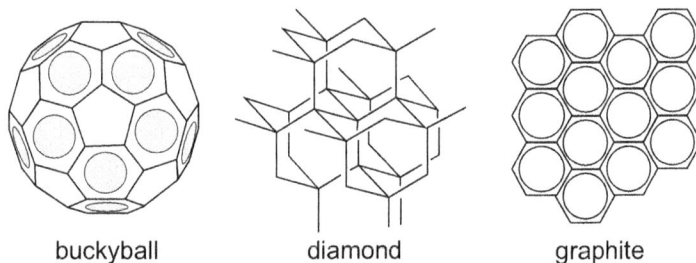

| buckyball | diamond | graphite |

$$H_f^\circ \text{ for } C_{60} (\textit{buckminsterfullerene or buckyball}) = 38.0 \ \frac{kJ}{mol}$$

$$H_f^\circ \text{ for } C \ (\textit{diamond}) = 1.88 \ \frac{kJ}{mol}$$

$$H_f^\circ \text{ for } C \ (\textit{graphite}) = 0 \ \frac{kJ}{mol}.$$

Chapter 5.4.3 Heat of reaction

The enthalpy change for a reaction is commonly called the **heat of reaction**. When a chemical reaction takes place, the enthalpies of the products will differ from the enthalpies of the reactants. The energy change for the reaction, ΔH_{rxn}, is determined by **the sum of the products minus the sum of the reactants**:

$$H_{rxn} = \left(H_{product\ 1} + H_{product\ 2} + H_{product\ 3} \cdots \right) - \left(H_{reactant\ 1} + H_{reactant\ 2} + H_{reactant\ 3} \cdots \right)$$

If the enthalpies of the products are greater than the enthalpies of the reactants then ΔH_{rxn} **is positive** and the reaction is **endothermic**. Endothermic reactions **absorb heat** from their surroundings. The simplest endothermic reactions break chemical bonds.

If the enthalpies of the products are less than the enthalpies of the reactants then ΔH_{rxn} is negative and the reaction is **exothermic**. Exothermic reactions **release heat** into their surroundings. The simplest exothermic reactions form new chemical bonds.

In a closed system, energy is always conserved. Thus for an exothermic reaction, in which the potential energy is lower in the products than the reactants, the products have a higher kinetic energy and are at a higher temperature. Conversely for a endothermic reaction, in which potential energy increases, the products have a lower kinetic energy and are at a lower temperature.

The heat absorbed or released by a chemical reaction often has the impact of changing the temperature of the reaction vessel and of the chemicals themselves. The measurement of these heat effects is known as **calorimetry**. **Calorimetry** will be discussed in more detail in the next section.

The enthalpy change of a reaction ΔH_{rxn} **is equal in magnitude but has the opposite sign to the enthalpy change for the reverse reaction**. If a series of reactions lead back to the initial reactants then the net energy change for the entire process is zero.

When a reaction is composed of substeps, the **total enthalpy change will be the sum of the changes for each step**. Even if a reaction in reality contains no substeps, we may still write any number of reactions in series that lead from the reactants to the products and their sum will be the heat of the overall reaction of interest. The ability to add together these enthalpies to form ultimate products from initial reactants is known as **Hess's Law**. It is used to determine one heat of reaction from others:

Hess' law states that energy changes are state functions. The amount of energy depends only on the states of the reactants and the state of the products, but not on the intermediate steps. Energy (enthalpy) changes in chemical reactions are the same, regardless of whether the reactions occur in one or several steps. The total energy change in a chemical reaction is the **sum of the energy changes** in its many steps leading to the overall reaction. When a reaction is reversed, the sign of the enthalpy of the reaction is changed; when two (or more) reactions are summed to obtain an overall reaction, the enthalpies of reaction are summed to obtain the net enthalpy of reaction.

"The net energy change during a reaction is the sum of the energy required to break the bonds in the reactant molecules and the energy released in forming the bonds of the product molecules. The net change in energy may be positive for endothermic reactions where energy is required, or negative for exothermic reactions where energy is released."

Chapter 5.4.4 Activation Energies _____

In order for one species to be converted to another during a chemical reaction, the reactants must collide. The collisions between the reactants determine how fast the reaction takes place. However, during a chemical reaction, only a fraction of the collisions between the appropriate reactant molecules result in conversion into product molecules. This occurs for two reasons:

- Not all collisions occur with **sufficiently high energy** for the reaction to occur.
- Not all collisions **orient the molecules properly** for the reaction to occur.

The **activation energy** (E_a) of a reaction is the **minimum energy needed to overcome the barrier to the formation of products** and allow the reaction to occur.

At the scale of individual molecules, a reaction typically involves a very small period of time when old bonds are broken and new bonds are formed. During this time, the molecules involved are in a **transition state** between reactants and products. A threshold of maximum energy is crossed when the arrangement of molecules is in an unfavorable intermediate state between reactants and products known as the **activated complex**. Formulas and diagrams of activated complexes are often written within brackets to indicate they are transition states that are present for extremely small periods of time.

The **activation energy (E_a)** is the difference between the energy of the reactants and the energy of the activated complex. The energy change during the reaction (ΔE) is the difference between the energy of the products and the energy of the reactants. The activation energy of the reverse reaction is $E_a \rightarrow \Delta E$. These energy levels are represented in an **energy diagram** such as the one shown below for $NO_2 + CO \rightarrow NO + CO_2$. This is an exothermic reaction because the products are lower in potential energy than the reactants. An endothermic reaction will go "uphill" because the products are higher in potential energy than the reactants.

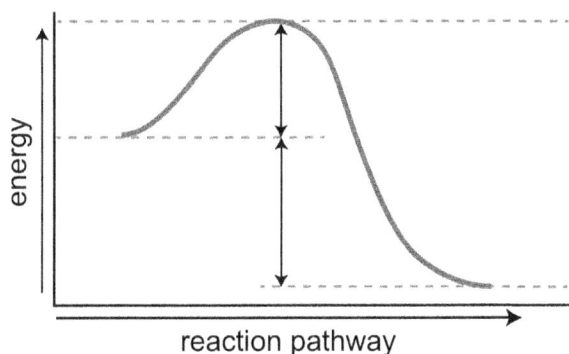

An energy diagram is a conceptual tool, so there is some variability in how its axes are labeled. The y-axis of the diagram is usually labeled energy (E), but it is sometimes labeled "enthalpy (H)". There is an even greater variability in how the x-axis is labeled. The terms "reaction pathway," "reaction coordinate," "course of reaction," or "reaction progress" may be used on the x-axis.

It is generally the case that exothermic reactions are more likely to occur spontaneously than endothermic reactions. Molecules usually seek the lowest potential energy state. However, entropy (reviewed below) also plays a critical role in determining whether a reaction occurs.

Chapter 5.5 Entropy

Entropy (S) may be thought of as **the disorder in a system** or as a measure of the **number of states a system may occupy**. Changes due to entropy occur in one direction with no driving force. For example, a small volume of gas released into a large container will expand to fill it, but the gas in a large container never spontaneously collects itself into a small volume. This occurs because a large volume of gas has more disorder and has more places for gas molecules to be. Entropy has units of J/K.

If two different chemicals are at the same temperature, in the same state of matter, and they have the same number of molecules, their entropy difference will depend mostly on the number of ways the atoms within the two chemicals can rotate, vibrate, and flex. Most of the time, **the more complex molecule will have the greater entropy** because there are more energetic and spatial states in which it may exist.

In the solid phase, each molecule may vibrate, but it is otherwise locked into place in an ordered position and may only be in a relatively small number of locations. In the gas phase, however, each molecule could be almost anywhere and there is greater disorder. Liquids are intermediate in terms of movement of molecules. Therefore, **the entropy of a material increases during a phase change that raises the freedom of molecular motion and decreases during a phase change that prevents molecular motion**. Entropy also increases with temperature because molecules experience more disorder when they have a wider range of energy states to occupy.

Based on this principle, gases have greater entropy than liquids, liquids have greater entropy than solids, and matter in the same state increases in entropy with temperature. Entropy is also an extensive property of matter. **A greater number of moles will have a larger entropy.**

Entropy also increases when there is an increase in volume (at constant temperature). The gas molecules are now able to move around a larger space. Review the gas laws in chapter 2 for more details.

The standard molar entropy, $S°$, is the absolute entropy of a chemical at 1 atm and 25 °C. The molar entropy of a substance is expressed in units of J/mol-**K**.

At zero Kelvin (0 K), there is no energy available for a chemical to sample different states. The **absolute entropy**, S, of a pure crystalline solid at 0 K is zero. Absolute entropy may be measured and calculated for different substances at different temperatures.

The entropy change of a reaction, ΔS, is given by the sum of the absolute entropies of all the products multiplied by their stoichiometric coefficients minus the sum of all the reactants multiplied by their stoichiometric coefficients:

For the reaction: $aA + bB \leftrightarrow pP + qQ$

$$\Delta S = pS(P) + qS(Q) - aS(A) - bS(B)$$

Chapter 5.6 Spontaneity and Gibbs free energy

Spontaneity

A reaction with a **negative ΔH and a positive ΔS** causes a decrease in energy and an increase in entropy. (For the purposes of the AP Chemistry exam, internal energy and enthalpy can be considered interchangeable.) **These reactions are thermodynamically favored (also known as "spontaneous")**. (Remember, however, that even if a reaction is thermodynamically favored, it will not necessarily proceed at a measurable rate.) A reaction with a positive ΔH and a negative ΔS causes an increase in energy and a decrease in entropy. These reactions do not occur to an appreciable extent, in the absence of external energy input, because the reverse reaction takes place spontaneously.

Whether reactions with the remaining two possible combinations (ΔH and ΔS both positive or both negative) occur depends on the temperature. In these cases we must look at the **Gibbs Free Energy, ΔG.**

ΔG = ΔH–TΔS where T is the temperature in K.

Gibb's Free Energy is the maximum amount of energy released from a system that is available to do work on the surroundings. If **ΔG** is **negative, the reaction is spontaneous and will take place in the forward direction**. If **ΔG** is positive, the reaction will not occur to an appreciable extent. If **$\Delta G = \Delta H - T\Delta S = 0$** exactly, then the system is at **equilibrium** and there there will be 50% reactants and 50% products.

A spontaneous reaction is called *exergonic*. A non-spontaneous reaction is known as *endergonic*.

Example: Will graphite and hydrogen gas react to form C_2H_4 at 25 °C and 100 kPa?

Solution: (part 1) Determine the standard heat of formation ΔH_f° for ethylene:

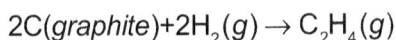

$$2C(graphite) + 2H_2(g) \rightarrow C_2H_4(g)$$

Use the heat of combustion for ethylene:

$$\Delta H_c^{\circ} = 1411.2 \frac{kJ}{mol\ C_2H_4} \quad \text{for } C_2H_4(g) + 3O_2(g) \rightarrow 2CO_2(g) + 2H_2O(l)$$

and the following two heats of formation for CO_2 and H_2O:

$$\Delta H_f^{\circ} = -393.5 \frac{kJ}{mol\ C} \quad \text{for } C(graphite) + O_2(g) \rightarrow CO_2(g)$$

$$\Delta H_f^{\circ} = -285.9 \frac{kJ}{mol\ H_2} \quad \text{for } H_2(g) + \frac{1}{2}O_2(g) \rightarrow H_2O(l).$$

"Some physical or chemical processes involve both a decrease in the internal energy of the components ($\Delta H^{\circ} < 0$) under consideration and an increase in the entropy of those components ($\Delta S^{\circ} > 0$). These processes are necessarily "thermodynamically favored" ($\Delta G^{\circ} < 0$)."

"If a chemical or physical process is not driven by both entropy and enthalpy changes, then the Gibbs free energy change can be used to determine whether the process is thermodynamically favored."

Also, find the standard change in entropy $\Delta S°$ for the formation of C_2H_4, given:

$$S°(C(graphite)) = 5.7\,\frac{J}{mol\,K} \quad S°(H_2(g)) = 130.6\,\frac{J}{mol\,K} \quad S°(C_2H_4(g)) = 219.4\,\frac{J}{mol\,K}$$

Example: (part 2): Will graphite and hydrogen gas react to form C2H4 at 25 °C and 100 kPa?

Solution: Use Hess's Law after rearranging the given reactions so they cancel to yield the reaction of interest. Combustion is exothermic, so ΔH for this reaction is negative. We are interested in C2H4 as a product, so we take the opposite (endothermic) reaction. The given ΔH are multiplied by stoichiometric coefficients to give the reaction of interest as the sum of the three:

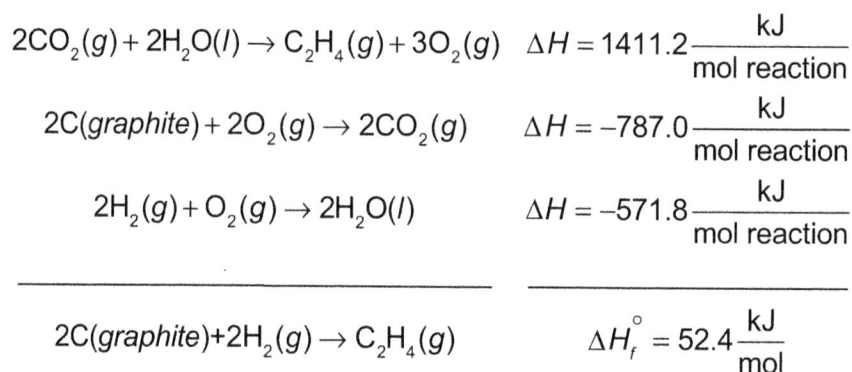

$$2CO_2(g) + 2H_2O(l) \rightarrow C_2H_4(g) + 3O_2(g) \quad \Delta H = 1411.2\,\frac{kJ}{mol\,reaction}$$

$$2C(graphite) + 2O_2(g) \rightarrow 2CO_2(g) \quad \Delta H = -787.0\,\frac{kJ}{mol\,reaction}$$

$$2H_2(g) + O_2(g) \rightarrow 2H_2O(l) \quad \Delta H = -571.8\,\frac{kJ}{mol\,reaction}$$

$$2C(graphite) + 2H_2(g) \rightarrow C_2H_4(g) \quad \Delta H_f^° = 52.4\,\frac{kJ}{mol}$$

Because 2 moles react we multiply 393.5 kJ/mol by 2 to get -787.0 kJ/mol. The same is true for the -571.8 kJ/mol; 2 mol react so it becomes 2 x -285.9 kJ/mol. The value for the first equation is not multiplied by 2 because the ΔH is for the equation as it is written.

The entropy change is found from:

$$\Delta S° = S°(C_2H_4) - 2S°(C) - 2S°(H_2) = 219.4\,\frac{J}{mol\,K} - 2 \times 5.7\,\frac{J}{mol\,K} - 2 \times 130.6\,\frac{J}{mol\,K}$$

$$= -53.2\,\frac{J}{mol\,K}.$$

This reaction is endothermic, with a decrease in entropy, so it is endergonic (not thermodynamically favored). Therefore, graphite and hydrogen gas will not react to form C_2H_4 spontaneously.

   The heat of combustion ΔH_c (also called enthalpy of combustion) is the heat of reaction when a chemical **burns in O_2** to form completely oxidized products such as **CO_2 and H_2O**. It is also the heat of reaction for **nutritional molecules that are metabolized** in the body. The burning of such molecules, known as catabolism, is a type of harnessed combustion reaction, mediated by enzymes to release the energy of the reaction slowly.

The standard heat of combustion $\Delta H_c°$ takes place at 25 °C and 100 kPa. **Combustion is always exothermic**, so the negative sign for values of ΔH_c is often omitted.

Light energy can also be used to drive a process that is not thermodynamically favored. An example of this would be the conversion of carbon dioxide to glucose through photosynthesis.

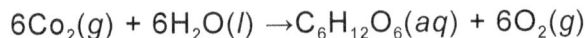

$$6Co_2(g) + 6H_2O(l) \rightarrow C_6H_{12}O_6(aq) + 6O_2(g)$$

This reaction has a $\Delta G = +2880kJ/mol$, and takes place through a multistep process that is initiated by the absorption of several photons in the range of 400-700nm. By using sunlight to drive the thermodynamically unfavorable conversion of carbon dioxide to glucose, plants are converting energy in the form of light into energy in the form of chemical potential energy, a form of energy they can use to power the reactions of living.

Electrolysis, described in Chapter 3, is another example of a reaction with a positive ΔG that proceeds via the input of energy. **Electrolytic cells use electricity** to force thermodynamically unfavorable redox reactions to occur.

In addition to using an external energy source such as electricity or light, a thermodynamically favorable reaction can be coupled with a thermodynamically unfavorable reaction in order to drive the unfavorable reaction forward. An important example of this is the conversion of ATP to ADP in biological systems. In this way, biological systems use ATP as a kind of "energy currency", producing ATP from combustion of food molecules or photosynthesis and then coupling it with thermodynamically unfavorable reactions to move them forward.

In calculating ΔG, be careful with the units; T must be in K and S is given in J/mol-K, while H is typically in kJ/mol-K.

ΔG is a state function like ΔH and ΔS. That means:

$$\Delta G_{rxn} = \Sigma \Delta G_{(products)} - \Sigma \Delta G_{(reactants)}$$

Example:

Sample Problem C:

Find the free energy of formation for the oxidation of water to produce hydrogen peroxide.

$$2\ H_2O_{(l)}\ +\ O_{2(g)}\ \rightarrow\ 2\ H_2O_{2(l)}$$

Given the following information:

	$\Delta G°_f$
$H_2O_{(l)}$	-56.7 kcal/mol
$O_{2(g)}$	0 kcal/mol
$H_2O_{2(l)}$	-27.2 kcal/mol

"External sources of energy can be used to drive change in cases where the Gibbs free energy change is positive."

Plugging all of the values you were given into the equation (remember that elements have a $\Delta G_f°$ of 0), you get

$$[2(-27.2)] - [2(-56.7) + 1(0)] = 59.0 \text{ kcal/mol}$$

Kinetic vs. thermodynamic control

While thermodynamically favored processes are commonly called "spontaneous", there are processes that are favored but do not occur to any appreciable extent. This may be due to kinetic factors such as high activation energies or unstable products. Such reactions are said to be under kinetic, as opposed to thermodynamic, control. Such reactions are not at equilibrium, because some factor, such as activation energy, is preventing them from proceeding to their most thermodynamically favorable state. Kinetic factors are reviewed in Chapter 4.

"A thermodynamically favored process may not occur due to kinetic constraints (kinetic vs. thermodynamic control)."

Chapter 5.7 Measurement of the transfer of thermal energy —

Phase changes occur when the relative importance of kinetic energy and intermolecular forces is altered sufficiently for a substance to change its state. Raw phase change data is often charted by recording the temperature over time when heat is added at a constant rate. A diagram for water at 1 atm from −50°C to 150°C is shown below. Note that temperature does not change during melting and boiling. Also note the difference in the length of time required for melting compared to boiling. This is a result of greater energy requirements to boil a substance than to melt it.

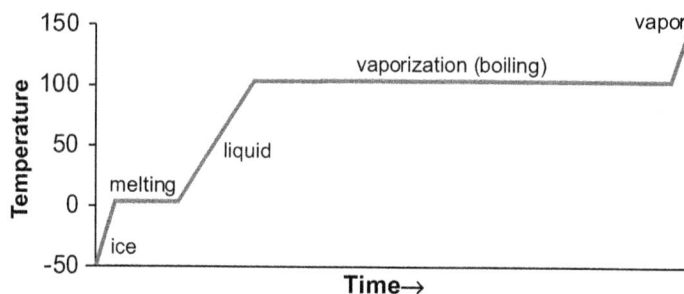

Phases and Phase changes are discussed in more detail in Chapter2.

Calorimetry

Calorimetry is an experimental technique that uses temperature to determine the amount of heat exchanged/transferred in a chemical system during processes such as phase transitions, physical changes, and chemical reactions. The apparatus that is used to measure the heat of reaction is called a calorimeter (from the words "calorie meter", although we now express the heat unit in joules rather than calories). A calorimeter is a well-insulated device in which a reaction takes place. No heat can leave or enter the system. The system is put in thermal contact with a bath composed of a substance with a known heat capacity. Examples include water, which is an excellent absorber of heat and has one of the highest specific heats (4.184 J/g°C). Since the heat capacity of the bath substance is known, changes in temperature are utilized to estimate amount of heat exchanged between the system and the heat bath. If the **heat bath increases** in temperature then the **energy of the system has decreased**. If the **heat bath decreases** in temperature, it has decreased in energy—it has transferred energy to the chemical system being measured, and **the energy of the system has increased** by the same amount.

Calorimetry is based on the following principles:

1. The heat energy of a system increases or decreases by heating or cooling processes, respectively.

2. The specific heat capacity of a substance is defined as the quantity of energy required to heat 1 gram of that substance by 1 °C.

3. Phase transitions from solid-liquid or liquid-gas consume energy, while the reverses processes release energy.

4. The enthalpy of vaporization is the amount of energy required to vaporize one mole of pure substance, and the reverse process (condensation) releases energy with an equal magnitude.

5. In an exothermic reaction, the energy released by the system is gained by the surrounding environment, and in an endothermic process, the system receives energy from the surrounding environment, in the form of heat.

6. In a chemical reaction performed at constant pressure, the change in enthalpy refers to the energy released (negative) or absorbed (positive).

Applications of calorimetry

Any process that results in a change in heat, either generated or consumed, is a candidate for a calorimetric study. Hence, calorimetry has a very broad range of applicability, including in drug design in the pharmaceutical industry, quality control of process streams in the chemical industry, and the study of metabolic rates in biological systems. Calorimetry is currently performed by means of a number of advanced methods such as differential Thermal Gravimetric Analysis (TGA), Differential Thermal Analysis (DTA), and Differential Scanning Calorimetry (DSC).

There are several different designs for calorimetric apparatuses. Among the most simple is one called a bomb calorimeter, shown in the figure below. This kind of calorimeter is normally used to study exothermic reactions that do not begin until they are initiated by applying heat – reactions such as the combustion of CH_4 or the reaction of H_2 with O_2. The apparatus consists of a strong steel container (the bomb) into which the reactants are placed. The bomb is then immersed in an insulated bath that is fitted with a stirrer and a thermometer. The initial temperature of the bath is measured and then the reaction is set off by a small heater wire within the bomb. The heat given off in the reaction is absorbed by the bomb and the bath, causing the temperature of the entire apparatus to rise.

"Calorimetry is an experimental technique that is used to determine the heat exchanged/transferred in a chemical system."

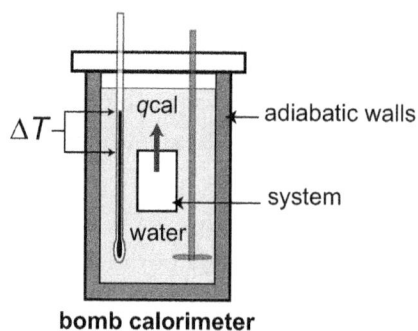

bomb calorimeter

Example: Measuring the heat of reaction with a bomb calorimeter

0.200g of H_2 and 1.600 of O_2 were compressed in an experiment into a 1.00 L bomb, which then was placed into a calorimeter with the capacity of 1.816×10^5 J/°C. The initial calorimeter temperature was measured to be 26°Cand after the reaction take place, the calorimeter's final temperature 26,155. Calculate the amount of heat given off in the reaction of H_2 and O_2 to form H_2O, expressed (a) in kilojoules and (b) in kilojoules per mole of H_2O formed.

Solution:

(a) To calculate the heat evolved, multiply the heat capacity (1.816×10^5 J/°C) by the temperature change in oC. The temperature change is 0.155. Therefore,

$(1.816 \times 10^5 \text{ J}/1 \text{ °C}) \times 0.155 \text{ °C} = 2.82 \times 10^4 \text{J} = 28.2 \text{ kJ}$. So, this is the amount of heat given off by the reactants.

(b) The amount of heat released in a reaction depends on the amount of the reactants used. Large amounts of reactants produce large energy changes. The heat of reaction for different reactions is usually expressed on a "per mole" basis.

The reaction between H_2 and O_2 to form water is as follows

$2 H_2 + O_2 \rightarrow 2 H_2O$

From 2 moles of H_2, 2 moles of H_2O is produced.

First calculate the number of moles of H_2

moles of H_2 = 0.200 g / 2g/mol H_2 s_2

Therefore, 0.1 moles of water will be produced. To get units of kilojoules per mole of water, calculate the ratio of the number of kilojoules to the number of moles.

28.2 kJ/0.1 moles of water =282 kJ/mol H_2O

A simpler calorimeter, often used in the chemistry lab, is a styrofoam cup that has good insulated walls to prevent heat exchange with the environment.

Chapter 5.8 Physical vs. chemical processes _____

Substances can change via physical and chemical processes. **Physical processes** involve **intermolecular forces**, including ion-dipole interactions, hydrogen bonding, and dipole interactions. These intermolecular forces are discussed in detail in Chapter 2. **Chemical processes** involve changes of the molecules themselves, via the breaking and/or formation of bonds. Both physical processes and chemical processes proceed only if they are thermodynamically favorable; that is, if Gibbs free energy is negative.

Hydrogen bonding can be a very important interaction. While the hydrogen "bond" is much weaker than a true chemical bond, it is a very strong intermolecular force. It is responsible for many of the unusual properties of water including a higher than expected boiling point and a high surface tension.

Hydrogen bonding is also very important for maintaining the structure of large biomolecules such as cellulose, proteins, RNA, and the double-helix DNA structure. Many biochemical processes involve "on-off switches" for biological functions that are controlled by the shape of large molecules like enzymes. The shapes of the molecules change via changes in intermolecular forces, most often hydrogen bonding, *within* a molecule. That is, one part of a very large molecule (such as a protein) forms an intermolecular bond with another part of the molecule and in this way folds into a functional shape. If a thermodynamically unfavorable shape change is needed, the change is often coupled to the release of a phosphate from the molecule ATP, a highly thermodynamically favorable process. In this way, the total ΔG for the required shape change is negative and the biological process can proceed. The formation of ATP is in turn formed by coupling with the thermodynamically favorable combustion of organic molecules such as glucose.

Chemical processes describe the change of a substance or substances into a new substance or substances, via the breaking and/or formation of chemical bonds. In contrast to physical processes, which involve only changes in intermolecular interactions, such as phase changes, chemical processes change the fundamental nature of the substance.

Both physical and chemical processes obey the same thermodynamic laws—if ΔG is negative, the process will proceed spontaneously. If ΔG is positive, the process will only proceed if energy is input from the surroundings, and will reverse to favor reactants if the energy is no longer supplied to drive the reaction forward.

Chemical changes are more commonly known as chemical reactions. These include combining or re-combining to create compounds, burning, decomposition, and the reactions of redox chemistry (reviewed in Chapter 3) and acid-base chemistry (reviewed in Chapter 6).

"Electrostatic forces exist between molecules as well as between atoms or ions, and breaking the resultant intermolecular interactions requires energy."

"Noncovalent and intermolecular interactions play important roles in many biological and polymer systems."

"At the particulate scale, chemical processes can be distinguished from physical processes because chemical bonds can be distinguished from intermolecular interactions."

A gray area exists between these two extremes, however. For instance, the dissolution of a salt in water involves the breaking of ionic bonds and the formation of intermolecular interactions between ions and solvent. It fulfills the criterion for being a physical process since evaporation will once again restore the salt as a solid held together by ionic bonds. However, from a thermodynamic point of view, since the magnitude of the intermolecular interactions between the salt ions and water molecules can be comparable to covalent bond strengths, a plausible argument can be made for classifying dissolution of a salt as either a physical or chemical process.

Sample Test Questions (Big Idea #5) _____

Multiple Choice Questions

Instructions: This section consists of some practice multiple choice questions as well as one Free Response question related to this chapter. You may NOT use a calculator for the Multiple Choice. However calculators are permitted for the Free Response.

1. If the average kinetic energy of the particles in the sample doubles, then

    A. The potential energy of the system is reduced

    B. The temperature in K also doubles

    C. The number of collisions is reduced.

    D. The volume of the system will double.

2. When 10 g samples of the following materials absorb 100 J of heat, the temperature increase indicated is observed. Which material has the largest specific heat?

    A. water 2 K

    B. wood 6 K

    C. copper 26 K

    D. cement 11 K

3. Which if the following is true regarding the C-C (triple bond) when compared to C=C
    I.   The triple bond is shorter than a double bond
    II.   The double bond is lower in energy that the triple bond
    III. The triple bond does not occur naturally
    IV.   In a triple bond one electron comes from one atom and the other two come from the other atom

    A. Only I & II are True

    B. Only III is true

    C. I, II & III are true

    D. Only I & IV are true

4. An ice cube whose mass is 0.080 kg is taken from a freezer where its temperature was -12.0 °C and dropped into a glass of water at 0.0 °C. If no heat is gained or lost from outside, how much water freezes onto the cube? The latent heat of fusion of water is 33.5 x 104 J/kg and the specific heat capacity of ice is 2000 J/kg °C.

    A. 24000g

    B. 2.4g

    C. 71.1g

    D. 7.1g

5. The heat capacity of lead is 0.13 J/g° C. How many joules of heat would be required to raise the temperature of 15 g of lead from 22° C to 37° C?

A.   2.0 J

B.   -0.13 J

C.   5.8 X10-4 J

D.   29 J

6. What is the final temp when 10grams of water at 0° C is added to 100 grams of water at 75° C?

A.   10.0° C

B.   68.1° C

C.   14° C

D.   70.5° C

7. 25.0 g of ice is warmed from -25.0° C to 0.0° C by placing it in a water bath that is 100° C , but does not melt. Which of the following statements are true.

A.   The water gained the same amount of heat energy as the bath lost.

B.   The water temperature changed more than the temperature of the water bath.

C.   The total energy of the system has decreased.

D.   None of the above.

8. Using the following data (at 25° C)

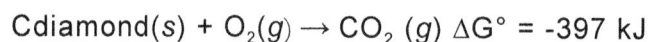

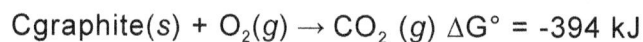

$$Cdiamond(s) + O_2(g) \rightarrow CO_2 (g) \quad \Delta G° = -397 \text{ kJ}$$

$$Cgraphite(s) + O_2(g) \rightarrow CO_2 (g) \quad \Delta G° = -394 \text{ kJ}$$

Calculate $\Delta G°$ for the reaction

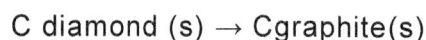

$$C \text{ diamond (s)} \rightarrow Cgraphite(s)$$

A. $\Delta G° = 791$kJ

B. $\Delta G° = -3$ kJ

C. $\Delta G° = 3$ kJ

D. $\Delta G° = 0$ kJ

9. Which of the following statements are false regarding ΔH and ΔS

A.   if ΔH is negative and ΔS is positive then the reaction is spontaneous at all temperatures

B.   if ΔH is negative and ΔS is negative spontaneous at low temperatures

C.   if ΔH is positive and ΔS is negative not spontaneous, ever

D.   none of the above

10. How much heat is required to warm 20g of liquid water from 25° C to the boiling point (Given

Specific Heats $H_2O(l)$ = 4.18 J/gC, $H_2O(s)$ = 2.0 J/gC and $H_2O(g)$ = 2.0J/gC)

$$H_2O(l, 25° C) \rightarrow H_2O(l, 100°C) \Delta H =$$

A. ΔH = 6.3 kJ

B. ΔH = -6.3kJ

C. ΔH = 3kJ

D. ΔH = 313kJ

11. According to the kinetic theory of gases which of the following are true

I. As you add heat to ice the temperature increase

II. As you add heat to ice the kinetic energy of the molecules increase

III. As you add heat to ice at 0° C you will increase the temperature to 1° C

IV. As you remove heat from liquid water at 0° C the temperature will be reduced to -1° C

A. All of sthe above are TRUE

B. Statements I and II are true

C. Statements II and IV are TRUE

D. Statements I, III and IV are True.

12. In which of the following reactions would ΔHrxn NOT be equal to ΔHf for the product?

A. $C(graphite) + 2H_2(g) \rightarrow CH_4(g)$

B. $CO(g) + \frac{1}{2} O_2(g) \rightarrow CO_2(g)$

C. $H2(g) + \frac{1}{2} O_2(g) \rightarrow H_{20}(g)$

D. $2Fe(s) + \frac{3}{2} O_2(g) \rightarrow Fe_2O_3(s)$

13. Given the following equation,

$$2 H_2(g) + O_2(g) \rightarrow 2 H_2O(l) \Delta HO = -571.6 \text{ kJ/mol.}$$

which statement is incorrect?

A. Δ Ho for the reverse reaction is +571.6 kJ/mol.

B. Δ Ho for forming one mole of liquid water is -285.8 kJ/mol.

C. Δ Ho for the reaction of one-half mole of oxygen is -285.8 kJ/mol. T

D. The standard heat of formation of liquid water is -571.6 kJ/mol.

14. What would be the total number of joules of heat absorbed by 30 grams of water when it is heated from 30 °C to 40 °C?

    A. 15

    B. 630

    C. 140

    D. 1260

15. The following questions refer to the graph below

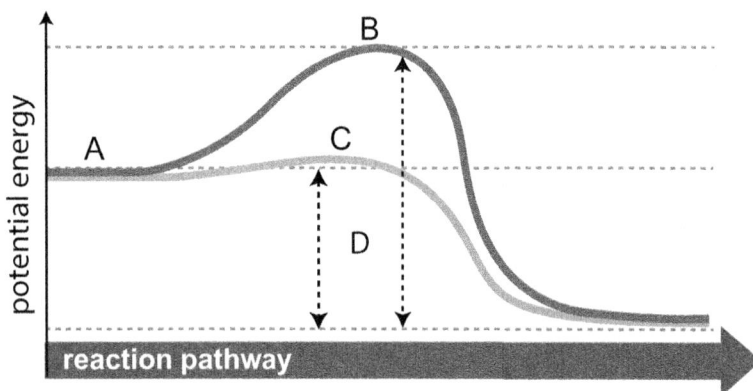

From this graph it can be determined that

I. The reaction is exothermic

II. A catalyst can lower the activation energy

III. The reaction is spontaneous at all temperatures

    A. I only

    B. II only

    C. I and II

    D. I,II and III are true

16. Which of the following statements are TRUE

    I. The enthalpy of a reaction does not depend on the elementary steps, but on the final state of the products and initial state of the reactants.

    II. Enthalpy is an extensive property and hence changes when the size of the sample changes.

    III. The sign of the reaction enthalpy changes when a process is reversed.

    IV. enthalpy is a state function, it is path independent.

    A. I, III, IV are TRUE

    B. I and II are TRUE

    C. II and III are TRUE

    D. All are TRUE

17. Which of the following is NOT a feature of the bomb calorimetry apparatus used to measure the internal energy of a reaction?

A.   The thermometer is inserted directly into the reaction vessel to measure ΔT of there-action.

B.   The large heat capacity of water is beneficial in measuring heat released by combustion reactions.

C.   The volume of the reaction vessel is held constant to eliminate energy released as work.

D.   The heat capacity of the calorimetershould be known to accurately correct for any heat lost to it.

18. Consider two solids S1 and S2 that are in contact with each other. In which of the following propositions do you think heat will flow spontaneously from S1 to S2?

A.   S1 and S2 are both at 30° C

B.   S1 is at 15° C and S2 is 10° C

C.   S1 is -15° C and S2 is -7° C

D.   S1 is 27° C and S2 is 37° C

19. What would occur if 1 mL of sulfuric acid is dissolved in a 100 mL beaker of water, indicating that the process is exothermic?

A.   Gas will be evolved from the solution.

B.   Decrease in temperature of the solution will be observed.

C.   An increase in temperature of the solution will be noticed.

D.   The solution changes in color.

20. An ice cube whose mass is 0.080 kg is taken from a freezer where its temperature was -12.0° C and dropped into a glass of water at 0.0° C. If no heat is gained or lost from outside, how much water freezes onto the cube? The latent heat of fusion of water is 33.5 x 104 J/kg and the specific heat capacity of ice is 2000 J/kg° C.

A.   0.19g

B.   5.73g

C.   0.005g

D.   3.35g

Free Response Question (Big Idea #5)

1.   Calculate the standard Gibbs free energy at 298K for the reaction

$$4NH_3(g) + 7O_2(g) \rightarrow 4NO_2(g) = 6H_2O(l)$$

Species	ΔH_f° kJ/mol	S° J/mol · K
$NH_3(g)$	-46.11	192.3
$O_2(g)$	0.0	205.0
$NO_2(g)$	+33.2	240.0
$H_2O(l)$	-285.8	69.91

1. **Answer B.**

   If the average kinetic energy of the particles in the sample doubles, then the temperature in K also doubles, since the temperature is the average kinetic energy of the system.

2. **Answer A.**

   The specific heat is the amount of heat required to heat 1gr of a substance one degree. In this case we are heating 10g of a substance and absorb 100J of heat.

   $$\text{Total Heat (q)} = gr * C * \text{Specific Heat}$$

   $$100J = 10g * \Delta T. * \text{Specific Heat}$$

   $$SH = 100 J / (10g * \Delta T)$$

   A   SH = 100/ 20. = 5

   B.   SH = 100/60 = 1.6

   C.   SH = 100/ 260 = 0.38

   D.   SH = 100/110 = 0.9

3. **Answer A.**

   The triple bond is shorter and has more energy than a double bond.

4. **Answer D.**

   Solution:

   1) Calculate amount of heat to raise ice to zero Celsius:
   $$q = (0.10 \text{ kg}) (12.0 \text{ °C}) (2000 \text{ J/kg °C}) = 2400 \text{ J}$$

   2) Calculate amount of water that freezes when 1920 J is removed:
   $$2400 \text{ J} = (x) (33.5 \times 104 \text{ J/kg})$$
   $$x = 7.1g$$
   (watch that the units are consistent)

5. **Answer D.**

   $$q = (15 \text{ g}) (15 \text{ °C}) (0.13J \text{ g}^{-1} \text{ °C}^{-1}) = 29J$$

6. **Answer B.**

   10 gram of water gains the heat of 10(t-0)s which is equal to the heat lost by the 100g of water = 100s(75-t).

   $$10s(t - 0) = 100s (75-t)$$
   $$10st = 100s (75 -t)$$
   $$10t = 7500 - 100t$$
   $$110t = 7500$$
   $$t = 7500/110 = 68.18 \text{ C}.$$

7. **Answer A.**

All others violate conservation of energy laws.

8. **Answer B.**

$$C_{diamond(s)} + O_2(g) \rightarrow CO_2(g) \quad \Delta G° = {}^-397 \text{ kJ}$$

$$C_{graphite(s)} + O_2(g) \rightarrow CO_2(g) \quad \Delta G° = {}^-394 \text{ kJ}$$

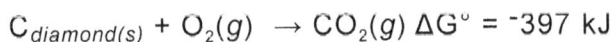

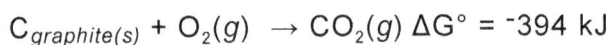

Need the reverse of the second equation so

$$\Delta G° = {}^-397 - ({}^-394) = {}^-3 \text{kJ}$$

9. **Answer D.**

None of the above statements are false.

10. **Answer A.**

$$q = \text{mass } H_2O \times \text{specific heat } H_2O \times \Delta T.$$
$$= 20g * 4.18 \text{ J/gC} * 75C$$
$$= 6270 \text{ J} = 6.3 \text{kJ}$$

11. **Answer B.**

Only Statements I and II are True.

Recall

Effect at 1 atm of the addition of heat to:	0 = no change, + = increase, ++ = strong increase			
	Temperature	Average speed of molecules	Average translational kinetic energy of molecules	Intermolecular freedom of motion
Ice at less than 0 °C	+	+	+	+
Ice at 0 °C	0	0	0	++ (melting)
Liquid water at 0 °C	+	+	+	+
Liquid water at 100 °C	0	0	0	++ (boiling)
Water vapor at 100 °C	+	+	+	0 (complete freedom for an ideal gas)

12. **Answer B.**

$\Delta H_{formation}$ of an element is always 0. Therefore all the others have the heat of formation of the reactants equal to zero – however for B

$$\Delta H_{rxn}^0 = \sum n \, \Delta H_{f \, prod}^0 - \sum n \, \Delta H_{f \, rct}^0$$

$$\sum_{CO_2, \, so} n \, \Delta H_{f \, rct}^0 = 0 \text{ for all except } CO + \frac{1}{2} O_2 \rightarrow$$

$$\Delta H_{rxn}^0 = \sum n \, \Delta H_{f \, prod}^0 \, .$$

But for $CO + \frac{1}{2} O_2 \rightarrow CO_2$,

$$\Delta H_{rxn}^0 = \Delta H_{f \, CO_2}^0 - H_{f \, CO}^0$$

13. **Answer D.**

D is incorrect because -571.6 kJ/mol represents the formation of 2 moles of water. The standard Heat of formation would be for one mole or 285kJ/mol.

14. **Answer D.**

Explanation: Heat capacity = mΔT, where m is the mass and ΔT is difference in temperature.
Heat capacity = 30 g x 4.2 J/g°C x 10°C = 1260 joules.

15. **Answer C.**

Only I and II are true. III may not be true at ALL temperatures.

16. **Answer D.**

All of the statements are TRUE

17. **Answer A.**

A is not true, **A.** The thermometer is not inserted directly into the reaction vessel but into the water bath.

18. **Answer B.**

The direction of heat flow is always from higher to lower temperature. Thus, in B, S1 is warmer than S2 and heat will flow from S1 to S2.

19. **Answer C.**

Explanation: Exothermic reactions cause heat to be released. Sulfuric acid needs heat to dissolve and releases this heat energy into water. As the water receives the heat energy, its temperature will increase.

20. **Answer A.**
    Solution:

1) Calculate amount of heat to raise ice to zero Celsius:

$$q = (0.080 \text{ kg}) (12.0 \text{ °C}) (2000 \text{ J/kg °C}) = 1920 \text{ J}$$

2) Calculate amount of water that freezes when 1920 J is removed:

$$1920 \text{ J} = (x) (33.5 \times 10^4 \text{ J/kg})$$

$$x = 5.73 \times 10^{-3} \text{ kg} = 5.73 \text{ g}$$

Answer to Free Response (Big Idea #5)

$$\Delta H^0_{rxn} = \sum n \, \Delta H^0_{f \, prod} - \sum n \, \Delta H^0_{f \, rct}$$
$$= \Big[6(-285.8 \text{ kJ/mol})$$
$$+ 4(33.2 \text{ kJ/mol}) \Big]$$
$$- 4(-46.11 \text{ kJ/mol})$$
$$= -1397.56 \text{ kJ/mol}$$

$$\Delta S^0_{rxn} = \sum n \, \Delta S^0_{f \, prod} - \sum n \, \Delta S^0_{f \, rct}$$
$$= \Big[6(69.91 \text{ J/mol} \cdot \text{K})$$
$$+ 4(240.0 \text{ J/mol} \cdot \text{K}) \Big]$$
$$- \Big[7(205.0 \text{ J/mol} \cdot \text{K})$$
$$+ 4(192.3 \text{ J/mol} \cdot \text{K}) \Big]$$
$$= -824.74 \text{ J/mol} \cdot \text{K}$$
$$= -0.82474 \text{ kJ/mol} \cdot \text{K}$$

$$\Delta G^0 = \Delta H^0 - T \Delta S^0$$
$$= -1397.56 \text{ kJ/mol}$$
$$- (298 \text{ K}) (-0.82474 \text{ kJ/mol} \cdot \text{K})$$
$$= -1151.79 \text{ kJ/mol}$$

Chapter 6: Big Idea #6
Equilibrium: Acids, Bases, and Solubility

Big Idea 6: Any bond or intermolecular attraction that can be formed can be broken. These two processes are in a dynamic competition, sensitive to initial conditions and external perturbations.

What you should already understand:

- concentration, mass, volume, and reaction rates
- calculation using molarity, molality, normality, weight percentage, and mole fraction
- definition of homogenous mixture, heterogenous mixture, and solution

What you will learn:

- chemical equilibrium
- Le Châtelier's principle
- solubility and factors affecting dissolution
- calculations of molar solubility, concentrations of solutes, and enthalpy of solution
- historic, Arrhenius, and Brønsted-Lowry concepts of acids and bases
- dynamic equilibrium, pH, and titration
- acid/base nomenclature
- strong and weak acids/bases, and conjugate acid/base pairs
- quantitative acid-base chemistry
- buffers, pH, and pK_a/pK_b
- effect of complexation on solubility
- equilibrium, temperature, and Gibbs free energy

Chapter 6.1. Solvents, solutes, and solutions _____

While any two phases may form a solution, liquid solutions, and especially aqueous ones, are especially important in chemistry. When a pure liquid and a gas or solid form a liquid solution, the pure liquid is called the solvent and the non-liquid is called the solute. When all components in the solution were originally liquids, then the one present in the greatest amount is called the solvent and the others are called solutes. The amount of solute in a solvent is called its concentration. A solution with a small concentration of solute is called dilute, and a solution with a large concentration of solute is called concentrated.

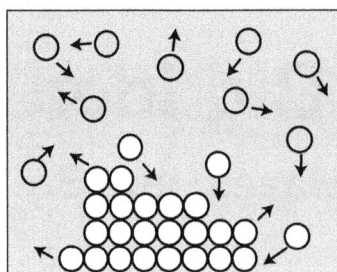

As more solid solute particles (circles in the figure to the left) dissolve in a liquid solvent (grey background), the concentration of solute increases, and the chance that dissolved solute (grey circles) will collide with the remaining undissolved solid (white circles) also increases. A collision may result in a dissolved solute particle reattaching itself to the solid. This process is called crystallization, and it is the opposite of the solution process. Particles in the act of dissolving or crystallizing are half-shaded in the figure.

Chapter 6.2. Solubility and Chemical Equilibrium _____

Equilibrium occurs when no additional solute will dissolve because the rates of crystallization and solution are equal. It is important to remember that this process is entirely reversible, and is dynamic—in a solution at equilibrium, the processes of solution and crystallization are still taking place, but since they are in equilibrium, the solution appears to be static.

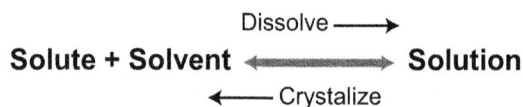

Dissolve ⟶

Solute + Solvent ⟷ **Solution**

⟵ Crystalize

A solution at equilibrium with undissolved solute is a **saturated** solution. The amount of solute required to form a saturated solution in a given amount of solvent is called the **solubility** of that solute. If less solute is present, the solution is called **unsaturated**. Under certain conditions, it is also possible to have more dissolved solute than the equilibrium amount, resulting in a solution that is termed **supersaturated**.

The **solubility-product constant**, K_{sp}, is the equilibrium constant for an ionic solid (salt) in contact with a saturated aqueous solution. The reaction quotient for such a dissolution, Q, is called the solubility product. At equilibrium, two processes are taking place at equal rates (in this case, dissolution and crystallization). The value of K_{sp} represents the ratio of products (solution) to reactants (solute + solvent) at equilibrium. The reaction quotient, Q, represents the ratio of products to reactants at a given instant. Therefore, when K = Q, the reaction has come to equilibrium. If Q is less than K, there are more reactants than products, and the reaction will proceed in the forward direction to reach equilibrium. If Q is greater than K, there are more products than reactants, and the reaction will go "backwards"—it will shift to the left-- to reach equilibrium. A useful shortcut for checking yourself when solving problems is to imagine the words "Quick checK" but substitute all but the first and last letters with >, =, or <. Then, imagine the < or > symbols as the open mouth of an alligator—whichever direction the alligator is chomping in is the direction the reaction will go. If Q < K, the alligator is chomping forwards; the reaction will go forwards. If Q > K, the alligator is chomping backwards; the reaction will go backwards. This quick check can save you from losing a point on the AP Exam due to a simple error.

"The magnitude of the equilibrium constant, K, can be used to determine whether the equilibrium lies toward the reactant side or product side."

The solubility of a substance can be calculated from the K_{sp} of the dissolution process.

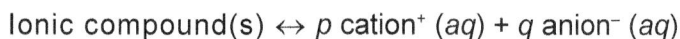

$$\text{Ionic compound(s)} \leftrightarrow p \text{ cation}^+ (aq) + q \text{ anion}^- (aq)$$

$$K_{sp} = \left[\text{cation}^+ \right]^p \left[\text{anion}^- \right]^q$$

Whereas K_{sp} is an equilibrium constant, solubility is the maximum mass of solid that is able to dissolve in a given quantity of water at equilibrium.

Example: At room temperature, solid lead chloride $PbCl_2$ is allowed to dissolve in pure water until equilibrium has been reached and the solution is saturated. The concentration of Pb^{2+} is 0.016 M. What is K_{sp} for $PbCl_2$?

Solution: The reaction and solubility product are shown below:

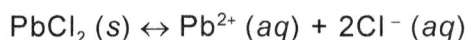

$$PbCl_2 (s) \leftrightarrow Pb^{2+} (aq) + 2Cl^- (aq)$$

$$K_{sp} = \left[Pb^{2+} \right] \left[Cl^- \right]^2$$

The only source of both ions in solution is $PbCl_2$, so the concentration of Cl^- must be twice that for Pb^{2+}, or 0.032 M. Therefore,

$$K_{sp} = 0.016 \text{ M} * (0.032 \text{ M})^2 = 1.6 \times 10^{-5}$$

To calculate the molar solubility (s) of a substance, use the K_{sp}:

$$K_{sp} = [Pb^{2+}][Cl^-]^2$$

The molar solubility of $PbCl_2$ will be equal to the molar solubility of Pb^{2+} because lead is the limiting factor. $[Cl^-] = 2[Pb^{2+}]$ so, rewriting this equation in terms of molar solubility,

$$K_{sp} = (s)(2s)^2 = 4s^3$$

From the K_{sp} value calculated above,

$$1.6 \times 10^{-5} / 4 = s^3 = 4 \times 10^{-6}$$

Taking the cube root of 4×10^{-6},

$$s = 0.016 \text{ M}$$

The maximum solubility of lead chloride at room temperature for the experiment in the example is 0.016 moles per liter of water.

Note: you would not see this problem on the AP Chemistry exam in the multiple choice section, although you may see it in the free response section, since it requires a calculator. You will, however, be expected to reason qualitatively about the solubility of a salt, given a certain K_{sp}. For example, you may be asked to compare two salts or be given a range of values that would require you to know how solubility is calculated, without requiring actual calculation of a precise value.

"The current state of a system undergoing a reversible reaction can be characterized by the extent to which reactants have been converted to products. The relative quantities of reaction components are quantitatively described by the reaction quotient, Q."

Chemical equilibrium applies not just to liquid solutions. Gasses are also in equilibrium, and this equilibrium can be affected not just by concentration but by partial pressures and temperature as well. In addition, different phases of the same substance can be in equilibrium—for example, liquid water and ice will be in equilibrium at 0°C. The water is freezing and melting, but at equilibrium the amount of liquid water that is freezing is equal to the amount of ice that is melting, so that, over time, there is no net change in the amount of ice and liquid water.

Chapter 6.3 Le Châtelier's principle

Chemical reactions often do not "go to completion." Instead, products are generated from reactants up to a certain point when the reaction no longer seems to occur, leaving some reactant unaltered. At this point, the system is in a state of **chemical equilibrium** because **the rate of the forward reaction is equal to the rate of the reverse reaction**. An observer at the macroscopic scale might say that no reaction is taking place at equilibrium, but at a microscopic scale, both the forward and reverse reactions are occurring at the same rate. **A reaction at equilibrium contains a constant ratio of chemical species.** The nature of these ratios is determined by an **equilibrium constant**.

Homogeneous equilibrium refers to a chemical equilibrium among reactants and products that are all in the same phase of matter. **Heterogeneous equilibrium** takes place between two or more chemicals in different phases.

In either case, if a reaction at equilibrium is disturbed, changes occur to reestablish equilibrium. **Le Châtelier's principle** states that the equilibrium will be reestablished in a manner that counteracts the effect of the initial disturbance.

A system at equilibrium is in a state of balance because forward and reverse processes are taking place at equal rates. If equilibrium is disturbed by a change in concentration, pressure, or temperature, the state of balance is upset for a period of time before the equilibrium shifts to achieve a new state of balance. **Le Châtelier's principle states that equilibrium will shift to partially offset the impact of an altered condition.**

For example, **if** a chemical reaction is at equilibrium, Le Châtelier's principle predicts that **adding a substance**—either a reactant or a product—will shift the reaction so **a new equilibrium is established by consuming some of the added substance**. Removing a substance will cause the reaction to move in the direction that forms more of that substance.

Example: The reaction $CO + 2H_2 \leftrightarrow CH_3OH$ is used to synthesize methanol. Equilibrium is established, and then additional CO is added to the reaction vessel. Predict the impact on each reaction component after CO is added.

Solution: Le Châtelier's principle states that the reaction will shift to partially offset the impact of the added CO. Therefore, the concentration of CO will decrease, and the reaction will "shift to the right." The concentration of H_2 will also decrease, and the concentration of CH_3OH will increase.

If a chemical reaction is at equilibrium in the gas phase, Le Chatelier's principle predicts that **an increase in pressure** will shift the reaction so **a new equilibrium is established by decreasing the number of moles of gas present**. A decrease in the number of moles partially offsets this rise in pressure. Decreasing pressure will cause the reaction to move in the direction that forms more moles of gas. These changes in pressure might result from altering the volume of the reaction vessel at constant temperature.

Example: The reaction, $N_2 + 3H_2 \leftrightarrow 2NH_3$ is used to synthesize ammonia. Equilibrium is established. Next, the reaction vessel is expanded at constant temperature. Predict the impact on each reaction component after this expansion occurs.

Solution: The expansion will result in a decrease in pressure (Boyle's Law). Le Châtelier's principle states that the reaction will shift to partially offset this decrease by increasing the number of moles present. There are 4 moles of gas on the left side of the equation and 2 moles of gas on the right, so the reaction will shift to the left. N_2 and H_2 concentration will increase. NH_3 concentration will decrease.

Similarly, Le Châtelier's principle predicts that **when heat is added** at constant pressure to a system at equilibrium, **the reaction will shift in the direction that absorbs heat** until a new equilibrium is established. For an endothermic process, the reaction will shift to the right towards product formation. For an exothermic process, the reaction will shift to the left towards reactant formation. If you understand the application of Le Châtelier's principle to concentration changes, then writing "heat" on the appropriate side of the equation will help you understand its application to changes in temperature.

Example: $N_2 + 3H_2 \rightarrow 2NH_3$ is an exothermic reaction.

First, equilibrium is established and then the temperature is decreased. Predict the impact of the lower temperature on each reaction component.

Solution: Since the reaction is exothermic, we may write it as:

$$N_2 + 3H_2 \rightarrow 2NH_3 + heat$$

To find the impact of temperature on equilibrium processes, we may consider heat as if it were a reaction component. Le Châtelier's principle states that after a temperature decrease, the reaction will shift to partially offset the impact of a loss of heat. Therefore more heat will be produced, and the reaction will shift to the right. N_2 and H_2 concentration will decrease. NH_3 concentration will increase.

"In many classes of reactions, it is important to consider both the forward and reverse reaction."

Chapter 6.4. Factors affecting solubility and dissolution rate

Fundamentally, solubility is related to Gibbs free energy (G°), which was extensively reviewed in Chapter 5. The change in free energy (ΔG°) of all of the intramolecular bonds in the solid that are broken and all of the intermolecular solid-solvent bonds that are formed together determine solubility. If the total ΔG° is negative, the dissolution will proceed spontaneously. Remember that ΔG° encompasses both ΔH (the change in enthalpy, or heat) and ΔS (the change in entropy, or randomness). In this light, Le Chatelier's principle can be seen as the response of a system to maintain an ideal balance of heat and entropy. Any factor affecting the heat or entropy of the system will affect whether dissolution proceeds. In other words, any factor changing the heat or entropy of a system is changing a system in which Q = K into a system in which Q differs from K. The system will then change until Q is once again equal to K.

Intermolecular forces in the solution process

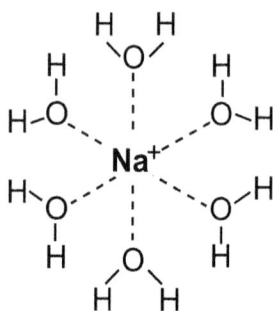

Solutions tend to form when the intermolecular attractive forces between solute and solvent molecules are about as strong as those that exist in the solute alone or in solvent alone. NaCl dissolves in water for two reasons:

1. The water molecules interact with the Na+ and Cl– ions with sufficient strength to overcome the attraction between them in the crystal.

2. Na+ and Cl– ions interact with the water molecules with sufficient strength to overcome the attraction water molecules have for each other.

The intermolecular attraction between solute and solvent molecules is known as solvation. When the solvent is water, it is known as hydration. This figure shows a hydrated Na+ ion.

Polar and nonpolar solutes and solvents

The only intermolecular bonds in a nonpolar liquid like heptane (C_7H_{16}) are relatively weak London dispersion forces. Heptane is immiscible in water because the attraction that water molecules have for each other via hydrogen bonding is too strong. Unlike Na^+ and Cl^- ions, heptane molecules cannot break these bonds. Because bonds of similar strength must be broken and formed for solvation to occur, nonpolar substances tend to be soluble in nonpolar solvents, and ionic and polar substances are soluble in polar solvents like water. Polar molecules are often called **hydrophilic** ("water-loving") and non-polar molecules are called **hydrophobic** ("water-hating"). This observation is often stated as "**like dissolves like.**" Network solids (e.g., diamond) are soluble in neither polar nor nonpolar solvents because the covalent bonds within the solid are too strong to break by solvation.

Solubility rules for ionic compounds

The AP Chemistry exam does not emphasize memorizing solubility rules, although you will be expected to know that all sodium, potassium, ammonium, and nitrate salts are soluble in water. Instead, it is important to be able to reason about the likelihood of dissolution, in line with Le Chatelier's principle. For example, a salt will be less soluble in a solution that has an ion in common with the salt. The acidity of a solution (described in depth below) will also affect the solubility of a salt if one of the salt's ions is acidic or basic. Again, a good understanding of Le Chatelier's principle will allow you to logically predict the effects of acidity on the solubility of a specific salt.

Another application of Le Chatelier's principle is temperature. For an endothermic dissolution, increasing temperature will increase solubility. However, **solubility will decrease with increasing temperature for an exothermic reaction**. For example, this is the case for cerium (III) sulfate, $Ce_2(SO_4)_3$, dissolving in water. Remember that the energy change associated with dissolution is the net result of two processes: 1) the energy required to break the solute-solute bonds, called the **crystal lattice energy** if the solute is an ionic compound, and 2) the energy released when the solute particles bond with the solvent molecules, called the **enthalpy of solution**.

Example: What is the enthalpy of solution for KCl in water?

Solution: The crystal lattice energy of KCl, the energy necessary to break apart the KCl crystal lattice and form free ions, is represented by:

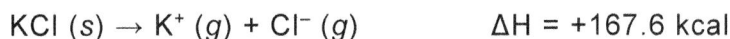

$$KCl\ (s) \rightarrow K^+\ (g) + Cl^-\ (g) \qquad \Delta H = +167.6\ kcal$$

The heat of hydration of KCl, the energy released when the free ions are hydrated, is represented by:

$$K+\ (g) + Cl^-\ (g) \rightarrow K^+\ (aq) + Cl^-\ (aq) \qquad \Delta H = -163.5\ kcal$$

The overall reaction is endothermic, and the heat of solution is positive since more energy is required in the first step than is released by the second step.

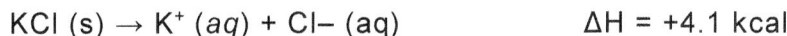

$$KCl\ (s) \rightarrow K^+\ (aq) + Cl-\ (aq) \qquad \Delta H = +4.1\ kcal$$

Solubility is one of the major realms in which the principle of dynamic equilibrium can be applied. Another major application of equilibrium is in the interaction between acids, bases, and water.

Chapter 6.5 Acidity and alkalinity _____

Practical understanding of acids and bases goes back to ancient times, when ancient Greek and Egyptian chemists recognized certain substances, characterized by a sour taste, and other, bitter tasting substances that felt slippery. From these early times through the

"The solubility of a substance can be understood in terms of chemical equilibrium."

Middle Ages, acids and bases have been put to practical use, such as in dying cloth and separating gold from silver, without really understanding what they were or how they worked.

Then, in the mid 17th century, **Robert Boyle**, famous for his gas laws, also moved the understanding of acids and bases forward through careful experimentation with dyes, which he discovered could be used as indicators of acid or base nature. In 1661, he summarized the properties of acids and bases (then called alkalies) as follows:

1. Acids have a sour taste.
2. Acids are corrosive.
3. Acids change the color of certain vegetable dyes, such as litmus, from blue to red.
4. Acids lose their acidity when they are combined with bases.
5. Bases feel slippery.
6. Bases change the color of litmus from red to blue.
7. Bases become less alkaline (basic) when they are combined with acids.

Boyle realized that bases "consume", or neutralize, acids. He found that color changes induced in his indicators by acids could be reversed by adding bases, and vice versa.

Chapter 6.6 Arrhenius and Bronsted-Lowry concepts of acids and bases

Two centuries after Boyle's methodical observations, **Svante Arrhenius** put forth the first post-periodic table theory of the nature of acids and bases. In 1884, he proposed that acids are neutral compounds that ionize when dissolved in water into a positive hydrogen ion and a corresponding negative ion. For HCl, this reaction is:

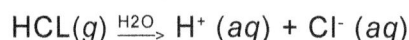

$$HCL(g) \xrightarrow{H2O} H^+ (aq) + Cl^- (aq)$$

We now know that in reality, the H+ ion does not float by itself in the aqueous solution; it is bound to a water molecule to form H_3O^+, known as a **hydronium ion**, so this reaction is more accurately depicted as:

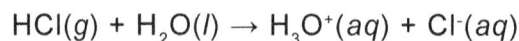

$$HCl(g) + H_2O(l) \rightarrow H_3O^+(aq) + Cl^-(aq)$$

Fundamentally, an acid is a compound with a weakly bound hydrogen atom that is subject to leave the compound when dissolved in a polar liquid like water.

For bases, Arrhenius proposed that bases are neutral compounds that ionize when dissolved in water into a negative OH- ion and a corresponding positive ion. For NaOH, this reaction is:

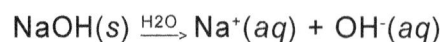

$$NaOH(s) \xrightarrow{H2O} Na^+(aq) + OH^-(aq)$$

Acids and bases that fit these definitions-- acids that dissolve into a positive H+

and a negative corresponding ion, and bases that dissolve into a negative OH⁻ and a corresponding positive ion-- are called **Arrhenius Acids** and **Arrhenius Bases.**

Arrhenius' theory was a very important advance in the understanding of acids and bases, but there were some "holes" in it; data it could not account for. Chiefly, Arrhenius' theory could not account for compounds which do not donate an OH⁻ ion, but are nevertheless basic, such as NH_3 or Na_2CO_3.

In the 1920s, Johannes **Brønsted** and Thomas **Lowry** addressed these issues with a slight, but important, change in Arrhenius' theory. Bronsted maintained Arrhenius' definition of acids as proton donors, but instead of bases being hydroxide ion donors, he proposed that bases were proton acceptors. For example, the equation for NH_3 is:

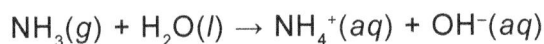

$$NH_3(g) + H_2O(l) \rightarrow NH_4^+(aq) + OH^-(aq)$$

The **Bronsted-Lowry** view of acids and bases makes it clear that acid-base reactions are reversible. Looking at the reverse of the reaction above:

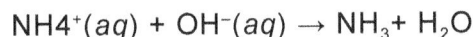

$$NH4^+(aq) + OH^-(aq) \rightarrow NH_3 + H_2O$$

Since a Bronstead acid is a proton donor, and a Bronstead base is a proton acceptor, **Brønsted acids and bases exist in conjugate pairs** that neutralize each other when they react. In the example above, NH_4^+ is the conjugate acid of NH_3- Acids that are able to transfer more than one proton are called **polyprotic acids**.

Examples:

1) In the reaction:

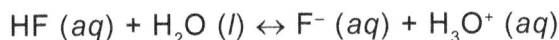

$$HF\ (aq) + H_2O\ (l) \leftrightarrow F^-\ (aq) + H_3O^+\ (aq)$$

HF transfers a proton to water. Therefore, HF is the Brønsted acid and H_2O is the Brønsted base. However, in the reverse direction, the hydronium ion transfers a proton to the fluoride ion. H_3O^+ is the conjugate acid of H_2O because it has an additional proton, and F^- is the conjugate base of HF because it lacks a proton.

2) In the reaction:

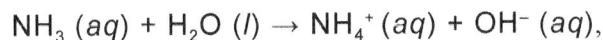

$$NH_3\ (aq) + H_2O\ (l) \rightarrow NH_4^+\ (aq) + OH^-\ (aq),$$

water transfers a proton to ammonia. H_2O is the Brønsted acid and OH^- is its conjugate base. NH_3 is the Brønsted base and NH_4^+ is its conjugate acid.

3) In the reaction:

$$H_3PO_4 + HS^- \leftrightarrow H_2PO_4^- + H_2S$$

$H_3PO_4/H_2PO_4^-$ is one conjugate acid-base pair and H_2S/HS^- is the other.

4) H_3PO_4 is a polyprotic acid. It may further dissociate to transfer more than one proton, so that the same compound can act as an acid or a base:

$$H_3PO_4 \leftrightarrow H_2PO_4^- + H+$$

$$H_2PO_4^- \leftrightarrow HPO_4^{2-} + H^+$$

$$HPO_4^{2-} \leftrightarrow PO_4^{3-} + H^+$$

A practical definition of acids and bases in water is the **Operational Definition** of acids and bases. In this view, an acid is any substance that, when dissolved in water, increases the concentration of H+ ions in the water (lowering the pH; more about this below). A base is any substance that, when dissolved in water, increases the concentration of OH- ions in the water (raising the pH, more about this below). Using this definition, it is easy to test whether a substance is basic or acidic; simply dissolve it in water and test the water, using litmus paper or a tool such as a pH meter, to see whether there is an excess of H+ ions or OH- ions.

Chapter 6.7 Dynamic equilibrium and pH

When an acid or a base dissolves in water, it is really reaching a **dynamic equilibrium** with the molecules of the water. Water molecules are **amphoteric**, meaning that they can both accept protons and donate protons. So, when an acid is dissolved in water, the water molecules are acting as a base by accepting the protons from the acid. A water molecule that has accepted a proton from an acid becomes a **hydronium ion** (H_3O^+). When a base is dissolved, the water molecules act as an acid by donating protons to the base. A water molecule that has donated a proton to a base becomes a **hydroxide ion** (OH-).

pH is a measure of the concentration of H+ ions, or more correctly in water, hydronium ions (H_3O^+). Roughly equal to the negative log10 of the concentration of H^+ ions, measured in moles/liter, pH values span a scale of 1 to 14 that is easy to use. Values less than 7 are acidic, more than 7 are basic, and equal to 7 are neutral. pOH is a measure of the concentration of OH- ions. In any solution, pH + pOH = 14, so pH = 14 - pOH, and pOH = 14 - pH. While pH meters provide a precise measure of this value, it can also be obtained approximately by using pH indicator paper.

Water itself is in fact in a state of dynamic equilibrium — pure water is a mix of H_2O, H_3O^+, and OH-, with the number of H_3O^+ ions being exactly equal to the number of OH- ions, giving a pH of 7.

Chapter 6.8 Acid-base nomenclature _____

Acids

There are special naming rules for acids that correspond with the **suffix of their corresponding anion** if hydrogen were removed from the acid (which is their conjugate base). Anions ending with –*ide* correspond to acids with the prefix *hydro*– and the suffix –*ic*. Anions ending with –*ate* correspond to acids with no prefix that end with –*ic*. Oxoanions ending with –*ite* have associated acids with no prefix and the suffix –*ous*. The *hypo*– and *per*– prefixes in the anions are retained in the acid. Some examples are shown in the following table (strong acids are in bold font):

anion	anion name	acid	acid name
Cl⁻	**chloride**	**HCl**	**hydrochloric acid**
CN^-	cyanide	HCN	hydrocyanic acid
CO_3^{2-}	carbonate	H_2CO_3	carbonic acid
SO_3^{2-}	sulfite	H_2SO_3	sulfurous acid
SO_4^{2-}	**sulfate**	**H_2SO_4**	**sulfuric acid**
ClO^-	hypochlorite	HClO	hypochlorous acid
ClO_2^-	chlorite	$HClO_2$	chlorous acid
ClO_3^-	chlorate	$HClO_3$	chloric acid
ClO_4^-	perchlorate	$HClO_4$	perchloric acid
Br⁻	bromide	HBr	hydrobromic acid
I⁻	iodide	HI	hydroiodic acid
NO_3^-	nitrate	HNO_3	nitric acid

Example: What is the molecular formula of phosphorous acid?

Solution: If we remember that the –*ous* acid corresponds to the –*ite* anion, and that the –*ite* anion has one less oxygen than (or has an oxidation number 2 less than) the –*ate* form, we only need to remember that phosphate is PO_4^{3-}. Then we know that phosphite is PO_3^{3-} and phosphorous acid is H_3PO_3.

Organic acids

Organic acids, which make up the majority of weak acids, are named for their **acyl group.** The acyl group, one or more carbons with covalently attached hydrogens, forms the backbone of many organic molecules, such as ketones, alcohols, esters, nitriles, and amides, as well as acids. In acids, a **carboxyl group** (COOH) is attached to the acyl group. They are acids because the electronegativity of the oxygens in this configuration makes it easy for these molecules to donate the end hydrogen atom as a proton. Organic acids are also known as **carboxylic acids.** To name these acids, the suffix "–oic" and the word "acid" is added to the name of the acyl group.

Small acyl groups are named with specific prefixes, such as "meth-" for one carbon and "eth-" for two carbons. Confusingly, there are at least two ways of naming each acyl group—the systematic (agreed-upon by IUPAC) way of naming chemicals, and the traditional way, which is generally not systematic, but may be more common. (For example, "acetic acid" is much more commonly used than "ethanoic acid".) You will need to know both, since both are commonly used in chemistry. The following table lists systematic and traditional names for small acyl groups and their corresponding acid names:

Number of carbons	Systematic prefix for acyl group	Systematic acid name	Traditional acid name	Formula
1	methan–	methanoic acid	formic acid	$HCOOH$
2	ethan–	ethanoic acid	acetic acid	CH_3COOH
3	propan–	propanoic acid	propanoic acid	CH_3CH_2COOH
4	butan–	butanoic acid	butyric acid	$CH_3CH_2CH_2COOH$
5	pentan–	pentanoic acid	valeric acid	$CH_3CH_2CH_2CH_2COOH$
6 and above	benzo-	benzoic acid	benzoic acid	$CH_3(CH_2)_4COOH$

Bases

The two main ways of naming bases reflect two main categories of bases: those that include hydroxide ions and those that include nitrogen atoms. The **hydroxide bases** consist of a metal and one or more hydroxide ions. To name these bases, simply name the metal and add the word "hydroxide". So, $NaOH$ is called sodium hydroxide, $Sr(OH)_2$ is called strontium hydroxide, and $Al(OH)_3$ is called aluminum hydroxide. The number of hydroxide ions does not change the name. If you are reading the name and looking for the formula, look at the periodic table—if the metal is in column 1, it will have one $OH-$; if it is in column 2, it will have two OH's, etc.

Nitrogen bases include ammonia ($NH3$) and organic compounds, also known as amines, with an ammonia-like group attached. These types of compounds, called amines, follow the same pattern as organic acids until there are six carbons. At that point, a benzene group forms, and the names change to be less systematic. The names of seven nitrogen bases are shown in the following table:

Number of carbons	Systematic prefix for acyl group	Base name	Formula
1	meth–	methylamine	CH_2NH_2
2	eth–	ethylamine	$CH_3CH_2NH_2$
3	prop–	propylamine	$CH_3CH_2CH_2NH_2$
4	but–	butylamine (4 isomers)	$CH3(CH_2)_3NH_2$
5	pent–	pentylamine	$CH_3(CH_2)_4NH_2$
6	benz-	Aniline, phenylamine, or aminobenzene	$C_6H_5NH_2$
7	benz-	benzylamine	$C_6H_5(CH_2)NH_2$

In addition to these, other nitrogenous bases can be formed by substituting more than one H on the nitrogen. If two hydrogens are substituted with one methyl group each, the base is called dimethylamine. It looks like this:

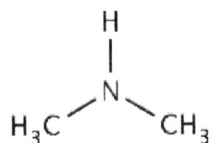

If three hydrogens are substituted, the base is called trimethylamine. It looks like this:

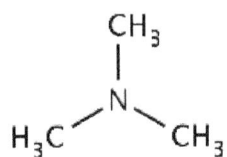

Extending the acyl groups, systematic names follow the same basic patterns as in the table above. For example, diethylamine has two ethyl groups (each two carbons in length) extending from the central nitrogen. In the case of two of more different groups, they are listed sequentially; for example, if one methyl group and one ethyl group extends from the N, the name is methylethylamine.

Chapter 6.9 Strong and weak acids and bases _____

Strong and weak acids

Both strong acids and weak acids transfer protons to water, but strong acids dissociate completely, transferring all of their H+ ions to water, whereas weak acids dissociate partially, transferring only some of their H+ ions to water. This means that strong acids are completely ionized, and (outside of cases of an acid being relatively insoluble, such as in a saturated solution) the concentration of H_3O^+ in a strong acid solution is equal to the initial concentration of the strong acid. Strong acids include binary acids of highly electronegative ions, like Cl-, Br-, and I-, and oxoacids. Superacids such as carborane $[H(CHB_{11}Cl_{11})]$, the strongest acid in the world, derive their strength from multiple electronegative central atoms.

Organic acids and acids like H_2SO_4 are **oxoacids**; the hydrogen that they will donate as a proton is attached to an oxygen. The more oxygen atoms, the stronger the acid for the same base atom. For example, sulfuric acid (H_2SO_4) is stronger than sulfurous acid (H_2SO_3). This is because the electrons are being pulled by all those electronegative oxygen atoms away from the hydrogen at the end, making it easier for the hydrogen to break off as a proton.

The other factor in oxoacid strength is the electronegativity of the central atom, which is reflected in the position of the central atom in the periodic table. For acids with the same number of oxygen atoms, acids with more electronegative central atoms will be stronger acids-- the higher the row in the periodic table and the farther right in the periodic table, the stronger the acid. The reason for this is again because more electronegative central atoms "hog" electrons, pulling them towards themselves and making it easier for the hydrogen to break of as a proton.

Six common strong acids, in order from strongest to weakest, are: perchloric acid, hydroiodic acid, hydrobromic acid, hydrochloric acid, sulfuric acid, and nitric acid. All six are stronger acids than the hydronium ion of water, and fully dissociate in water.

Weak acids are only partially ionized in water. They include the organic (carboxylic) acids and mineral acids with less-electronegative central atoms and/or oxoacids with fewer oxygen atoms. The vast majority of weak acids are organic acids. Some common non-organic weak acids include: HF, H_3PO_4, H_3AsO_4, $HClO_3$, $HClO_2$, and HClO. Organic acids are stronger if they have a highly electronegative atom attached. For example, adding chlorine atoms steadily increases the strength of ethanoic acid (acetic acid, the weak acid in vinegar), so that chloroethanoic acid ($CH_2ClCOOH$) is stronger than ethanoic acid (CH_3COOH), dichloroethanoic acid (CH_3Cl_2COOH) is stronger than chloroethanoic acid, and trichloroethanoic acid (CCl_3COOH) is the strongest of the four.

When a weak acid is added to water, some of the water molecules will accept a proton from the weak acid, but some of the weak acid molecules will accept a proton from the water molecules:

weak acid + H_2O ⇔ conjugate weak base + H_3O^+

"Common strong acids include HCl, HBr, HI, $HClO_4$, H_2SO_4, and HNO_3. The molecules of strong acids completely ionize in solution to produce hydronium ions."

The result will be reflected in a single quantitative value, the pH, but its nature is a dynamic equilibrium, just as solubility is, and is subject to Le Chatelier's principle in the same way. The ratio is determined by the strength of the weak acid; in other words, the equilibrium, and therefore the pH, is determined by the pK_a of the weak acid.

The pK_a of an acid is determined by the structure of the acid; by factors such as bond strength and electronegativity. The strongest acid in a polyprotic (more than one acidic hydrogen) series is always **the acid with the most protons** (e.g. H_2SO_4 is a stronger acid than HSO_4^-). The strongest acid in a series with the same central atom is always **the acid with the central atom at the highest oxidation number** (e.g. $HClO_4 > HClO_3 > HClO_2 > HClO$). The strongest acid in a series with different central atoms at the same oxidation number is usually **the acid with the central atom at the highest electronegativity** (e.g. the K_a of HClO > HBrO > HIO). This electronegativity trend stretches across the periodic table for oxides.

Strong and weak bases

Metal hydroxides are strong bases, but most of these are not very soluble, and weakly soluble strong bases have only mild effects. (However, 100% of the *dissolved* base is ionized.) Ionic oxides containing a large cation with a low charge (Rb_2O, for example) are most soluble and form the strongest bases.

Common strong bases that are soluble (and therefore have strong effects) include sodium hydroxide (NaOH), potassium hydroxide (KOH), cesium hydroxide (CsOH), and lithium hydroxide (LiOH). Atoms with lower electronegativity make stronger bases.

All bases related to ammonia, including organic bases, known as amines, are weak bases. Like weak acids, they are partially ionized in solution, and in dynamic equilibrium with water. Just as the pH of a weak acid solution can be determined by pKa, the pH of a weak base solution can be determined through pK_b .

weak base + H₂O <=> conjugate weak acid + OH-

Acid and base **strength is not necessarily related to safety**. Weak acids like HF may be extremely corrosive and dangerous. Conversely, carborane $[H(CHB_{11}Cl_{11})]$, the strongest acid in the world, is relatively non-corrosive.

The **common strong acids and bases** are reviewed here:

Strong acid		Strong base	
HCl	Hydrochloric acid	LiOH	Lithium hydroxide
HBr	Hydrobromic acid	NaOH	Sodium hydroxide
HI	Hydroiodic acid	KOH	Potassium hydroxide
HNO_3	Nitric acid	$Ca(OH)_2$	Calcium hydroxide
H_2SO_4	Sulfuric acid	$Sr(OH)_2$	Strontium hydroxide
$HClO_4$	Perchloric acid	$Ba(OH)_2$	Barium hydroxide

"Weak acid molecules react with water to transfer a proton to the water molecule. However, weak acid molecules only partially ionize in this way. The equilibrium constant for this reaction is Ka, often reported as pKa The pH of a weak acid solution can be determined from the initial acid concentration and the pKa."

"Common strong bases include group I and II hydroxides. When dissolved in solution, strong bases completely dissociate to produce hydroxide ions."

Chapter 6.10 Conjugate acid/base pairs

Conjugate acids and bases are simply the same molecule with a proton on or off, respectively; a conjugate acid is its conjugate base with a proton. A conjugate acid/base pair in solution can be represented as:

$$\text{conjugate acid} \leftrightarrow \text{conjugate base} + H^+$$

In general, **if two acid/base conjugate pairs are present, the stronger acid will transfer a proton to the conjugate base of the weaker acid.**

In water, the vast majority of weak acid molecules remain un-ionized. However, it is important to remember that this is not a static endpoint-- a solution of a weak acid is a dynamic equilibrium between the un-ionized acid and its conjugate base. Though it is convenient to think of it as a solution of fixed concentration, it is in fact a constantly changing mix of protons off and protons on, since the water is acting as a conjugate base or conjugate acid to the dissolved acid or base. The concentration and pH remain constant because, at equilibrium, the proton transfers in the forward direction are balanced by those in the reverse direction.

The concentration of H_3O^+, which determines the acidity of the solution, depends on the **aqueous dissociation constants,** K_a and K_b, which quantify acid and base strength, respectively. At 25°C, for any conjugate acid-base pair, $pK_a + pK_b = 14$. K_a and K_b values have been compiled in tables, allowing easy calculation of the predicted acidity of a solution. K_w is the **autoionization constant of water**, and is equal to $[H^+][OH^-]$ (1.0 x 10^{-14} for pure water at 25°C). Just as we use pH, the negative log of the concentration of protons, because it is more manageable for calculations, we also use pK_a, pK_b, and pK_w in quantifying acid and base strength since it is less awkward and more intuitive to work with the negative log of these constants. The negative log of K_a, pK_a, can be used to calculate pH.

Using ethanoic acid as an example, the dissociation of a weak acid in water looks like this:

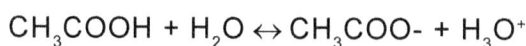

$$CH_3COOH + H_2O \leftrightarrow CH_3COO\text{-} + H_3O^+$$

$CH_3COO\text{-}$ is the conjugate base of CH_3COOH,

and H_3O^+ is the conjugate acid of H_2O.

Whereas, for weak acids, you need to know the K_a of the acid to calculate the concentration of H_3O^+, because strong acids completely dissociate in water, the molar concentration of H_3O^+ is simply equal to the initial concentration of the acid.

Like strong acids, strong bases completely dissociate in water. They dissociate into their conjugate acids and hydroxide ions. (However, remember that many strong bases are not very soluble in water.)

Like weak acids, weak bases only partially dissociate. They form a mixture of their conjugate acids and conjugate bases in water, with the composition of the mixture determined by the K_b of the base.

"When an acid molecule loses its proton, it becomes a base, since the resultant ion could react with water as a base. The acid and base are referred to as a conjugate acid-base pair."

"For pure water, pH = pOH, and this condition is called "neutrality". At 25°C, pK_w = 14 and thus pH and pOH add to 14. In pure water at 25°C, pH = pOH = 7."

For example, the dissociation of the weak base NH_3 in water looks like this:

$$NH_3 + H_2O \leftrightarrow NH_4^+ + OH^-$$

where NH_4^+ is the conjugate acid of NH_3.

It is important to remember that pH depends on both the strength of the acid or base and its concentration. A given acidic pH can be obtained by either a strong acid or a weak acid. However, since the weak acid will be only partially dissociated, more of it will be required. To achieve a given acidic pH, the concentration of a weak acid will have to be much greater than the concentration of a strong acid capable of achieving the same pH. Given the right concentrations, though, both can be used to create solutions with the same $[H_3O^+]$.

Chapter 6.11 Reactions of acids and bases

The chemical reaction between an acid and a base is called **neutralization**. The products of neutralization reactions are neither acids nor bases; the H^+ and OH^- combine to form water, and the cation of the base and anion of the acid combine to form a **salt**. The essential neutralization reaction itself, regardless of the type of acid and base, is:

$$H_3O^+ + OH^- => H_2O$$

When a strong acid reacts with a strong base, the reaction goes to completion. This means that all of the limiting reagent will be used up in the reaction, and the end product will depend on whichever reactant is in excess.

For example, **HCl and NaOH** will react completely to produce the salt **NaCl** and water:

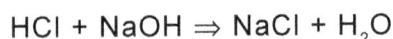

$$HCl + NaOH \Rightarrow NaCl + H_2O$$

So, if there is more HCl to start with, the ending solution will have an excess of H_3O^+ ions, and will be acidic. If there is more NaOH to start with, the end solution will have an excess of OH^- ions, and will be basic. If equal molar amounts of the strong base and the strong acid are added, the resulting solution will be neutral.

The anion of the salt (Cl^- in the example above) is the conjugate base of the acid, and the cation of the salt (Na^+ in the example above) is the conjugate acid of the base. In this example, a strong acid is reacting with a strong base, so equal molar amounts of each result in a neutral solution. However, if the acid is a weak acid, its conjugate base will react with water, increasing the concentrations of OH^- ions and making the solution more basic. If the cation of a salt is the conjugate acid of a weak base, it will react with water to increase the concentration of H_3O^+ ions, making the solution more acidic.

When a strong base is added to a weak acid, the essential reaction is:

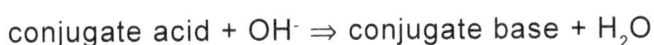

$$\text{conjugate acid} + OH^- \Rightarrow \text{conjugate base} + H_2O$$

"A solution of a weak base involves an equilibrium between an un-ionized base and its conjugate acid. The equilibrium constant for this reaction is K_b, often reported as pK_b. The pH of a weak base solution can be determined from the initial base concentration and the K_b."

"In a neutralization reaction, protons are transferred from an acid to a base."

When a strong acid is added to a weak base, the essential reaction is:

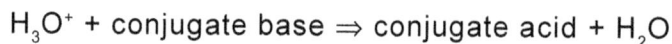

$$H_3O^+ + \text{conjugate base} \Rightarrow \text{conjugate acid} + H_2O$$

To summarize, the effect of a salt on a water solution will be as follows. When both the acid and the base forming the salt are strong, the salt solution will be neutral. When a strong acid reacts with a weak base to create the salt, the salt solution will be acidic. When a weak acid reacts with a strong base to create the salt, the salt solution will be basic. When a weak acid and a weak base react, whether the resulting salt solution will be acidic or basic depends on the aqueous dissociation constants, K_a and K_b. If K_a is bigger, that means that the acid is the stronger of the two, and the salt solution will be acidic. If K_b is bigger, that means that the base is stronger, and the salt solution will be basic.

Ironically, a weak acid solution at a given pH will resist pH change by a strong base more than a strong acid at the same pH. This is because, in order to create a solution with the same pH, much more weak acid will be required. That means, more acid is present to react with the strong base and neutralize it. We will talk more about this below, when we discuss buffers and titration.

Chapter 6.12 Quantitative Acid-Base Chemistry _____

Dissociation

As mentioned above, water itself can be considered a weak acid, with H_2O the conjugate base of H_3O^+, and the equilibrium constant for the dissociation of water is K_w, the **autoionization constant of water**.

$$K_w = [H^+][OH^-] = 1.0 \times 10^{-14}$$

The negative log of this constant, pK_w, is equal to pH + pOH, which is equal to 14.

Pure water at equilibrium has an equal concentration of the two ions. Therefore,

$$K_w = [H^+]^2 = 1.0 \times 10^{-14}$$

Solving for $[H^+]$ yields $[H^+] = 1.0 \times 10^{-7}$ M for pure water.

Remember, the concentration of H^+ (*aq*) ions is often expressed in terms of pH. **The pH of a solution is the negative base-10 logarithm of the hydrogen-ion molarity:**

$$pH = -\log[H^+] = \log(1/[H^+])$$

A ten-fold increase in [H+] decreases the pH by one unit. [H+] may be found from pH using the expression:

$$[H^+] = 10^{-pH}$$

Because $[H^+] = 10^{-7}$ M for pure water, the **pH of a neutral solution is 7**. In an

acidic solution, $[H^+] > 10^{-7}$ M and pH < 7. In a basic solution, $[H^+] < 10^{-7}$ M and pH > 7.

Example: An aqueous solution has an H^+ ion concentration of 4.0×10^{-9}. Is the solution acidic or basic? What is the pH of the solution?

Solution: The solution is basic because $[H^+] < 10^{-7}$ M.

$$pH = -\log[H^+] = -\log 4 \times 10^{-9} = 8.4$$

The pH and [H+] of a solution containing a strong acid or strong base may be found using stoichiometry alone for a strong acid, and stoichiometry together with K_w for a base.

The **acid-dissociation constant**, K_a, is the equilibrium constant for the ionization of a weak acid to a hydrogen ion and its conjugate base.

$$HX\ (aq) \leftrightarrow H^+\ (aq) + X^-\ (aq)$$

Polyprotic acids have unique values for each dissociation: K_{a1}, K_{a2}, etc.

Example: Hydrofluoric acid is dissolved in pure water until $[H^+]$ reaches 0.006 M. What is the concentration of undissociated HF? K_a for HF is 6.8×10^{-4}.

Solution: $HF\ (aq) \quad H^+\ (aq) + F^-\ (aq), \quad K_a = \dfrac{[H^+][F^-]}{[HF]}$

The principle source of both ions is dissociation of HF (autoionization of water is negligible). Therefore $[F-] = [H+] = 0.006$ M, and

$$[HF] = \frac{[H^+][F^-]}{K_a} = \frac{(0.006)^2}{6.8 \times 10^{-4}} = 0.05\ M$$

"The **base-dissociation constant**, K_b, is the equilibrium constant for the addition of a proton to a weak base by water to form its conjugate acid and an OH^- ion. In these reactions, it is water that is dissociating as a result of reaction with the base:

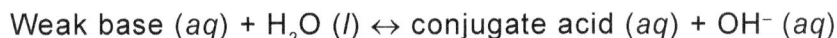

$$Weak\ base\ (aq) + H_2O\ (l) \leftrightarrow conjugate\ acid\ (aq) + OH^-\ (aq)$$

$$K_b = \frac{[conjugate\ acid][OH^-]}{[weak\ base]}$$

The concentration of water is nearly constant and is incorporated into the dissociation constant.

For ammonia (the most common weak base), the equilibrium reaction and base-dissociation constant are:

The pH of an acid solution depends on both the strength of the acid and the concentration of the acid. To achieve solutions of equal pH, the weak acid solution must be at a much greater concentration than the strong acid solution."

$$NH_3 \ (aq) + H_2O \ (l) \leftrightarrow NH_4^+ \ (aq) + OH^- \ (aq)$$

$$K_b = \frac{\left[NH_4^+\right]\left[OH^-\right]}{\left[NH_3\right]}$$

Example: K_b for ammonia at 25 °C is 1.8 X 10^{-5}. What is the concentration of OH^- in an ammonia solution at equilibrium containing 0.2 M NH_3 at 25°C?

Solution: Let x = $[OH^-]$. The principle source of both ions is NH_3 (autoionization of water is negligible).

Therefore: x = $[OH^-]$ = $[NH_4^+]$.

$$K_b = \frac{\left[NH_4^+\right]\left[OH^-\right]}{\left[NH_3\right]} = \frac{x^2}{0.2} = 1.8 \times 10^{-5}$$

Solving: for x yields: $x = \sqrt{(0.2)\left(1.8 \times 10^{-5}\right)} = 0.002 \ M \ OH^-$

The product of K_a for an acid and K_b for its conjugate base will always be K_w. This is demonstrated below for the weak acid, HF and its conjugate base, F^-:

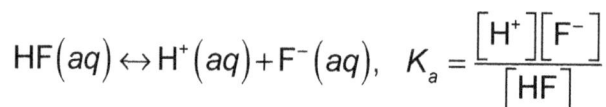

$$HF(aq) \leftrightarrow H^+(aq) + F^-(aq), \quad K_a = \frac{\left[H^+\right]\left[F^-\right]}{\left[HF\right]}$$

$$F^- \ (aq) + H_2O \ (l) \leftrightarrow HF \ (aq) + OH^- \ (aq) \quad Kb = [HF] \ [OH-] \ / \ [f-]$$

Multiplication of K_a and K_b yields:

$$Ka * Kb = [H+] \ [F-] \ [HF] \ [OH-] \ / \ [HF] \ [F-] = [H+] \ [OH-] = K_w$$

pH from strong acids and bases

Since strong acids completely dissociate, the $[H^+]$ of a strong acid is equal to the molar concentration of the acid. The pH of a strong acid is therefore:

$$pH = -\log [H^+] = -\log M_{strong \ acid}$$

For strong bases, the $[OH^-]$ is equal to the molar concentration of the base multiplied by the number of $[OH^-]$ ions per mole of the base. So, the concentration of OH^- ions will be twice as high for the strong base $Ba(OH)_2$ as for the strong base $NaOH$. (This is different from polyprotic strong acids because the second hydrogen in a polyprotic acid will not dissociate completely. Remember, while H_2SO_4 is a strong acid, HSO_4^- is not, so it would be a mistake to think that H_2SO_4 results in twice the $[H^+]$ as a monoprotic strong acid like HCl.)

$$pOH = -\log [OH-] = -\log M_{strong \ base} \times number \ of \ OH- \ ions \ per \ mole$$

After using [OH-] to calculate pOH, simply subtract this from 14 to get the pH. To calculate the pH of a strong base:

pH = 14 − pOH = 14 − (-log M$_{strong\ base}$ × number of OH- ions per mole)

pH from weak acids and bases

To calculate the pH of solutions from weak acids and weak bases, we need to use K_a and K_b. Consider the example of acetic acid (ethanoic acid; CH_3COOH).

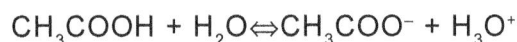

$$CH_3COOH + H_2O \Leftrightarrow CH_3COO^- + H_3O^+$$

The concentrations for this reaction will be related to K_a:

$$K_a = [H^+][CH_3COO^-] / [CH_3COOH]$$

Looking up K_a in a table, we find that the K_a for acetic acid is 1.8×10^{-5}. If we add one mole of acetic acid to one liter of water, we have an initial concentration of CH_3COOH of 1 mol/L, an initial concentration of CH_3COO^- of 0 mol/L, and an initial concentration of H_3O^+ of 0 mol/L. The loss of protons by the acid will be equal to the gain of protons by the base, and the concentration of conjugate base will also be equal to the protons gained by the base. In table form, this looks like:

	CH_3COOH	H^+	CH_3COO^-
Initial Concentration	1 mol/L	0	0
Change	-x	+x	+x
Equilibrium Concentration	1 − x	0 + x	0 + x

Tables like this are commonly referred to as ICE tables to represent the Initial, Change, and Equilibrium states of the system. These tables appear often in the Free-Response section of the exam due to the intensity of the required math. (See the section on math review for an ICE Table calculation that requires knowledge of the quadratic formula.)

So, to solve for K_a, plug the values from the table into the equation above:

$$K_a = [H^+][CH_3COO^-] / [CH_3COOH]$$

$$1.8 \times 10^{-5} = (0 + x)(0 + x) / (1 − x)$$

For weak acids, x can be thought to be negligible in the initial concentration term if x is less than one tenth of the initial concentration. To see if this might be true, we can run the calculation as if it were true:

$$1.8 \times 10^{-5} = x^2 / 1$$

$$x = \sqrt{(1.8 \times 10^{-5})} = .00424$$

Since the square root of 1.8×10^{-5} is less than one tenth of 1, the change in the concentration of acid can indeed be considered negligible, so we can use this value for x. Solving for x using this caveat:

	CH_3COOH	H^+	CH_3COO^-
Initial Concentration	1 mol/L	0	0
Change	-x	+x	+x
Concentration at Equilibrium	0.996 mol/L	.00424 mol/L	.00424 mol/L

If the concentration of H+ is .0042 mol/L, then:

$$pH = -log [H^+] = -log (.00424) = 2.37$$

So, using the K_a, we can predict that adding 1 mole of acetic acid to 1 liter of water will yield a pH of 2.37.

For bases, the problem can be set up the same way, except that water must be included when considering weak bases:

	NH_3	H_2O	NH_4^+	OH^-
Initial Concentration	1 mol/L	1 L	0	0
Change	-x	-x	+x	+x
Concentration at Equilibrium	1 − x	~1	0 + x	0 + x

Henderson-Hasselbalch equation

An easier way to do these kinds of problems quickly is to use the Henderson-Hasselbalch equation. This is just a rearrangement of the calculations above:

$$pH = pK_a + log([HA]/[A-])$$

and

$$pOH = pK_b + log([BH+]/[B])$$

This equation shows clearly that, when the concentrations of conjugate acid and conjugate base are the same (i.e., the solution is neutral), the pH is equal to the pK_a and the pOH is equal to the pK_b.

Calculating the pH of salt solutions

Remember, the acidity of a salt solution depends on the nature of the acids and bases that created it. To review:

strong acid + strong base $\Rightarrow$ neutral salt solution
strong acid + weak base $\Rightarrow$ acidic salt solution
weak acid + strong base $\Rightarrow$ basic salt solution
weak acid + weak base $\Rightarrow$ depends on K_a and K_b

To calculate the precise pH of basic and acidic salts, we use the K_a and K_b of the conjugate acid or base that makes up the salt, and the concentration of the salt. (Calculating the pH of mixed salts is beyond the scope of the AP exam, so we will focus on the pH of single salts only.) Derived from the equation $K_w = (K_a)(K_b)$ are two simplified equations to calculate the pH and pOH of acidic and basic salts, respectively. C_s is the concentration of the salt.

Acidic salts:

$$[H+] = \sqrt{(K_w/K_b) * C_s}$$

Basic salts:

$$[OH-] = \sqrt{(K_w/K_a) * C_s}$$

Both of these equations can be used as long as the concentration of the salt is greater than $100(K_w/K_b)$.

For example, let us consider NaF, a basic salt, dissolved in water at a concentration of 1 mol/L. (We already know it is basic, since the parent base, NaOH, is a strong base, while the parent acid, HF, is a weak acid.) The Na+ can be ignored, since it will not act as an OH- acceptor. The Ka of HF is 7.2×10^{-4}. The reaction we will consider, then, is:

$$F- + H_2O \Leftrightarrow HF + OH-$$

Using the above equation, $[OH-] = \sqrt{(K_w/K_a) * C_s}$,

$$[OH-] = \sqrt{(1 \times 10^{-14} / 7.2 \times 10^{-4}) * 1 \text{ mol/L}}$$

$$[OH-] = \sqrt{1.4 * 10^{-11}}$$

$$[OH-] = 3.7 * 10^{-6}$$

$$pOH = -\log(3.7 * 10^{-6}) = 5.4$$

$$pH = 14 - 5.4 = 8.6$$

Remember, you will not be asked to solve any problems that require a calculator on the multiple choice section of the AP Chemistry exam, but you will be expected to understand this material qualitatively. You may be asked to solve a problem that has been set up so that a calculator is not required, or choices may be given that are in ranges such that you can choose an answer without having calculated the exact right answer. However, questions such as the ones detailed above frequently appear as Free Response questions so be sure to know how to set up an ICE table and perform the appropriate algebraic manipulations. (See the chapter on math review for examples.)

Chapter 6.13 Titration

Standard titration

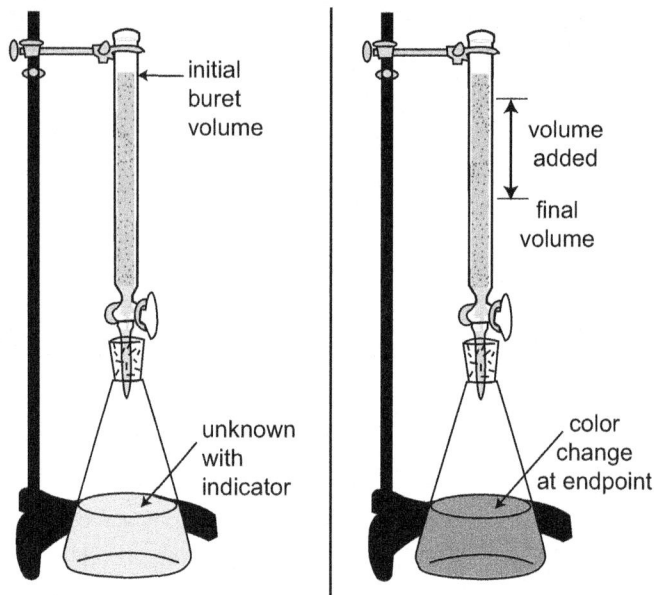

In a typical acid-base titration, an acid-base indicator (such as phenolphthalein) or a pH meter is used to monitor the course of a neutralization reaction. The goal of titration is to determine an unknown concentration of an acid (or base) by neutralizing it with a known concentration of base (or acid).

To measure the concentration of acid or base in an unknown solution, the **titrant**, the strong acid or base of known concentration, is poured into a **buret** (also spelled burette) until it is nearly full, and an initial buret reading is taken. Buret numbering is close to zero when nearly full. A known volume of the solution of unknown concentration is added to a flask and placed under the buret. An indicator is added (or the pH meter probe is inserted).

The buret stopcock is opened, and titrant is slowly added until the pH meter reads 7 (for a strong acid and strong base) or the solution permanently changes color, which indicates a rapid change in pH. This is the titration **endpoint** if we are doing a neutralization titration, and a final buret reading is made. This buret reading tells us how much of the known-concentration titrant was required to neutralize the unknown solution. The endpoint occurs when the number of **acid and base equivalents in the flask are identical**.

The endpoint is also known as the titration **equivalence point**.

Titration data typically consist of:

- Initial buret volume

- Final buret volume

- Concentration of known solution

- Volume of unknown solution

To determine the unknown concentration, first find the volume of titrant added at the known concentration by subtracting the initial buret volume from the final buret volume. We will call this $V_{titrant}$.

The AP exam will only consider acids and bases with one equivalent; e.g. monoprotic acids and bases. At the equivalence point, **the number of moles of the unknown solution will equal the moles of the added titrant**. Since the number of moles is equal to the molar concentration times volume,

$$(C_{unknown})(V_{unknown}) = (C_{titrant})(V_{titrant})$$

Therefore,

$$C_{unknown} = (C_{titrant})(V_{titrant}) / V_{unknown}$$

Units of molarity may be used for concentration, since the AP exam considers only monoprotic acids and bases in titration calculations.

Example: A 20.0 mL sample of an HCl solution is titrated with 0.200 M NaOH. The initial buret volume is 1.8 mL and the final buret volume at the $\text{titration}_{endpoint}$ is 29.1 mL. What is the molarity of the HCl sample?

Solution: Two solution methods will be used.

A. Calculate the moles of the known substance added to the flask:

$$0.200 \frac{mol}{L} \times \frac{1\ L}{1000\ mL} \times (29.1\ mL - 1.8\ mL) = 0.00546\ mol\ NaOH.$$

At the endpoint, this base will neutralize 0.00546 mol HCl. Therefore, this amount of HCl must have been present in the sample before the titration.

$$\frac{0.00546\ mol\ HCl}{0.0200\ L} = 0.273\ M\ HCl.$$

B. Utilize the formula:

$$C_{unknown} = (C_{titrant})(V_{titrant}) / V_{unknown}$$

$$C_{HCl} = C_{NaOH} * V_{NaOH} / V_{HCl}$$

$$C_{HCl} = 0.200\ M * 27.3\ mL / 20.0\ mL$$

$$C_{HCl} = 0.273\ M$$

Titrating with the unknown

In a common variation of standard titration, an unknown is added to the buret as a titrant, and the reagent of known concentration is placed in the flask. The chemistry involved is the same as in the standard case, and the mathematics is also identical except for the identity of the two volumes.

"At the equivalence point, the moles of titrant and the moles of titrate are present in stoichiometric proportions. In the vicinity of the equivalence point, the pH rapidly changes. This can be used to determine the concentration of the titrant."

Interpreting titration curves

A **titration curve** is a plot of a solution's **pH charted against the volume of an added acid or base**. Titration curves are obtained if a pH meter is used to monitor the titration, instead of an indicator.

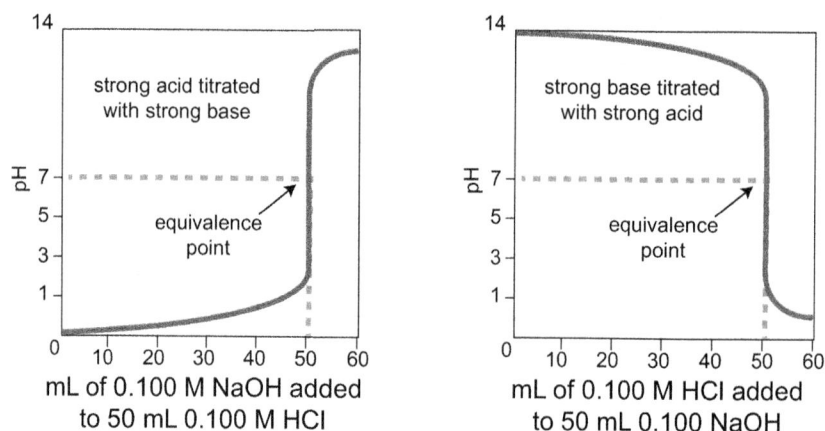

strong acid titrated
with strong base

equivalence
point

mL of 0.100 M NaOH added
to 50 mL 0.100 M HCl

strong base titrated
with strong acid

equivalence
point

mL of 0.100 M HCl added
to 50 mL 0.100 NaOH

In the case of strong acids and bases, as the titration proceeds, the pH barely changes, until the equivalent point, so the shape of the titration curve will be very flat-- 10-90% of the titrant will be added to produce a pH change of less than 1.5. Then, at the equivalence point, the titration curve becomes nearly vertical, with the pH changing very rapidly with just a small amount of added titrant. In addition to determining the equivalence point, the **shape of titration curves** may be interpreted to determine **acid/base strength and the presence of a polyprotic acid.**

The pH at the equivalence point of a titration is the **pH of the salt solution obtained when the amount of acid is equal to amount of base**. For a strong acid and a strong base, the equivalence point occurs at the neutral pH of 7. For example, an equimolar solution of HCl and NaOH will contain NaCl(aq) at its equivalence point.

The salt solution at **the equivalence point of a titration involving a weak acid or base will not be at neutral pH**. For example, an equimolar solution of NaOH and hypochlorous acid HClO at the equivalence point of a titration will be a base because it is indistinguishable from a solution of sodium hypochlorite. A pure solution of NaClO(aq) will be a base because the ClO^- ion is the conjugate base of HClO, and it consumes $H^+(aq)$ in the reaction: $ClO + H^+ => HClO$.

In a similar fashion, an equimolar solution of HCl and NH_3 will be an acid because a solution of $NH_4Cl(aq)$ is an acid. It generates $H^+(aq)$ in the reaction: $NH_4^+ \Rightarrow NH_3 + H^+$.

Contrast the following **titration curves for a weak acid or base** with those for a strong acid and strong base above:

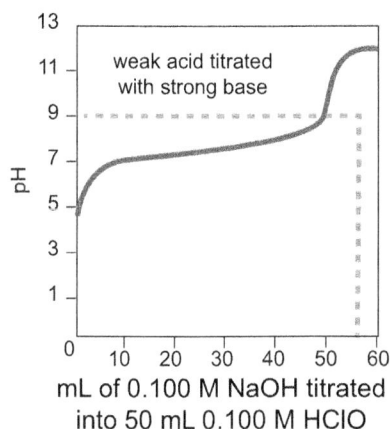

mL of 0.100 M NaOH titrated into 50 mL 0.100 M HClO

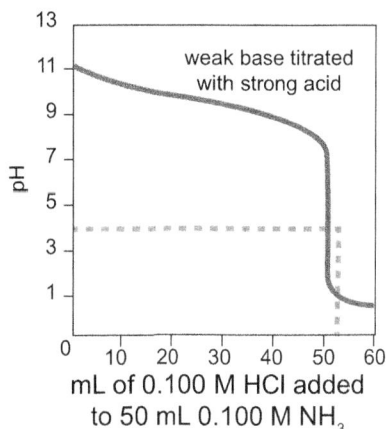

mL of 0.100 M HCl added to 50 mL 0.100 M NH$_3$

Titration of **polyprotic acid** results in **multiple equivalence points** and a curve with more "bumps" as shown below for sulfurous acid and the carbonate ion:

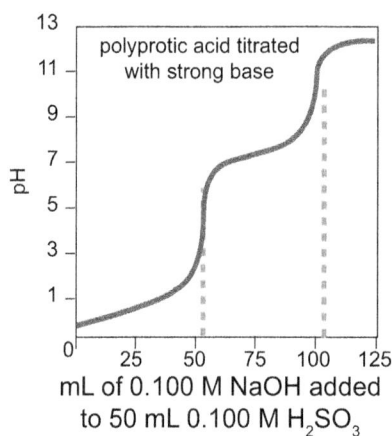

mL of 0.100 M NaOH added to 50 mL 0.100 M H$_2$SO$_3$

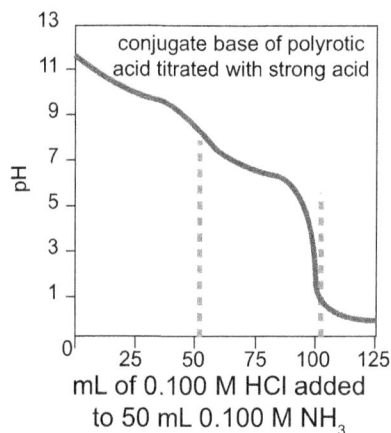

mL of 0.100 M HCl added to 50 mL 0.100 M NH$_3$

Note: On the AP exam, you will not be asked to compute the concentration of each species present in the titration curve for polyprotic acids. However, you will be expected to be able to determine the number of labile protons, and you will be expected to be able to see which species are present in large concentrations in any given location on the curve.

Half-equivalence points

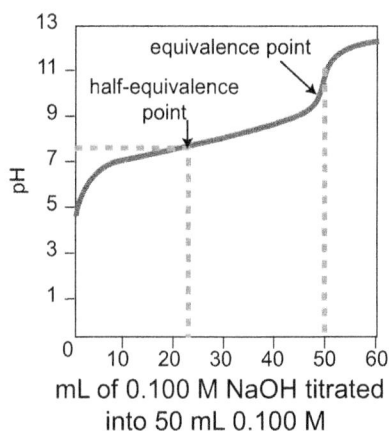

mL of 0.100 M NaOH titrated into 50 mL 0.100 M

The volume at the half-equivalence point, that is, half of the volume needed for the titration to reach the equivalence point, can be used to calculate the pK_a of an acid. For monoprotic acids, the pK_a is equal to the pH at the half-equivalence point. This pH can be determined experimentally by pH meter, or can be approximated by looking at a titration curve. For the titration curve to the left, for the weak acid HClO, the pH at the half-equivalence point is approximately 7.5. This indicates that HClO is a very weak acid.

The chemical species present at the half-equivalent point differ between strong and weak acids. At the half-equivalence point of a strong acid, the main species present will be H_3O^+, the anion of the acid, and the cation of the base. At the half-equivalence point of a weak acid, the main species will be H_3O^+, the anion of the acid (conjugate base of the acid), the cation of the base, and the undissociated acid (HA), with the undissociated acid and its conjugate base present at equal concentrations.

In both cases, the total positive charge is equal to the total negative charge.

Chapter 6.14 Buffers, pH, and pK$_a$

As a measure of the tendency of a solution to accept or donate a proton, pH is an important feature of a solution. Above, we have discussed how pH can be lowered by adding acid, raised by adding base, or neutralized when an acid and base react. For many applications, and in living systems, it is vital that the pH not change, or change very little.

A **buffer solution** is a solution that **resists a change in pH** upon the addition of small amounts of an acid or a base. Buffers are important when a change in pH would be detrimental, for example, in cell growth medium. Buffer solutions require a large concentration of an acid to neutralize any added base and a large concentration of base to neutralize any added acid. These two components present in the buffer also must not neutralize each other. The best way to achieve these aims is a **conjugate acid-base pair** of a weak acid or weak base. Buffers are prepared by mixing **a weak acid or base** with a **salt of the acid or base**, providing the conjugate.

Consider the buffer solution prepared by mixing together acetic acid ($HC_2H_3O_2$) and sodium acetate ($C_2H_3O_2^-$) and containing Na^+ as a spectator ion. The equilibrium reaction for this acid/conjugate base pair is:

$$HC_2H_3O_2 \leftrightarrow C_2H_3O_2^- + H^+$$

If H^+ ions from a strong acid are added to this buffer solution, Le Châtelier's principle predicts that the reaction will shift to the left and much of this H^+ will be consumed to create more $HC_2H_3O_2$ from $C_2H_3O_2^-$. If a strong base that consumes H^+ is added to this buffer solution, Le Châtelier's principle predicts that the reaction will shift to the right and much of the consumed H^+ will be replaced by the dissociation of $HC_2H_3O_2$. The net effect is that **buffer solutions prevent large changes in pH that occur when an acid or base is added to pure water** or to an unbuffered solution. Natural buffers such as blood and seawater contain several conjugate acid-base pairs to buffer the solution's pH and decrease the impact of acids and bases on living things.

The amount of acid or base that a buffer solution can neutralize before large pH change begins to occur is called its **buffering capacity**. While calculation of exact pH changes in buffers is beyond the scope of the AP Chemistry exam, it is important to remember that buffering capacity is a function of the absolute concentrations of the acid and base forms. Weak acids and bases make better buffers because a much higher concentration of the acid or base is required to achieve a given pH.

"For a weak acid solution and a strong acid solution with the same pH, it takes much more base to neutralize the weak acid solution because the initial acid concentration is much larger. Therefore, a weak acid solution resists changes in pH for a much greater amount of added base."

To have a high enough concentration of each conjugate (acid and base), it is important to choose weak acids or bases with pK_a values near the desired pH of the buffer you are designing. The pK_a does not have to exactly equal the desired pH; the solution can be adjusted slightly by carefully adding drops of strong base or strong acid, using the same technique and equipment as described for titration.

If you know the pH of a buffer solution and the pK_a of the buffer, you can determine the protonation state of the acid; that is, the ratio of conjugate acid to conjugate base. You will not need to do this exact calculation, but you will be expected to conceptually understand this and use this understanding to answer questions based on reasoning.

An excellent flash animation with audio to explain the action of buffering solutions is found at http://www.mhhe.com/physsci/chemistry/essentialchemistry/flash/buffer12.swf.

Lewis acids and bases: complexation reactions, ligands/complexing agents/chelates/sequestering agents

Traditionally, acid-base reactions are only considered in aqueous solutions. However, there is another concept of acids and bases. Lewis bases are considered electron-pair donors and Lewis acids are considered electron-pair acceptors. If protons are available for combination or substitution, then they are likely being transferred from an acid to a base. An unshared electron pair on one of the reactants may form a bond in a Lewis acid-base reaction.

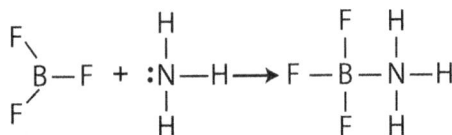

This type of reaction is called a **complexation reaction.** The Lewis base in this reaction is the NH_3 and the Lewis acid is the BF_3. Lewis bases are also known as **ligands, complexing agents, chelates,** and **sequestering agents.** Many Lewis acids are metal ions. The number of electron pairs they can accept is given by their **coordination number.** The silver ion (Ag^+) has a coordination number of 2, and tends to accept two electron pairs. Metal ions with coordination numbers of 4 (accepting 4 electron pairs) include Au^{3+} (gold), Cu^{2+} (copper), Zn^{2+} (zinc), and Pt^{2+} (platinum). Metal ions with coordination numbers of 6 (accepting six electron pairs) include Fe^{2+} and Fe^{3+} (iron), Co^{3+} (cobalt), Ti^{4+} (titanium), Mn^{2+} (manganese), and Cr^{3+} (chromium).

Complexing reactions can be used to make insoluble metals soluble by converting them to metal ions in a soluble complex. This type of reaction is used in extracting metal from ore. Cyanide is used to extract gold this way; cyanide is the Lewis base, gold is the Lewis acid, and the complex they form is soluble. After washing out the gold from the ore in this way, the metal itself can be removed from the cyanide-gold complex using one of a number of processes, yielding pure gold.

"The pH is an important characteristic of aqueous solutions that can be controlled with buffers. Comparing pH to pKa allows one to determine the protonation state of a molecule with a labile proton."

Chapter 6.15 Equilibrium, temperature, and Gibbs free energy

The equilibrium constant, K, whether K_a, K_b, or K_{sp}, specifies the relative proportions of products and reactants, whether in a solubility or acid/base reaction, at equilibrium. K is the concentration(s) of the product(s) divided by the concentration(s) of the reactant(s), so if K = 1, the concentration of the products is equal to the concentrations of the reactants. If K is near 1, there will be significant quantities of both reactants and products at equilibrium. If K is greater than 1, the higher it is, the more products will be favored in the equilibrium; if K is less than 1, the lower it is, the more reactants will be favored in the equilibrium.

Equilibrium is related to temperature and Gibbs free energy, with any reaction wanting to go "downhill" to a lower $\Delta G°$. So, while K is a description of what will happen in an equilibrium, the *reason* for the equilibrium is $\Delta G°$, the change in free energy. Those species with lower $\Delta G°$ will have higher concentrations at equilibrium. If $\Delta G° < 0$ then K > 1, so products will be in greater concentration than reactants at equilibrium. If $\Delta G° > 0$ then K < 1, and reactants will be higher in concentration than products at equilibrium. If $\Delta G°$ is near 0, then K will be near 1, and there will be significant amounts of both products and reactants at equilibrium.

The equation relating K to $\Delta G°$ is:

$$K = e^{-\Delta G°/RT}$$

Using this equation, you can reason about whether products or reactants will be favored in an equilibrium. RT at room temperature is 2.4 kJ/mol. (However, on the AP Chemistry exam, you will not need to calculate an exact answer, since that would require a calculator.)

It is important to consider the difference between exothermic/endothermic and exergonic/endergonic. Whereas exothermic/endothermic refers to the energy released or consumed by a reaction, exergonic and endergonic refer to negative and positive changes in $\Delta G°$, respectively, and as such they take into account entropy:

$$\Delta G° = \Delta H - (T * \Delta S)$$

K is related to how exergonic a reaction is. From the equation above, it is clear that when temperature rises, the change in entropy is magnified. If ΔS is positive, raising the temperature can change a positive $\Delta G°$ into a negative one (K > 1), even if ΔH is positive enough to favor reactants (K < 1) at a lower temperature.

"The equilibrium constant is related to temperature and the difference in Gibbs free energy between reactants and products."

"When the difference in Gibbs free energy between reactants and products ($\Delta G°$) is much larger than the thermal energy (RT), the equilibrium constant is either very small (for $\Delta G° > 0$) or very large (for $\Delta G° < 0$). When $\Delta G°$ is comparable to the thermal energy (RT), the equilibrium constant is near 1."

Sample Test Questions (Big Idea #6) _____

Multiple Choice Questions

Instructions: This section consists of some practice multiple choice questions as well as one Free Response question related to this chapter. You may NOT use a calculator for the Multiple Choice. However calculators are permitted for the Free Response.

For questions 1-7, refer to the following titration diagram:

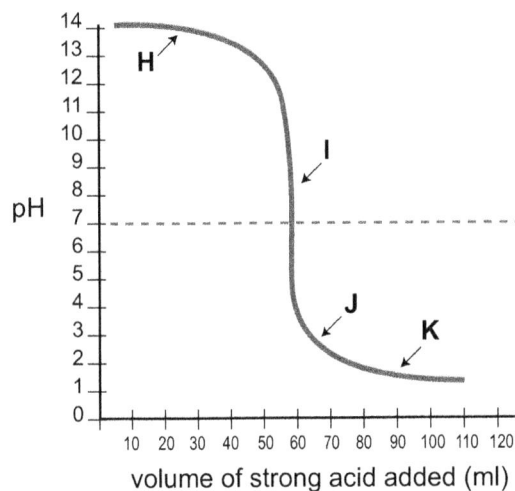

1. The diagram above describes adding strong acid to:

   A.   Strong base

   B.   Strong acid

   C.   Weak base

   D.   Weak acid

2. Which of the four labeled points is the equivalence point?

   A.   H

   B.   I

   C.   J

   D.   K

3. At what volume does pH = pOH?

   A.   25 ml

   B.   45 ml

   C.   55 ml

   D.   85 ml

4. Which of the following points fall within buffer regions?

   A. H and K

   B. I only

   C. J only

   D. none of these

5. After 100 ml of titrant have been added, what species has the highest concentration?

   A. OH-

   B. H_3O+

   C. HA

   D. BOH

6. At point K, if you added 20 ml of NaOH to the solution, what species would have the highest concentration?

   A. OH-

   B. H_3O+

   C. HA

   D. NaOH

7. At point K, if you added 40 ml of NaOH to the solution, what species would have the highest concentration:

   A. OH-

   B. H_3O+

   C. HA

   D. BOH

For questions 8-17, refer to the following titration diagram:

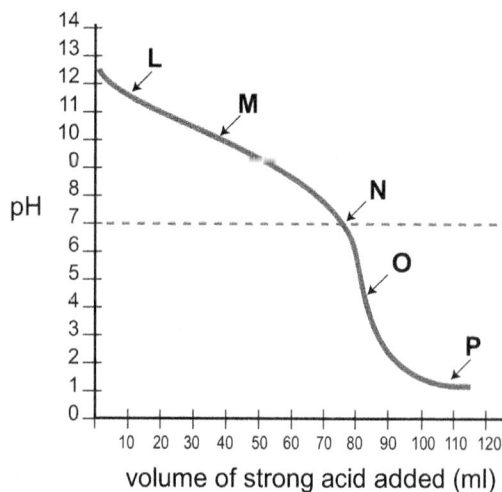

8. The diagram above describes adding strong acid to:

    A. Strong base

    B. Strong acid

    C. Weak base

    D. Weak acid

9. Which of the four labeled points is the equivalence point?

    A. M

    B. N

    C. O

    D. P

10. At what volume does pH = pOH?

    A. 35 ml

    B. 55 ml

    C. 75 ml

    D. 82 ml

11. Which of the following points falls within a buffer region?

    A. L

    B. O

    C. N

    D. none of these

12. After 100 ml of titrant have been added, what species has the highest concentration?

    A. OH-

    B. H_3O+

    C. HA

    D. BH

13. After 20 ml of titrant have been added, what species has the highest concentration?

    A. OH-

    B. H_3O+

    C. BH+

    D. Both A and C

14. At point O, if you added 20 ml of HCl to the solution, which of the following species would have the highest concentration?

   A.   OH-

   B. H_3O+

   C.   HCl

   D.   BH

15. At point M, which of the following are true:

   A. $[H_3O+] = [OH-}$

   B. $[HCl] = [BH+]$

   C. $[B-] = [BH+]$

   D.   All of the above

16. At point M, if you added 10 ml of HCl to the solution, which of the following species would have the highest concentration?

   A.   B-

   B. H_3O+

   C.   HCl

   D.   BH+

17. At point M, if you added 10 ml of NaOH to the solution, which of the following species would have the highest concentration?

   A.   B-

   B. H_3O+

   C.   NaOH

   D.   BH+

18. How much HCl is needed to turn Litmus, an indicator that changes color at pH 7, from blue to red?

   A.   ~82 ml

   B.   ~75 ml

   C.   50 ml

   D.   ~35 ml

19. What type of titration is this?

   A.   Weak acid titrated by a strong base

   B.   Weak acid titrated by a weak base

   C.   Strong base titrated by a strong acid

   D.   Strong acid titrated by a strong base

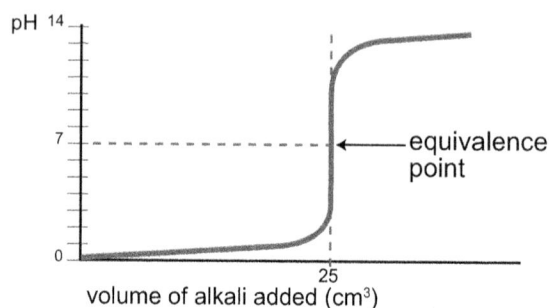

20. What type of titration is this?

    A.   Strong base titrated by a strong acid

    B.   Weak base titrated by a strong acid

    C.   Weak base titrated by a weak acid

    D.   Weak acid titrated by a strong base

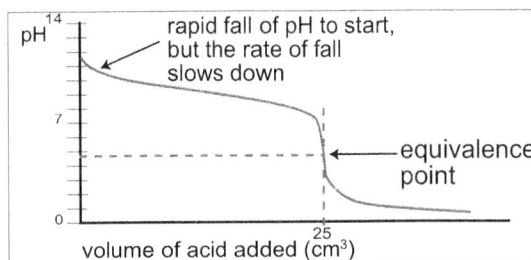

21. What type of titration is this?

    A.   Strong acid titrated by a strong base

    B.   Weak acid titrated by a weak base

    C.   Weak acid titrated by a strong base

    D.   Weak base titrated by a strong acid

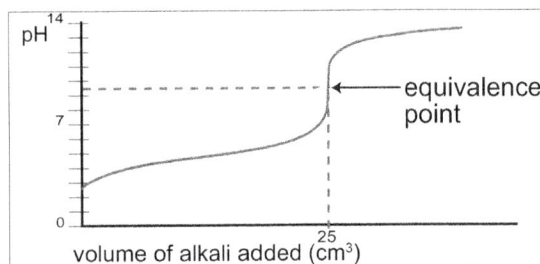

22. What type of titration is this?

    A.   Strong base titrated by a weak acid

    B.   Weak base titrated by a strong acid

    C.   Weak acid titrated by a weak base

    D.   Weak base titrated by a weak acid

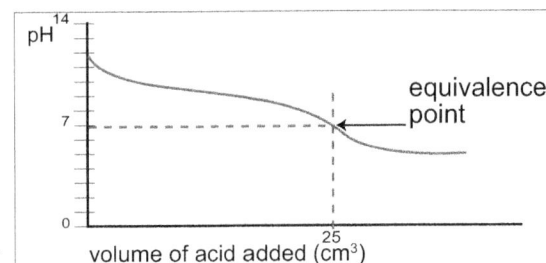

23. What type of titration is this?

    A.   Diprotic acid titrated with a strong base

    B.   Monoprotic acid titrated with a strong base

    C.   Diprotic base titrated with a strong acid

    D.   Monoprotic base titrated with a strong acid

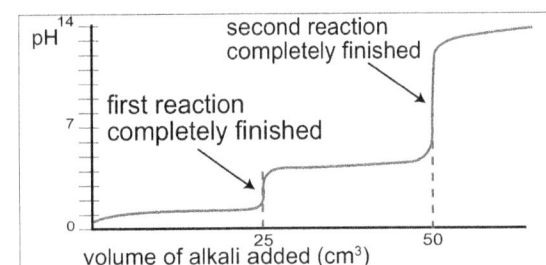

24. Which of the following will have a pH above 7 at the equivalence point:

    A.   Weak acid titrated with a strong base

    B.   Weak base titrated with a strong acid

    C.   Strong base titrated with a strong acid

    D.   Strong acid titrate with a strong base

25. Which of the following will have a pH at 7 at the equivalence point:

    A. Strong acid titrate with a strong base

    B. Weak base titrated with a strong acid

    C. Strong base titrated with a strong acid

    D. Both A and C

26. Litmus is red when the pH is below 5 and blue when the pH is above 8. Methyl orange is red when the pH is below 3.7 and yellow-orange when the pH is above 4.4. Phenolphthalein is bright pink when the pH is above 9.8 and colorless when the pH is below pH 8.2. For the titration to the right, which of these indicators provides an endpoint that reflects the equivalence point?

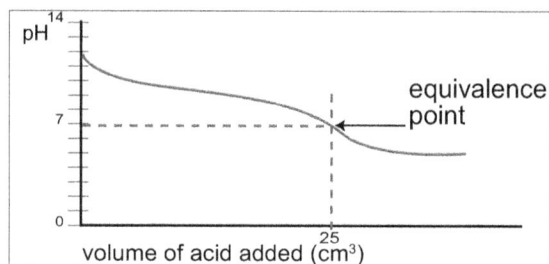

    A. Phenolphthalein and litmus

    B. Methyl orange only

    C. Phenolphthalein only

    D. Litmus only

Free Response Question (Big Idea #6)

One mole of acetic acid is dissolved in one liter of water, following the reaction below. K for this process, known as the "acid dissociation constant" for acetic aid, is about 1.8×10^{-5}. Give that the pH of a solution is defined by $pH = -\log_{10}([H_3O^+])$, what is the pH of this solution at equilibrium?

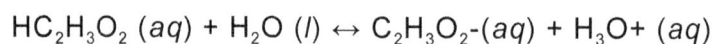

$$HC_2H_3O_2 \ (aq) + H_2O \ (l) \leftrightarrow C_2H_3O_2\text{-}(aq) + H_3O\text{+} \ (aq)$$

Answers to Sample Questions (Big Idea #6)

1. **Answer A.** The steep, nearly vertical slope of the curve as it changes from base to acid indicates the addition of a strong acid to a strong base.

2. **Answer B.** The equivalence point is where the amount of acid is equal to the amount of base. It can be found at the point at which the base changes to acid.

3. **Answer C.** pH equals pOH and pH = 7. The volume at this point is 55 ml.

4. **Answer D.** There are no buffer regions in this neutralization; since it is a strong acid being added to a strong base, the change happens very quickly, with no buffer region.

5. **Answer B.** After 100 ml of titrant were added, the pH is approximately 1. This is acidic, so the highest concentration species is H_3O+.

6. **Answer B.** Even if you added 20 ml of NaOH to the solution, it would be in the range of pH well below 7, and therefore would be acidic, with H_3O+ the predominant species.

7. **Answer A.** Adding 40 ml would be enough to raise the pH above 7, making the solution basic, and therefore OH- would be the species with the highest concentration.

8. **Answer C.** The gradual slope indicates that acid is being added to a weak base.

9. **Answer C.** The almost vertical slope at O indicates that this is the equivalence point.

10. **Answer C.** At pH 7, pH = pOH, and the volume at this point is 75 ml

11. **Answer A.** At point L, there is a large amount of conjugate acid and conjugate base of the weak base; it is a buffer region.

12. **Answer B.** H3O+. At this point, the pH is very low, around 1; acid is dominant.

13. **Answer D.** Both A and C. At this point, it is basic, but less basic than at the beginning of the titration, so there is both OH- and conjugate acid present.

14. **Answer B.** H_3O+. At that point, it was already acidic; adding more acid would make it more acidic.

15. **Answer C.** [B-] = [BH+] The conjugate acid and conjugate base are in approximately equal concentrations.

16. **Answer D.** BH+ At this point, although the pH is still basic, the dominant weak species is the conjugate acid of the weak base.

17. **Answer A.** B- At point M, if you added 10 ml of NaOH to the solution, this will make the solution basic enough that the conjugate weak base will dominate.

18. **Answer B.** Litmus will turn red when the pH becomes acidic—this change happens when ~75 ml of HCl is added.

19. **Answer D.** The low and high pH sections with a nearly vertical line in between show that this is a strong acid titrated by a strong base.

20. **Answer B.** The gradual slope from high pH to the equivalence point followed by a sudden drop to low pH indicates that this is a weak base titrated by a strong acid

21. **Answer C.** The gradual slope from low pH to the equivalence point followed by a sudden rise to high pH indicates that this is a weak acid titrated by a strong base

22. **Answer D.** The gradual fall from high pH to an equivalence point in the middle of a gradual slope, leading to a moderately low pH indicates that this is a weak base titrated by a weak acid

23. **Answer A.** The two vertical equivalence points indicate that two reactions have each come to completion; since the titration started at very low pH and ended at very high pH, this indicates a diprotic acid titrated with a strong base

24. **Answer A.** If the pH is above 7 at the equivalence point, the base in the titration is stronger than the acid. Of the four answers, only "weak acid titrated with a strong base" is correct.

25. **Answer D.** Both a strong acid titrated with a strong base and a strong base titrated with a strong acid would have a pH of 7 at the equivalence point. This is because both strong acids and strong bases are completely dissociated in water; at the equivalence point, the number of OH^- molecules and H_3O^+ molecules will be equal, giving a neutral pH of 7.

26. **Answer D.** Litmus only. The pH at the equivalence point is 7; of the three indicators, only litmus will change in the range of pH 7.

Answer to Free Response (Big Idea #6)

$HC_2H_3O_2$ (aq)	H_2O (l)	$C_2H_3O_2$ (aq)	H_3O+ (aq)
1	XXXXX	0	0
-X	XXXXX	+X	+X
1-X	XXXXX	X	X

$K = 1.8 \times 10^{-5} = x^2/(1-x)$

$(1.8 \times 10^{-5}) - (1.8 \times 10^{-5})x = x^2$

$x^2 + (1.8 \times 10^{-5})x - (1.8 \times 10^{-5}) = 0$

$x = 0.00423$ M

$pH = -\log_{10}([H_3O^+]) = -\log_{10}(x) = -\log_{10}(0.00423) = \underline{\mathbf{2.37}}$

SECTION V:
Appendix 1:

Math Review

Math Review

AP Chemistry is much more than a high school chemistry class. It will cover the material you would learn in a first year college chemistry course. The biggest differences between AP chem and a regular chem class will be in ability to extrapolate basic concepts and principles to describe and understand more complex chemical phenomena. In many cases this will involve using more Math than a basic chemistry class. Calculators are not permitted for the multiple choice questions on the AP test so the actual calculations will be easy but the basic concepts of units and conversions must be recalled. Calculators are allowed for the Free-Response section so those questions may require more in depth mathematical analysis.

Things you should know:
- Scientific Notation
- Using logs
- Using the Pythagorean Theorem.

Things this book will cover:
- Using the above concepts in chemical analysis.

A. The Celsius, Fahrenheit, and Kelvin temperature scales

The **Fahrenheit** (°F) non-metric temperature scale was proposed in 1724 by Gabriel Fahrenheit who was looking to improve upon Galileo's thermometer by changing from an enclosed gas to mercury. Mercury has a large uniform thermal expansion, does not adhere to the glass, and its silvery color makes it easy to read.

On his scale, he found the temperature of boiling water to be 212°. He adjusted the freezing point temperature for water to 32° so that the interval between freezing and boiling would be a rational number - 180°.

Anders Celsius proposed the **Celsius** scale which was designed to have 100 degrees between the boiling point temperature of water and its freezing point temperature at standard atmospheric pressure.

The use of the Celsius or Fahrenheit scale requires using negative numbers when measuring low temperatures. This proves to be inconvenient, so in the late 1800s Lord Kelvin suggested a new temperature scale. This temperature scale is based on absolute zero, the temperature at which a material has cooled to the point where it has no more heat to lose (or the point at which all molecular motion ceases). This temperature is the basis for the Kelvin scale and the value zero kelvin is assigned to it. At sea level, water freezes at around 273 K and boils at 373 K. Notice that there are still 100 degrees between the freezing and boiling point temperatures on the Kelvin scale. Temperatures on this scale are called **kelvins**, *not* degrees kelvin, kelvin is *not* capitalized, and the symbol (capital K) stands alone with no degree symbol.

The Celsius scale is now defined as follows:

1. The triple point of water is defined to be 0.01° C.

2. A degree Celsius equals the same temperature change as a degree on the ideal-gas scale, now called the **Kelvin scale**.

3. On the Celsius scale the boiling point of water at standard atmospheric pressure is 99.975° C in contrast to the 100° originally defined by the Centigrade scale.

Conversions between scales are simple but require a little math:

*   To convert from Celsius to Fahrenheit, multiply by 1.8 and add 32:
    ° F = 1.8 x ° C + 32

    From Fahrenheit to Celsius, subtract 32 and divide by 1.8:
    ° C = (°F - 32) / 1.8

*   To convert from Celsius to Kelvin, add 273 (273.15 to be more exact) to the Celsius temperature: K = ° C + 273

    From Kelvin to Celsius, subtract 273 from the Kelvin temperature:
    ° C = K - 273

Common temperature comparisons

temperature	degree Celsius	degree Fahrenheit
symbol	°C	°F
boiling point of water	100.	212.
average human body temperature	37.	98.6
average room temperature	20. to 25.	68. to 77.
melting point of ice	0.	32.

B. Unit/ Dimensional Analysis

There are a number of important physical quantities that have unique units that are important to remember and to convert between. Many of these will be given as part of the test. However, it might be necessary to convert between different units or to derive one unit from another.

Derived units measure a quantity that may be **expressed in terms of other units**. The derived units important for chemistry are:

Derived quantity	Unit name	Expression in terms of other units	Symbol (if any)
Area	square meter	m^2	
Volume	cubic meter	m^3	
	liter	$dm^3 = 10^{-3}\ m^3$	L or l
Mass	unified atomic mass unit	$(6.022 \times 10^{23})^{-1}\ g$	u or Da
Time	minute	60 s	min
	hour	60 min = 3600 s	h
	day	24 h = 86400 s	d
Speed	meter per second	m/s	
Acceleration	meter per second squared	m/s^2	
Temperature*	degrees Celsius	$K - 273.15°$	°C
Mass density	gram per liter	$g/L = 1\ kg/m^3$	
Amount-of-substance concentration (molarity†)	molar	mol/L	M
Molality‡	molal	mol/kg	m
Chemical reaction rate	molar per second†	$M/s = mol/(L{\cdot}s)$	
Force	newton	$m{\cdot}kg/s^2$	N
Pressure	pascal	$N/m^2 = kg/(m{\cdot}s^2)$	Pa
	standard atmosphere§	101325 Pa	atm
Energy, Work, Heat	joule	$N{\cdot}m = m^3{\cdot}Pa = m^2{\cdot}kg/s^2$	J
	nutritional calorie§	4184 J	Cal
Heat (molar)	joule per mole	J/mol	
Heat capacity, entropy	joule per kelvin	J/K	
Heat capacity (molar), Entropy (molar)	joule per mole kelvin	$J/(mol{\cdot}K)$	
Specific heat	joule per kilogram kelvin	$J/(kg{\cdot}K)$	
Power	watt	J/s	W
Electric charge	coulomb	$s{\cdot}A$	C
Electric potential, electromotive force	volt	W/A	V
Viscosity	pascal second	$Pa{\cdot}s$	
Surface tension	newton per meter	N/m	

*Temperature differences in kelvin are the same as in degrees Celsius. To obtain degrees Celsius from Kelvin, subtract 273.15 (see below).

†Molarity is considered to be an obsolete unit by some physicists.

‡**Molality, *m*, is often considered obsolete. Differentiate *m* and meters (m) by context.**

These are commonly used non-SI units.

Decimal multiples of SI units are formed by attaching a **prefix** directly before the unit and a symbol prefix directly before the unit symbol. SI prefixes range from 10^{-24} to 10^{24}. Only the prefixes you are likely to encounter in chemistry are shown below:

Factor	Prefix	Symbol	Factor	Prefix	Symbol
10^9	giga-	G	10^{-1}	deci-	d
10^6	mega-	M	10^{-2}	centi-	c
10^3	kilo-	k	10^{-3}	milli-	m
10^2	hecto-	h	10^{-6}	micro-	μ
10^1	deca-	da	10^{-9}	nano-	n
			10^{-12}	pico-	p

Example: 0.0000004355 meters is 4.355×10^{-7} m or 435.5×10^{-9} m. This length is also 435.5 nm or 435.5 nanometers.

Example: Find a unit to express the volume of a cubic crystal that is 0.2 mm on each side so that the number before the unit is between 1 and 1000.

Solution: Volume is length x width x height, so this volume is $(0.0002 \text{ m})^3$ or 8×10^{-12} m^3. Conversions of volumes and areas using powers of units of length must take the power into account. Multiply the factor in the chart above by the power of the unit to obtain the new exponent, as follows:

$$1 \text{ m}^3 = 10^3 \text{ dm}^3 = 10^6 \text{ cm}^3 = 10^9 \text{ mm}^3 = 10^{18} \text{ μm}^3$$

The length 0.0002 m is 2×10^2 μm, so the volume is also 8×10^6 μm^3. This volume could also be expressed as 8×10^{-3} mm^3, but none of these numbers are between 1 and 1000.

Expressing the volume in liters is helpful in cases like these. There is no power on the unit of liters, therefore:

$$1 \text{ L} = 10^3 \text{ mL} = 10^6 \text{ μL} = 10^9 \text{ nL}$$

Converting cubic meters to liters (canceling the meters) gives:

$$8 \times 10^{-12} \text{ m}^3 \times \frac{10^3 \text{ L}}{1 \text{ m}^3} = 8 \times 10^{-9} \text{ L}$$

The crystal's volume is 8 nanoliters (8 nL).

Dimensional analysis is a structured way to convert units. It involves a conversion factor that allows the units to be canceled out when multiplied or divided. A conversion factor is the same measurement written as an equivalency between two different units such as 1 meter = 100 centimeters.

One Dimension Unit Conversions

These are the steps to converting one dimension measurements.

1. Write the term to be converted, (both number and unit)
    $$6.0 \text{ cm} = ? \text{ km}$$

2. Write the conversion formula(s) needed for the involved units.
    $$100 \text{ cm} = 1 \text{ m}$$
    $$1000 \text{m} = 1 \text{ km}$$

3. Make a fraction of the conversion formula, such that:

    a. if the unit in step 1 is in the numerator, that same unit in this step must be

    in the denominator $\dfrac{0.001}{100} \dfrac{km}{cm}$

    b. if the unit in step 1 is in the denominator, that same unit in this

    step must be in the numerator $\dfrac{100}{0.001} \dfrac{cm}{km}$

Since both the numerator and denominator are equal in value, the fraction equals 1.

4. Multiply the term in step 1 by the correct fraction in step 3. Since the fraction equals 1, you can multiply by it without changing the size of the original term.

5. Cancel units : $6.0 \text{ c}\cancel{\text{m}} \times \left(\dfrac{0.00100}{100} \dfrac{km}{\cancel{c}\text{m}} \right)$

6. Perform the indicated calculation, rounding the answer to the correct number of significant figures.
    $$6.0 \times 1 / 100 = 0.060 \text{ km (or } 6.0 \times 10^{-2} \text{ km)}$$

Two or Three Dimension Unit Conversions

The process is nearly the same for two and three dimension conversions.

Example: How many cm^3 is 1 m^3?

Solution: Remember, 100 cm = 1 m and that 1 m^3 is really 1m x 1m x 1m. Substituting in the 100 cm for every meter the problem can be rewritten as

100 cm x 100 cm x 100 cm or $1m^3 = 1,000,000 \text{ cm}^3$

$1 m^3 \times 1,000,000 \text{ cm}^3/1m^3 = 1,000,000 \text{ cm}^3$ (or 1.0×10^6 cm)

Example: Convert 4.17 kg/m^2 to g/cm

Solution: First to convert from kg to g use 1000 g = 1.00 kg as the conversion factor.

$$4.17 \text{ } \cancel{\text{kg}}/\text{m}^2 \times 1000 \text{ g}/1.00 \text{ } \cancel{\text{kg}} = 4170\text{g} /\text{m}^2$$

Then use 1.00 m = 100 cm to convert the denominator. Remember that m² is m x m and replacing m with 100 cm, the denominator becomes 100 cm x 100 cm. or 10 000 cm² The conversion factor for the denominator becomes 1.00m² = 10 000 cm² (cancel m)

$$4170 \text{ g/m}^2 \times 1.00 \text{ m}^2/10{,}000 \text{ cm}^2 = 0.417 \text{ g/cm}^2$$

Dimensional analysis can also be used to help solve mathematical problems.

Example: The density of gold is 19.3 g/cm³. How many grams of gold would be found in 55 cm³?

Solution: Using dimensional analysis, some unit must be made to cancel. The answer needs to be in grams, so the cm³ needs to be canceled out. Multiply or divide the units so that the cm³ cancel. In this case the units part of the problem works out as follows:

$$\text{g/cm}^3 \times \text{cm}^3 = \text{g}$$

Now adding in the values from the problem, we get:

$$19.3 \text{ g/cm}^3 \times 55 \text{ cm}^3 = 1060 \text{ g of gold}$$

C. Calculating pH and pOH

The concentration of H+(aq) ions is often expressed in terms of pH. **The pH of a solution is the negative base-10 logarithm of the hydrogen-ion molarity.**

$$\text{pH} = -\log[\text{H}^+] = \log(1/[\text{H}^+])$$

A ten-fold increase in [H+] decreases the pH by one unit. [H+] may be found from pH using the expression:

$$[\text{H}^+] = 10^{-\text{pH}}$$

Because [H+] = 10-7 M for pure water, **the pH of a neutral solution is 7.**

In an **acidic solution,** [H+] > 10-7 M and pH < 7.

In a basic solution, [H+] < 10-7 M and pH > 7.

The negative base -10 log is a convenient way of representing other small numbers used in chemistry by placing the letter "p" before the symbol. Values of K_a are often represented as $\text{p}K_a$, with $\text{p}K_a = -\log K_a$. The concentration of OH⁻(aq) ions may also be expressed in terms of pOH, with pOH = -log [OH⁻].

The ion-product constant of water, $K_w = [\text{H+}][\text{OH-}] = 1.0 \times 10^{-14}$ at 25 °C. The value of K_w can used to determine the relationship between pH and pOH by taking the negative log of the expression:

$$-\log K_w = -\log\left[H^+\right] - \log\left[OH^-\right] = -\log\left(10^{-14}\right).$$

$$\text{Therefore: pH} + \text{pOH} = 14.$$

Example: An aqueous solution has an H^+ ion concentration of 4.0×10^{-9}. Is the solution acidic or basic? What is the pH of the solution? What is the pOH?

Solution: The solution is basic because $[H+] < 10^{-7}$ M.

$$pH = -\log\left[H^+\right] = -\log\ 4\times10^{-9}$$

$$= 8.4.$$

$$pH + pOH = 14.\ \ \text{Therefore } pOH = 14 - pH = 14 - 8.4 = 5.6.$$

D. Using the Quadratic Formula

The multiple choice part of the exam does not allow the use of calculators, however the Free Response section does. In this section you may come across an Equilibrium type question that may require solving the quadratic formula.

The Quadratic Formula is

$$x = \frac{-b \pm \sqrt{b^2 - 4ac}}{2a}$$

A typical Equilibrium Constant problem will give initial concentrations and ask you to find some later concentration. For example:

$$A(g) \rightarrow B\ (g) + C\ (g)\ \ Kc = 1.6\times10^{-2}$$

Given an initial concentration of 0.15mol of X create an ICE table (see Chapter 6.12)

	A	B	C
Initial	0.15M	0	0
C	-x	x	x
Equilibrium	0.15 -x	x	x

At equilibrium:

$$K_{eq} = \frac{[B]\,[C]}{[A]} = \frac{[x][x]}{0.15\ -x}$$

Rearranging the equation

$$(K_{eq}) * (0.15\ -x) = x^2$$

$$(K_{eq}) * 0.15) - (K_{eq} * x) = x^2$$

$$x^2 + 1.6\times10^{-2}\ x - (1.6\times10^{-2})(0.15) = 0$$

Use quadratic formula above where

A = 1

B=1.6×10^{-2} 1.6×10^{-2}

C=$(-0.150)(1.6 \times 10^{-2})=-2.4 \times 10^{-3}$

Solving for X gives two solutions. Intuition must be used in determining which solution is correct. If one gives a negative concentration, it can be eliminated, because negative concentrations are unphysical. In this case the solutions are: x = 0.0416, -0.0576. x = 0.0416 makes chemical sense and is therefore the correct answer.

SECTION VI:
Appendix 2:

Nomenclature and Organic Chemistry Review

Nomenclature and Organic Chemistry Review

Carbon

Carbon is among the most important elements in the periodic table. Its ability to form chains that can be long enough to fold into rings and other shapes allows the formation of the tools and structures of life—carbohydrates, proteins, and lipids. Carbon is unique among elements in this ability to form large, highly complex structures. Carbon can bond with 1, 2, 3, or 4 other atoms, using single, double, or triple bonds. In methane, CH_4, carbon is bonded to four hydrogens with single bonds. In carbon dioxide, CO_2, carbon is bonded to each of two oxygens with a double bond. In carbon monoxide, CO, carbon is bonded to a single oxygen with a triple bond.

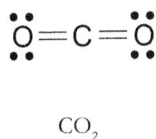

When carbon forms into chains lined with hydrogen atoms, these molecules are known as organic molecules. Organic molecules, which include oxygen, nitrogen, and other atoms that alter their chemical functioning, make up the field of **organic chemistry.**

The relatively large compounds that carbon can form lead to the possibility of various three-dimensional shapes, and these shapes have important effects. This can already be seen in the clear difference between diamond and graphite, but becomes even more important when considering organic molecules.

Alkanes

The simplest types of organic molecules, called hydrocarbons, contain only carbon and hydrogen. Even with only carbon and hydrogen, highly complex shapes can be created through branching and "kinks" introduced into the chains in the form of double or triple bonds, or the unique bonds seen in benzene rings. **Alkanes contain only single bonds.** Alkanes have the maximum number of hydrogen atoms possible for their carbon backbone, so they are called **saturated**. Alkenes, alkynes, and aromatics are **unsaturated** because they have fewer hydrogens.

Straight-chain alkanes are also called **normal alkanes**. These are the simplest hydrocarbons. They consist of a linear chain of carbon atoms. The names of these molecules contain the suffix –*ane* and a **root based on the number of carbons in the**

chain according to the table on the following page. The first four roots, *meth*–, *eth*–, *prop*–, and *but*– have historical origins in chemistry, and the remaining alkanes contain common Greek number prefixes. Alkanes have the general formula C_nH_{2n+2}. For example, pentane has five carbons and ten hydrogens:

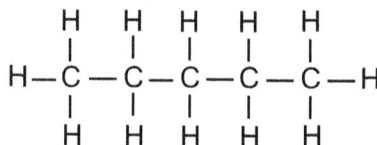

Alkanes are non-polar, and therefore do not have the dipole forces that hold small polar molecules like H_2O into liquids at room temperature. The only intermolecular attractive forces capable of holding them together in liquid or solid phase is Van der Waals dispersion forces. Since Van der Waals forces are quite weak for small molecules, alkanes smaller than pentane are gases at room temperature. (Van de Waals forces increase with increasing volume of the molecule, because there is more of a chance for electrons to momentarily be located unevenly, creating a temporary dipole.) The boiling point of pentane is 36.1°C, so even this relatively large molecule is still quite volatile.

A single molecule may be represented in multiple ways. Six ways that pentane might be represented are:

n-pentane (the *n* represents a *normal* alkane)

$CH_3CH_2CH_2CH_2CH_3$

$CH_3(CH_2)_3CH_3$

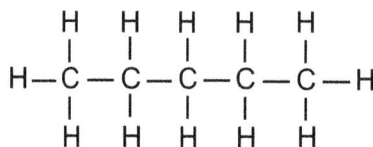

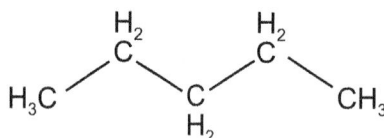

The table below shows the unbranched alkanes up to eight carbons in length. Methane and ethane in the table are shown as three-dimensional structures with dashed wedge shapes attaching atoms behind the page and thick wedge shapes attaching atoms in front of the page.

Number of carbons	Name	Formula	Structure
1	Methane	CH_4	
2	Ethane	C_2H_6	
3	Propane	C_3H_8	
4	Butane	C_4H_{10}	
5	Pentane	C_5H_{12}	
6	Hexane	C_6H_{14}	
7	Heptane	C_7H_{16}	
8	Octane	C_8H_{18}	

Branched alkanes are named using a four-step process:

1. Find the longest continuous carbon chain. This is the parent hydrocarbon.

2. Number the atoms on this chain beginning at the end near the first branch point, so the lowest locant numbers are used. Number functional groups from the attachment point. In the case of a branched alkane, the functional group is simply an attached straight-chain alkane. This type of group is called an alkyl group. To name alkyl groups, replace the –ane part of the alkane name with –yl.

3   Determine the numbered locations and names of the substituted alkyl groups. Use *di–*, *tri–*, and similar prefixes for alkyl groups represented more than once. Separate numbers by commas and groups by dashes.

4. List the locations and names of alkyl groups in alphabetical order by their name (ignoring the *di–*, *tri–* prefixes) and end the name with the parent hydrocarbon.

Example: Name the following hydrocarbon:

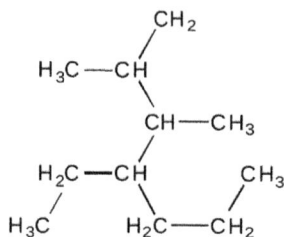

Solution:

1. The longest chain is seven carbons in length, as shown by the bold lines below. This molecule is a heptane.

2. The atoms are numbered from the end nearest the first branch as shown:

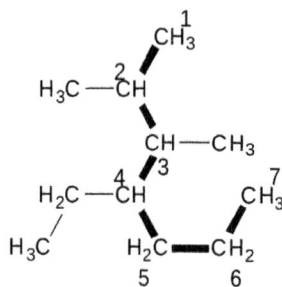

3. Methyl groups are located at carbons 2 and 3 (2,3-dimethyl), and an ethyl group is located at carbon 4.

4. "Ethyl" precedes "methyl" alphabetically. The hydrocarbon name is: 4-ethyl-2,3-dimethylheptane.

The following branched alkanes also have IUPAC-accepted common names:

Structure	Systematic name	Common name
	2-methylpropane	isobutane
	2-methylbutane	isopentane
	2,2-dimethylpropane	neopentane

The following alkyl groups have IUPAC-accepted common names. The systematic names assign a locant number of 1 to the attachment point:

Structure	Systematic name	Common name
	1-methylethyl	isopropyl
	2-methylpropyl	isobutyl
	1-methylpropyl	sec-butyl
	1,1-dimethylethyl	tert-butyl

Branched alkanes give rise to the possibility of two or more molecules with the same number of carbons and hydrogens but completely different shapes giving rise to different chemical properties. For example, consider these three molecules, all C_5H_{12}.

 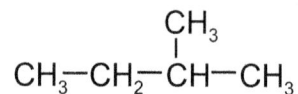

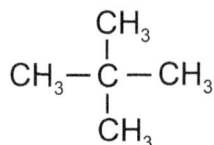

The first is pentane, the second 2-methyl butane, and the third, dimethyl propane. Different molecules with the same chemical formula are called **structural isomers.**

Alkenes

Alkenes contain one or more double bonds. Alkenes are also called olefins. Small alkenes tend to have lower boiling points than alkanes with the same number of carbons. This is because, with two fewer electrons per double bond, alkenes have a smaller total volume than alkanes, and the strength of Van der Waals forces are associated with the volume of the molecule.

In larger molecules, the double bonds serve to introduce "kinks" in the carbon chains. This geometric arrangement of the atoms affects properties of the molecules by affecting their intermolecular attraction. In other words, these shapes make it harder for alkenes to "stick together". This causes physical affects; for example, fats made mainly of straight-chain alkanes, known as **saturated** fats (e.g. butter), stack neatly together, and tend to be solid at room temperature. Fats that have double bonds, known as **unsaturated** fats, are prevented by their geometrical arrangement from stacking tightly together into a solid structure; these fats (e.g. vegetable oil) tend to be liquid at room temperature.

The suffix used in the naming of alkenes is –*ene*, and the number roots are those used for alkanes of the same length. A number preceding the name shows the location of the double bond for alkenes of length four and above. Alkenes with one double bond have the general formula C_nH_{2n}. Multiple double bonds are named using –*diene*, –*triene*, etc. The suffix –*enyl*– is used for functional groups after a hydrogen is removed from an alkene. Ethene and propene have the common names **ethylene** and **propylene**. The ethenyl group has the common name **vinyl** and the 2-propenyl group has the common name **allyl**.

Examples:

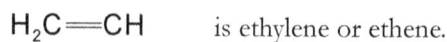   is ethylene or ethene.

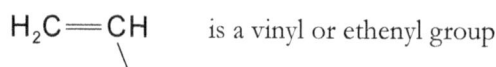   is a vinyl or ethenyl group

	is propylene or propene.
	is an allyl or 2- propenyl group.
	is 2-hexene.
	2-methyl-1,3-butadiene (common name: isoprene).

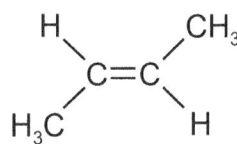

Whereas a single bond can freely rotate, a double bond cannot, so the same alkene can have different shapes based on the arrangement of atoms around the double bond. The prefix cis indicates that carbons bonded via a double bond are on the same side; trans means that they are on opposite sides. For example, consider the arrangement of atoms in cis-2-butene and trans-2-butene, below.

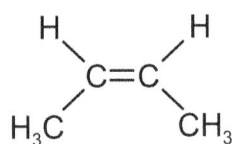

cis-2-butene   trans-2-butene

As seen in the above example, **cis-trans isomerism** is often part of the complete name for an alkene. Not as obvious from a two-dimensional picture is the affect of the cis or trans arrangement on the shape of the molecule. In three dimensions, these two versions of 2-butene are quite different in shape, and this will have important effects on their chemical properties. For example, *cis*-2-butene has a higher boiling point than *trans*-2-butene. Cis-trans isomers, also known as diasteriomers, are **stereoisomers** but they are not mirror images of each other.

Alkynes contain one or more triple bonds. This makes them smaller than alkanes, but, surprisingly, their boiling points are *higher* than alkanes of the same number of carbon atoms. The reason for this is pi bonds. Electrons in pi bonds are "looser"; farther from their nuclei than electrons in sigma bonds, and therefore more likely to randomly form instantaneous dipoles. In a triple bond, there are two pi bonds, and this is enough to overcome the effect of smaller volume on the Van der Waals forces. This illustrates the importance of multiple factors in the behavior of organic molecules, including volume, geometric shape, and types of bonds. While many factors affect boiling point, including isomerism, there are certain stable trends (e.g., larger unbranched alkanes have higher boiling points than smaller unbranched alkanes). One of these is:

Alkynes | Alkanes | Alkenes

One other big factor affecting intermolecular attraction of both alkynes and alkenes is the location of the double or triple bond. For example, while 1-butyne has a higher boiling point (8.1°C) than its corresponding alkanes and alkenes (ranging from -6.3°C to 3.7°C), 2-butyne has a much higher boiling point than 1-butyne: 27 °C. With larger molecules, geometric shape becomes an even more important factor in intermolecular forces.

Alkynes are named in a similar way to alkenes. The suffix used for alkynes is –*yne*. Ethyne is often called **acetylene**. Alkynes with one triple bond have the general formula C_nH_{2n-2}. Multiple triple bonds are named using –*diyne*, –*triyne*, etc. The infix –*ynyl*– is used for functional groups composed of alkynes after the removal of a hydrogen atom.

Hydrocarbons with **both double and triple bonds are known as alkenynes**. The locant number for the double bond precedes the name, and the locant for the triple bond follows the suffix –*en*– and precedes the suffix –*yne*.

Examples:

$HC \equiv CH$	is acetylene or ethyne.
	is 1-butyne.
$HC \equiv C - C \equiv C - CH_3$	is 1, 3-pentadiyne.
	is a 4-hexynyl group.
	is 1-buten-3-yne. This compound has the common name of vinylacetylene

Ring compounds: cycloalkanes, –enes, and –ynes _____

In addition to straight and branched chains, carbons can also form rings. The structures for these cyclic hydrocarbons are often written as if the molecule lay entirely within the plane of the paper even though in reality, these rings dip above and below a single plane. When there is more than one substitution on the ring, numbering begins with the first substitution listed in alphabetical order.

Hydrocarbons are divided into classes called **aliphatic** and **aromatic**. Aromatic hydrocarbons are related to benzene and are always cyclic (more about them below). Aliphatic hydrocarbons may be open-chain or cyclic. Aliphatic hydrocarbons include the alkanes, alkenes, and alkynes described above, and also cyclic hydrocarbons other than benzene rings. Aliphatic cyclic hydrocarbons are called **alicyclic**. **Alicyclic hydrocarbons use the prefix** *cyclo–* before the number root for the molecule. **Cis-trans isomerism** is also often part of the complete name for a cycloalkane.

Examples:

	cyclopropane
	methylcyclohexane 1,3-
	1,3-cyclohexadiene
	is 1-ethyl-3-propylcyclobutane.

Because single bonds allow rotation, cyclohexane freely alternates between different stereoisomeric conformations, the most stable of which are the "chair" form and the "boat" form.

planar conformation   boat conformation   chair conformation

A demonstration of how cyclohexane freely alternates between chair and boat shapes can be found here: https://www.youtube.com/watch?v=noqc6cgQOm4

Aromatic hydrocarbons

Aromatic hydrocarbons include benzene, its derivatives, and molecules that include one or more benzene rings. These molecules are called **arenes** to distinguish them from alkanes, alkenes, and alkynes. Benzene rings are written as if every other bond is a double bond, but in reality each benzene ring is a 6-carbon system with electrons in delocalized π orbitals, that move freely throughout the ring. The double bonds are said to be conjugated, and this type of structure is called a **conjugated system.** This creates an unusually stable structure; it is very difficult to break carbon-carbon bonds within a benzene ring. In contrast to the unconjugated 6-carbon ring cyclohexane, benzene rings are planar.

Substitutions onto the benzene ring are named in alphabetical order using the lowest possible locant numbers. The prefix *phenyl–* may be used for $C_6H_5–$ (benzene less a hydrogen) attached as a functional group to a larger hydrocarbon residue. Arenes in general form aryl functional groups. A phenyl group may be represented in a structure by the symbol Ø. The prefix *benzyl–* may used for $C_6H_5CH_2–$ (methylbenzene with a hydrogen removed from the methyl group) attached as a functional group.

Examples:

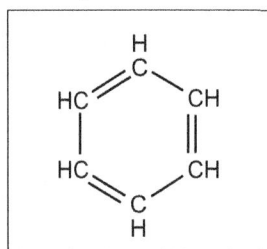 Or 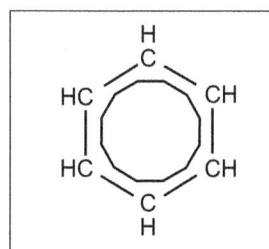 benzene

The most often used common names for aromatic hydrocarbons are listed in the following table. Naphthalene is the simplest molecule formed by fused benzene rings.

Structure	Systematic name	Common name
![toluene structure] HC—CH, HC, C—CH₃, HC—CH	methylbenzene	toluene
![o-xylene structure] HC═CH, HC, C—CH₃, HC—C, CH₃	1,2-dimethylbenzene	ortho-xylene or o-xylene

	1,3-dimethylbenzene	*meta*-xylene or *m*-xylene
	1,4-dimethylbenzene	*para*-xylene or *p*-xylene
	ethenylbenzene	styrene
		naphthalene

Side chains and functional groups

The functional groups in organic chemistry behave in characteristic ways because of the potential energy associated with particular geometric arrangements of atoms. Chemical behavior—whether compounds or atoms will bond and what types of bonds they will form-- arises from electrostatic interaction between electrons and nuclei. Larger charges tend to lead to larger strengths of interaction so polar groups are more reactive than simple alkanes. In addition, triple bonds are stronger than double or single bonds because they share more pairs of electrons. Groups of atoms in organic molecules that react in different ways than regular alkanes are called **functional groups**.

Hydrocarbons consist entirely of nonpolar C-H bonds with no unpaired electrons. These compounds are relatively unreactive. The substitution of one or more atoms with unpaired electrons into the hydrocarbon backbone creates a **hydrocarbon derivative**. The unpaired electrons result in polar or charged portions of these molecules. These atoms fall into categories known as **functional groups**, and they create **local regions of reactivity**. Alkyl, alkenyl, alkynyl, and aryl groups may also be considered functional groups in some circumstances as described in the previous skill.

The carboxyl group is an arrangement of -COOH atoms that makes an organic molecule into an organic **acid**. The hydrogen atom in the –COOH group can be removed (providing H^+ ions to a solution). When this occurs, the oxygen retains both the electrons it shared with the hydrogen and the molecule then has a negative charge.

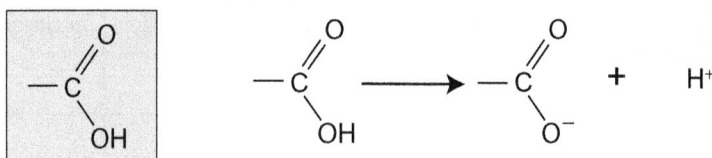

If polar or ionizing functional groups are attached to otherwise hydrophobic (fat soluble) molecules, the molecule may become hydrophilic (water soluble) due to the functional group. Some ionizing functional groups are: -COOH, -OH, -CO, and $-NH_2$.

Some other common functional groups include:

Hydroxyl group:

The hydroxyl group, -OH, is the functional group identifying alcohols. The hydroxyl group makes the molecule polar which tends to increase the solubility of the compound in polar solvents. (This is different from the –COOH group of acids; the double bonded oxygen attracts the electron cloud, pulling electrons away from the –OH to a great enough extent that the molecule lose an H+ and becomes negatively charged.)

Carbonyl group:

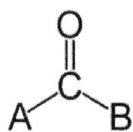

The carbonyl group is a -C=O group. It is found in aldehydes (carboxyl group is attached to another carbon and a hydrogen) and ketones (carboxyl group is attached to two other carbons).

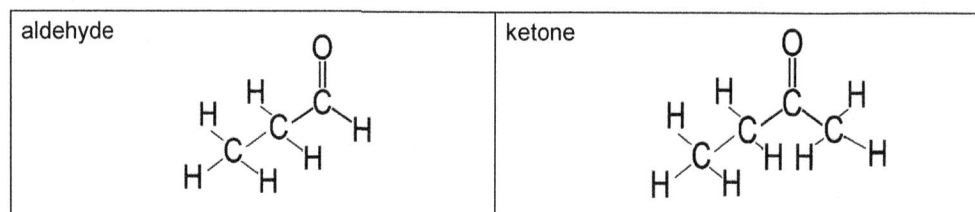

The double bonded oxygen atom is highly electronegative so it creates a polar molecule and will exhibit properties of polar molecules.

Amino group:

An amino group contains an ammonia-like functional group composed of a nitrogen and two hydrogen atoms covalently bonded. The nitrogen atom has unshared electrons and so can bind free H^+ ions (proton). This gives the molecule basic properties. An organic compound that contains an amino group is called an amine. The amines are weak bases because the unshared electron pair of the nitrogen atom can form a coordinate bond with a proton. Another group of molecules that contains the amino group are amino acids. Every amino acid contains both a $-NH_2$ group and a $-COOH$ group.

This is the amino acid glycine.

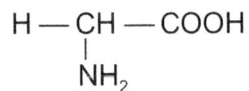

$$H - \underset{\underset{NH_2}{|}}{CH} - COOH$$

In the tables on the following pages, the symbols R and R' represent hydrocarbons in covalent linkage to the functional group. Many derivatives are named in a similar manner to alkenes and alkynes, but the location and suffix of the functional group is used in place of *–ene* and *–yne*.

The **acyl group** is:

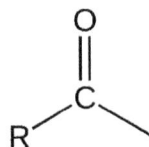

$$\underset{R}{\overset{\displaystyle O}{\underset{\displaystyle \|}{C}}}\diagdown$$

There are non-systematic number roots for hydrocarbon derivatives containing acyl groups or derived from acyl groups. These are the ketones, carboxylic acids, esters, nitriles, and amides:

Number of carbons (including the acyl carbon)	Systematic prefix	Accepted prefix for acyl and nitrile functional groups
1	meth–	form–
2	eth–	acet–
3	prop–	propion–
4	but–	butyr–
5	pent–	pent–
6 and above	same for larger numbers	

Class of molecule	Functional group	Structure	Affix	Example
Alcohol	Hydroxyl —OH	primary: R—CH₂—ÖH; secondary: R₁R₂CH—ÖH; tertiary: R₁R₂R₃C—OH	-ol	*ethanol* (H₃C—CH₂—OH)
Ether	Oxy (—O—)	R₁—Ö—R₂	-Oxy-	*methoxyethane OR ethyl methyl ether* (H₃C—O—CH₂—CH₃)
Aldehyde	Carbonyl	R—C(=O)—H	-al	*propionaldehyde OR propanal* (O=CH—CH₂—CH₃)
Ketone	Carbonyl	R₁—C(=O)—R₂	-one	*acetone* (H₃C—C(=O)—CH₃)
Carboxylic acid	Carboxyl	R—C(=O)—ÖH	-oic	*formic acid OR methanoic acid* (HC(=O)—OH)
Ester	Oxycarbonyl	R₂—C(=O)—Ö—R₁	-yl -oate	*methyl butyrate OR methyl butanoate* (CH₃—CH₂—CH₂—C(=O)—O—CH₃)

Class of molecule	Functional group	Structure	Affix	Example
Acid anhydride	Carbonyloxycarbonyl		-oic anhydride	
Nitrile	Cyanide $—C≡N$	$R—C≡N$	-nitrile	acetonitrile OR ethanonitrile $H_3C—C≡N$
Amine	Amino $—N—$	primary, secondary, tertiary	-amine	ethanamine
Amide	Aminocarbony 1	primary, secondary, tertiary	-amide	propionamide OR propanamide
Alkyl halide	Halide $—X$ (where X is F, Cl, Br, or I)	$R—X$	flouro- chloro- bromo- iodo-	2-bromobutane

Sulfhydryl group

A thiol is a compound that contains a functional group composed of a sulfur atom and a hydrogen atom (-SH). This functional group is referred to either as a *thiol group* or a *sulfhydryl group*. More traditionally, thiols have been referred to as *mercaptans*.

The small difference in electronegativity between the sulfur and the hydrogen atom produces a non-polar covalent bond. This in turn provides for no hydrogen bonding giving thiols lower boiling points and less solubility in water than alcohols of a similar molecular mass.

Phosphate group

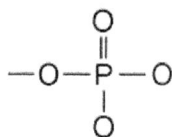

$$\begin{array}{c} \text{O} \\ \| \\ -\text{O}-\text{P}-\text{O} \\ | \\ \text{O} \end{array}$$

The phosphate ion as a functional group in a hydrocarbon chain is ideal for energy transfer reactions (such as those that occur with ATP) because of its symmetry and rotating double bond.

In biological systems, phosphates are most commonly found in the form of adenosine phosphates, (AMP, ADP and ATP) and in DNA and RNA.

Lewis electron dot structures, valence shell electron pair repulsion (VSEPR) theory, and molecular orbital (M/O) theory.

Lewis dot structures are used to keep track of each atom's valence electrons in a molecule. Drawing Lewis structures is a three-step process:

1. Determine the number of valence shell electrons for each atom. If the compound is an anion, add the charge of the ion to the total electron count because anions have "extra" electrons. If the compound is a cation, subtract the charge of the ion.

2. Write the symbols for each atom and show how the atoms within a molecule are bound to each other.

3. Draw a single bond (one pair of electron dots or a line) between each pair of connected atoms. Place the remaining electrons around the atoms as unshared pairs. If every atom has an octet of electrons except H atoms with two electrons, the Lewis structure is complete. Shared electrons count towards both atoms. If there are too few electron pairs to do this, draw multiple bonds (two or three pairs of electron dots between the atoms) until an octet is around each atom (except H atoms with two). If there are too many electron pairs to complete the octets with single bonds then the octet rule is broken for this compound.

Example: Draw the Lewis structure of HCN.

Solution: From the periodic table, we know that each atom contributes the following number of electrons: H - 1, C - 4, N - 5. Summing these numbers, we see that this molecule will have a total of 10 valence electrons (because it has no net charge we do not add to or subtract from this number).

1. The atoms are connected with C at the center and will be drawn as: H C N. Having H as the central atom is impossible because H has one valence electron and will always only have a single bond to one other atom. If N were the central atom then the formula would probably be written as HNC.

2. Connecting the atoms with 10 electrons in single bonds gives the structure to the right. H has two electrons to fill its valence subshells, but C and N only have six each. A triple bond between these atoms fulfills the octet rule for C and N and is the correct Lewis structure.

H : C̈ : N̈

H : C⋮⋮⋮N:

Molecular Orbital Theory

The electron configurations of isolated atoms are found using atomic orbitals; the configurations of atoms about to bond are represented by atomic and hybridized atomic orbitals; and the electron configurations of molecules are represented by molecular orbitals. Molecular orbital theory is an advanced topic, but it may be simplified to representing the bonds between atoms as overlapping electron density shapes from atomic orbitals. There are two typical locations for molecular orbitals.

The bonding sigma orbital (σ) surrounds a line drawn between the two atoms in a bond. At least one electron pair in every bond is in a bonding σ orbital. Sigma bonds get their name from s orbitals because the spherical electron density shapes of two s orbitals overlap to form a σ orbital. A drawing of this overlap and the resulting molecular orbital is shown to the right for H_2. Hybrid or p atomic orbitals also form a σ orbital when they overlap such that the axis between the bonded atoms runs through the center of the combined electron density.

The bonding pi orbital (π) occurs in the region of a line drawn between the two atoms in a bond. Two overlapping p orbitals will form π ⊂⊂ bonds to contain the additional shared electrons in molecules with double or triple bonds. π bonds prevent atoms from rotating about the central axis between them.

In CH_4, the electron density of the four $sp3$ orbitals of C each overlap with an s orbital of H to form four σ bonds. In C2H$_4$ (an alkene), two sp_2 orbitals on each C overlap with H, the remaining $sp2$ orbitals overlap with each other in a σ bond, and the p orbitals (drawn as shaded shapes) overlap with each other above and beneath the carbons in a π⊂⊂ bond (also drawn as shaded shapes). In CO_2, the C atom has two sp hybrid orbitals and two p

orbitals. These form one bond and one π_{CC} bond with the two unfilled p orbitals on each O atom. In $C2H_2$ (an alkyne), a triple bond forms with one C and two π_C.

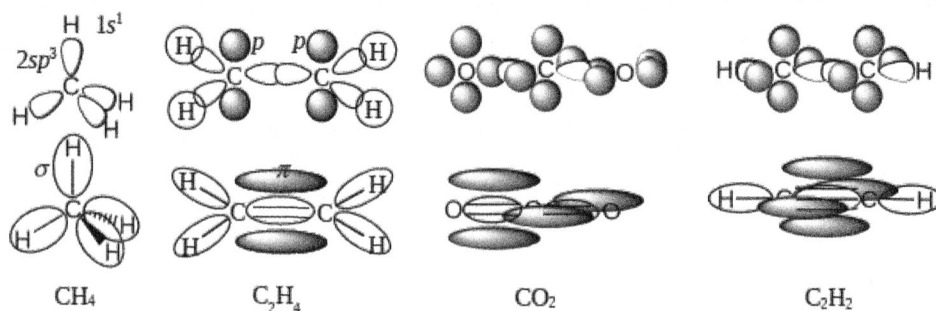

CH₄   C₂H₄   CO₂   C₂H₂

Molecules with double bonds next to each other and aromatic molecules based on benzene contain more than two π_C orbitals on adjacent atoms. The bonds and the entire molecules are described as being conjugated. Electrons in these molecules are free to move from one bond to the next on the same molecule and so are delocalized. Delocalized electrons are found throughout the entire substance in materials dealing with metallic bonds.

Molecular orbital theory also predicts **antibonding orbitals** that **prevent bonding** because they are at a higher energy level than the electrons in individual atoms. Antibonding electrons play a role in explaining why molecules like H_2 form while molecules like He_2 do not, but they are not required to predict molecular structures.

SECTION VII:
Appendix 3:

**Periodic Table
and Formula Sheet of
Equations and
Constants**

Periodic Table of the Elements

1	2	3	4	5	6	7	8	9	10	11	12	13	14	15	16	17	18
1 **H** 1.008																	2 **He** 4.00
3 **Li** 6.94	4 **Be** 9.01											5 **B** 10.81	6 **C** 12.01	7 **N** 14.01	8 **O** 16.00	9 **F** 19.00	10 **Ne** 20.18
11 **Na** 22.99	12 **Mg** 24.30											13 **Al** 26.98	14 **Si** 28.09	15 **P** 30.97	16 **S** 32.06	17 **Cl** 35.45	18 **Ar** 39.95
19 **K** 39.10	20 **Ca** 40.08	21 **Sc** 44.96	22 **Ti** 47.90	23 **V** 50.94	24 **Cr** 52.00	25 **Mn** 54.94	26 **Fe** 55.85	27 **Co** 58.93	28 **Ni** 58.69	29 **Cu** 63.55	30 **Zn** 65.39	31 **Ga** 69.72	32 **Ge** 72.59	33 **As** 74.92	34 **Se** 78.96	35 **Br** 79.90	36 **Kr** 83.80
37 **Rb** 85.47	38 **Sr** 87.62	39 **Y** 88.91	40 **Zr** 91.22	41 **Nb** 92.91	42 **Mo** 95.94	43 **Tc** (98)	44 **Ru** 101.1	45 **Rh** 102.91	46 **Pd** 106.42	47 **Ag** 107.87	48 **Cd** 112.41	49 **In** 114.82	50 **Sn** 118.71	51 **Sb** 121.75	52 **Te** 127.60	53 **I** 126.91	54 **Xe** 131.29
55 **Cs** 132.91	56 **Ba** 137.33	57 ***La** 138.91	72 **Hf** 178.49	73 **Ta** 180.95	74 **W** 183.85	75 **Re** 186.21	76 **Os** 190.2	77 **Ir** 192.2	78 **Pt** 195.08	79 **Au** 196.97	80 **Hg** 200.59	81 **Tl** 204.38	82 **Pb** 207.2	83 **Bi** 208.98	84 **Po** (209)	85 **At** (210)	86 **Rn** (222)
87 **Fr** (223)	88 **Ra** 226.02	89 **†Ac** 227.03	104 **Rf** (261)	105 **Db** (262)	106 **Sg** (266)	107 **Bh** (264)	108 **Hs** (277)	109 **Mt** (268)	110 **Ds** (271)	111 **Rg** (272)							

*Lanthanide Series

58 **Ce** 140.12	59 **Pr** 140.91	60 **Nd** 144.24	61 **Pm** (145)	62 **Sm** 150.4	63 **Eu** 151.97	64 **Gd** 157.25	65 **Tb** 158.93	66 **Dy** 162.50	67 **Ho** 164.93	68 **Er** 167.26	69 **Tm** 168.93	70 **Yb** 173.04	71 **Lu** 174.97

†Actinide Series

90 **Th** 232.04	91 **Pa** 231.04	92 **U** 238.03	93 **Np** (237)	94 **Pu** (244)	95 **Am** (243)	96 **Cm** (247)	97 **Bk** (247)	98 **Cf** (251)	99 **Es** (252)	100 **Fm** (257)	101 **Md** (258)	102 **No** (259)	103 **Lr** (262)

Formula Sheet of Equations and Constants

Throughout the exam the following symbols have the definitions specified unless otherwise noted.

L, mL = liter(s), milliliter(s)   mm Hg = millimeters of mercury
g   = gram(s)   J, kJ = joule(s), kilojoule(s)
nm   = nanometer(s)   V   = volt(s)
atm   = atmosphere(s)   mol   = mole(s)

ATOMIC STRUCTURE

$$E = h\nu$$
$$c = \lambda\nu$$

E = energy
ν = frequency
λ = wavelength

Planck's constant, $h = 6.626 \times 10^{-34}$ J s
Speed of light, $c = 2.998 \times 10^{8}$ m s^{-1}
Avogadro's number $= 6.022 \times 10^{23}$ mol^{-1}
Electron charge, $e = -1.602 \times 10^{-19}$ coulomb

EQUILIBRIUM

$$K_c = \frac{[C]^c[D]^d}{[A]^a[B]^b}, \text{ where } a\,A + b\,B \rightleftarrows c\,C + d\,D$$

$$K_p = \frac{(P_C)^c(P_D)^d}{(P_A)^a(P_B)^b}$$

$$K_a = \frac{[H^+][A^-]}{[HA]}$$

$$K_b = \frac{[OH^-][HB^+]}{[B]}$$

$$K_w = [H^+][OH^-] = 1.0 \times 10^{-14} \text{ at } 25°C$$
$$= K_a \times K_b$$

$$pH = -\log[H^+], \quad pOH = -\log[OH^-]$$

$$14 = pH + pOH$$

$$pH = pK_a + \log\frac{[A^-]}{[HA]}$$

$$pK_a = -\log K_a, \quad pK_b = -\log K_b$$

Equilibrium Constants

K_c (molar concentrations)
K_p (gas pressures)
K_a (weak acid)
K_b (weak base)
K_w (water)

KINETICS

$$\ln[A]_t - \ln[A]_0 = -kt$$

$$\frac{1}{[A]_t} - \frac{1}{[A]_0} = kt$$

$$t_{1/2} = \frac{0.693}{k}$$

k = rate constant
t = time
$t_{1/2}$ = half-life

GASES, LIQUIDS, AND SOLUTIONS

$$PV = nRT$$

$$P_A = P_{total} \times X_A, \text{ where } X_A = \frac{\text{moles A}}{\text{total moles}}$$

$$P_{total} = P_A + P_B + P_C + \ldots$$

$$n = \frac{m}{M}$$

$$K = \text{°C} + 273$$

$$D = \frac{m}{V}$$

$$KE \text{ per molecule} = \frac{1}{2}mv^2$$

Molarity, M = moles of solute per liter of solution

$$A = abc$$

P = pressure
V = volume
T = temperature
n = number of moles
m = mass
M = molar mass
D = density
KE = kinetic energy
v = velocity
A = absorbance
a = molar absorptivity
b = path length
c = concentration

Gas constant, R = 8.314 J mol^{-1}K^{-1}

$\qquad\qquad$ = 0.08206 L atm mol^{-1} K^{-1}

$\qquad\qquad$ = 62.36 L torr mol^{-1} K^{-1}

1 atm = 760 mm Hg = 760 torr

STP = 273.15 K and 1.0 atm

Ideal gas at STP = 22.4 L mol^{-1}

THERMODYNAMICS / ELECTROCHEMISTRY

$$q = mc\Delta T$$

$$\Delta S° = \sum S° \text{ products} - \sum S° \text{ reactants}$$

$$\Delta H° = \sum \Delta H_f° \text{ products} - \sum \Delta H_f° \text{ reactants}$$

$$\Delta G° = \sum \Delta G_f° \text{ products} - \sum \Delta G_f° \text{ reactants}$$

$$\Delta G° = \Delta H° - T\Delta S°$$

$$= -RT \ln K$$

$$= -nFE°$$

$$I = \frac{q}{t}$$

q = heat
m = mass
c = specific heat capacity
T = temperature
$S°$ = standard entropy
$H°$ = standard enthalpy
$G°$ = standard Gibbs free energy
n = number of moles
$E°$ = standard reduction potential
I = current (amperes)
q = charge (coulombs)
t = time (seconds)

Faraday's constant, F = 96,485 coulombs per mole
$\qquad\qquad\qquad\qquad$ of electrons

$$1 \text{ volt} = \frac{1 \text{ joule}}{1 \text{ coulomb}}$$

SECTION VIII:
Practice Test 1

**Plus Answers
and Rationale**

Practice Test 1

Part I - 60 Multiple Choice Questions

1: How many atoms of O are present in 15.2 g of Fe SO_4?

   A. 4.0

   B. 2.4 x 10^{24}

   C. 2.4 x 10^{23}

   D. 6.02 x 10^{23}

Questions 2-5 refer to the following five orbital diagrams labeled A, B, C, and D.

From the orbital diagrams select an example which demonstrates:

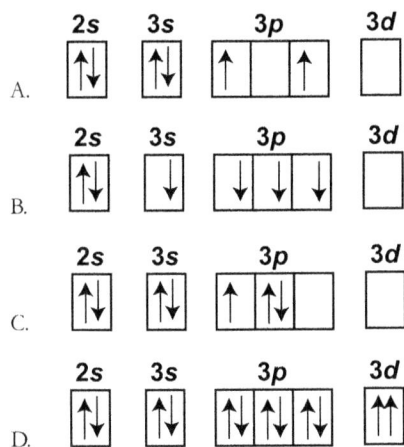

2. a violation of Hund's rule

3. a violation of the Pauli exclusion principle

4. a ground state orbital diagram

5. an excited state orbital diagram

6. What is the electron configuration for Ge?

   A. $1s^2\ 2s^2\ 2p^6\ 3s^2\ 3p^6$

   B. $1s^2\ 2s^2\ 2p^6\ 3s^2\ 3p^6\ 4s^1$

   C. $1s^2\ 2s^2\ 2p^6\ 3s^2\ 3p^6\ 4s^2$

   D. $1s^2\ 2s^2\ 2p^6\ 3s^2\ 3p^6\ 4s^2\ 3d^{10}\ 4p^2$

7. Which energy sublevel is being filled by the elements Sc through Zn?

A. 3d

B. 4s

C. 4p

D. 4d

8. Which of the following represents Titanium (Ti)

A. 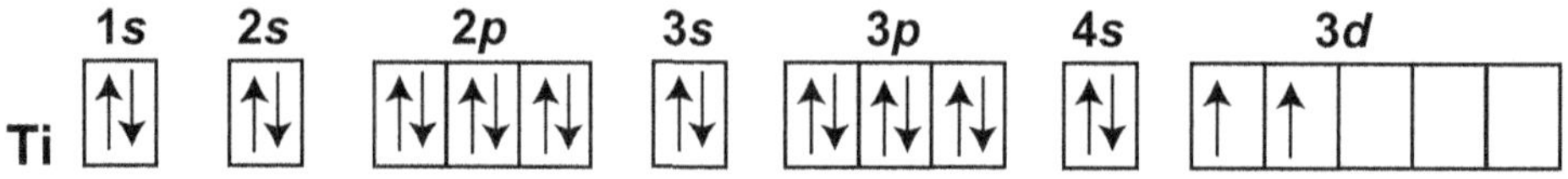 Ti — 1s 2s 2p 3s 3p 4s 3d

B. 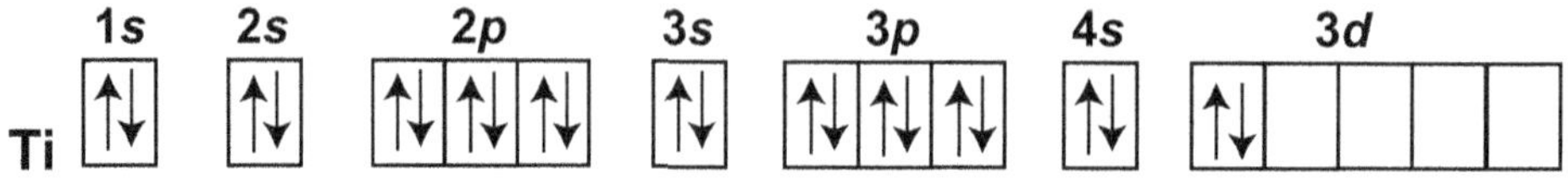 Ti — 1s 2s 2p 3s 3p 4s 3d

C. 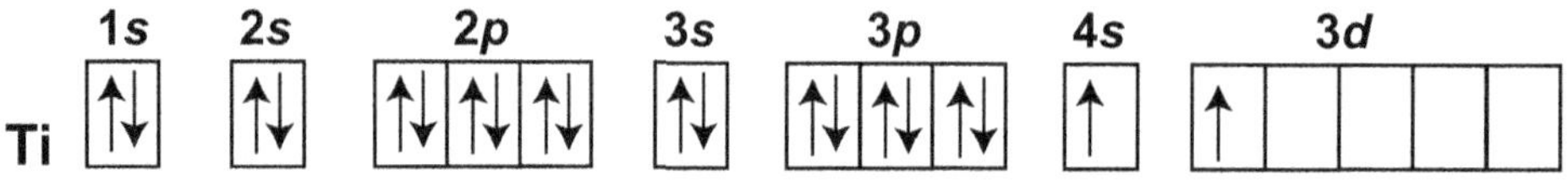 Ti — 1s 2s 2p 3s 3p 4s 3d

D. 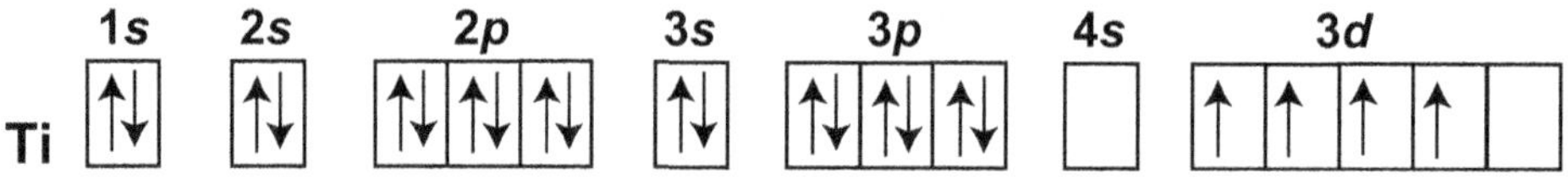 Ti — 1s 2s 2p 3s 3p 4s 3d

9. Predict which of the following has chemical properties most similar to arsenic.

A. Au

B. Cl

C. Ge

D. Sb

10. In each set of the following elements the size of atom increases from left to right except?

A. Li, Na, K

B. As, Ge, Ga

C. Br, Se, As

D. Al, Si, P

11. 100 μL of SO_2 contains 2.2 x 1017 molecules at STP. According to Avogadro's hypothesis, how many molecules are in 200 μL of methane gas, CH_4, at STP?

A. .2 x 10^{17}

B. 44 x 10^{17}

C. 4.4 x 10^{17}

D. 0.22 x 10^{17}

12. Which of the following graphs represents the plot of volume versus Kelvin temperature at constant pressure?

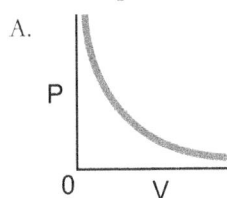

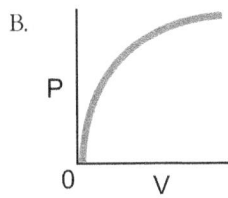

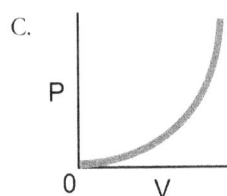

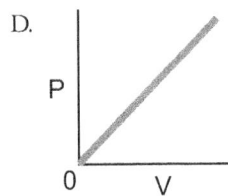

A.

B.

C.

D.

13. Phenacyl chloride is a riot control gas with the formula of C_7H_7OCl. What would be the percentage of carbon in the compound?

A. **59.2%**

B. 8.42%

C. 11.2%

D. 24.9%

14. Which of the following is not true according to the kinetic theory of gases?

A. molecules occupy a negligible volume

B. molecules are not attracted to one another

C. molecules have elastic collisions

D. molecules in the same container all have the same velocity

15. A sample of Argon gas is at $10°$ C. If both the volume and pressure double, what is the new Kelvin temperature?

A. 40 K

B. 313 K

C. 566 K

D. 1132 K

16. A container with two gases, helium and argon, is 30.0% by volume helium. Calculate the partial pressure of helium and argon if the total pressure inside the container is 4.00 atm.

A. PHe = 0.300 atm

B. PHe = 1.20 atm.

C. PHe = 2.8 at

D. PHe = 0.12 atm

17. Which of the following statements regarding molecular geometries are true ?

    I.   Double bonds are planar

    II.   A central atom with sp2 hybridization is trigonal

    III.   A central atom with sp2 hybridization is bipyramidal

    IV.   Single bonds are always planar

A.   I and II are true

B.   I and IV are true

C.   II and IV are true

D.   II and III are true

18. Find the mass of CO_2 produced by the combustion of 15 kg of isopropyl alcohol in the reaction

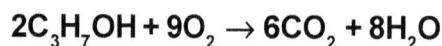

$$2C_3H_7OH + 9O_2 \rightarrow 6CO_2 + 8H_2O$$

A.   33 kg

B.   44 kg

C.   50 kg

D.   60 kg

19. What is the shape of an NH_3 molecule using the VSEPR model.

A.   Trigonal planar

B.   Tetrahedral

C.   Trigonal bipyramidal

D.   Trigonal pyramidal

20. How many mL of water is needed to add to 150 mL of 6M HNO_3 to prepare a solution of 0.6 M HNO_3?

A. 350 ml

B. 2050 ml

C. 1350 ml

D. 1200 ml

21. Given:

$E° = -2.37V$ for $\qquad Mg_2+ (aq) + 2e^- \leftrightarrow Mg (s)$

and

$E° = 0.80 V$ for $\qquad Ag+ (aq) + e- \leftrightarrow Ag (s)$

What is the standard potential of a voltaic cell composed of a piece of magnesium dipped in a 1 M Ag^+ solution and a piece of silver dipped in 1 M Mg^{2+}?

A. 0.77 V

B. 1.57 V

C. 3.17 V

D. 3.97 V

22. Consider the reaction between iron and hydrogen chloride gas:

$$Fe(s) + 2HCl(g) \rightarrow FeCl_2(s) + H_2(g)$$

7 moles of iron and 10 moles of HCl react until the limiting reagent is consumed. Which statements are true?

    I.   HCl is the excess reagent

    II.   HCl is the limiting reagent

    III.   7 moles of H_2 are produced

    IV.   2 moles of the excess reagent remain

A. I and III

B. I and IV

C. II and III

D. II and IV

23. **Which reaction is not a redox process?**

A. Combustion of octane: $2C_8H_{18} + 25O_2 \rightarrow 16XO_2 + 18H_2O$

B. Depletion of a lithium battery: $Li + MnO_2 \rightarrow LiMnO_2$

C. Corrosion of aluminum by acid: $2Al + 6HCl \rightarrow 2AlCl_3 + 3H_2$

D. Taking an antacid for heartburn: $CaCO_3 + 2HCl \rightarrow CaCl_2 + H_2CO_3 \rightarrow CaCl_2 + CO_2 + H_2O$

24. 32.0 g of hydrogen and 32.0 grams of oxygen react to form water until the limiting reagent is consumed. What is present in the vessel after the reaction is complete?

A. 16.0 g O_2 and 48.0 g H_2O

B. 24.0 g H_2 and 40.0 g H_2O

C. 28.0 g H_2 and 36.0 g H_2O

D. 28.0 g H_2 and 34.0 g H_2O

25. Balance the equation for the neutralization reaction between phosphoric acid and calcium hydroxide by filling in the blank stoichiometric coefficients.

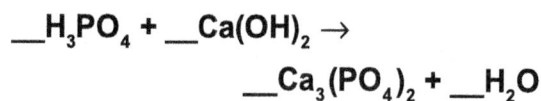

$$\_\_H_3PO_4 + \_\_Ca(OH)_2 \rightarrow$$
$$\_\_Ca_3(PO_4)_2 + \_\_H_2O$$

A. 4, 3, 1, 4

B. 2, 3, 1, 8

C. 2, 3, 1, 6

D. 2, 1, 1, 2

26. Consider the unbalanced chemical equation: $CO(g) + O_2(g) \rightarrow CO_2(g)$.

When this equation is completely balanced using the smallest whole numbers, what is the sum of the coefficients?

A. 4

B. 3

C. 6

D. 5

27. Consider the following electrochemical overall reaction.

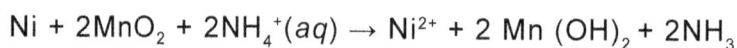

$$Ni + 2MnO_2 + 2NH_4^+(aq) \rightarrow Ni^{2+} + 2\,Mn\,(OH)_2 + 2NH_3$$

What phenomenon occurs during this reaction?

A. Ni is oxidized and its oxidation number increases

B. MnO_2 is oxidized and its oxidation number decreases

C. Ni is reduced and its oxidation number decreases

D. MnO_2 is reduced and its oxidation number increases

28. If yeast consumes glucose aerobically ($C_6H_{12}O_6$) until all the sugar is used up (18 grams), how many moles of carbon dioxide do they produce?

A. 0.6 M

B. 20 M

C. 2.6 M

D. 6 M

29. Iron(II) sulfide reacts with hydrochloric acid according to the reaction:

$$FeS(s) + 2HCl(aq) \rightarrow FeCl_2(s) + H_2S(g)$$

A reaction mixture initially contains 0.214 mol FeS and 0.664 mol HCl.

Once the reaction has occurred as completely as possible, what amount (in moles) of the excess reactant is left?

A.   0.43mol HCL left over

B.   0.24 mol HCl left over

C.   0.24 mol FeS left over

D.   0.43 mol FeS left over

30. Identify the type of reaction when $H_2CO_3 \rightarrow H_2O + CO_2$

A.   Synthesis reaction

B.   Decomposition reaction

C.   Single replacement reaction

D.   Double replacement reaction

31. Consider the energy diagram of the following reaction: A + B → C

1/Concentration versus Time

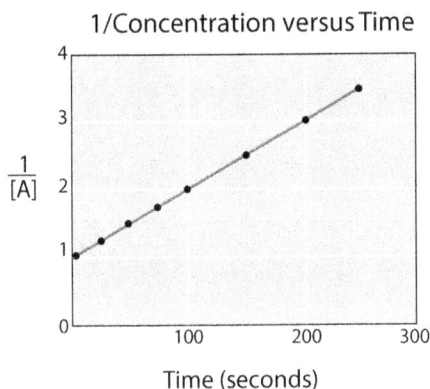

The rate of this reaction is

A.   Zero-order

B.   First Order

C.   Second Order

D.   Cannot be determined without additional information

32. For a certain reaction a plot of [A] vs. time yields a straight line with a slope of -5 and an intercept at 200 moles. From this information we know that:

    I.   The initial concentration of [A] = 200 moles.

    II.   The [A] is decreasing

    III.   The [A] is increasing

    IV.   This is a zero-order reaction

A.   I only

B.   I and II

C.   I and III

D.   III and IV

33. Which of the following statements are true about this reaction

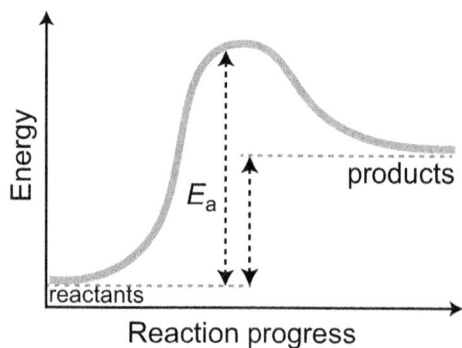

A. The reaction is endothermic

B. The reaction will proceed spontaneously

C. The activation energy is less than the energy of the products

D. The reaction will have $-\Delta G$

34. At a certain temperature, T, the equilibrium constant for the reaction
$$NO\ (g) \leftrightarrow N_2\ (g) + O_2\ (g)$$

is $K_{eq} = 2 \times 10^3$. If a 1.0 L container at this temperature contains 90 mM N_2, 20 mM O_2, and 5 mM NO, what will occur?

A. The reaction will make more N_2 and O_2.

B. The reaction is at equilibrium.

C. The reaction will make more NO.

D. The temperature, T, is required to solve this problem.

35. Which statements about reaction rates are true?

  I. Catalysts shift an equilibrium to favor product formation.

  II. Catalysts increase the rate of forward and reverse reactions.

  III. A greater temperature increases the chance that a molecular collision will overcome a reaction's activation energy.

  IV. A catalytic converter contains a homogeneous catalyst.

A. I and II

B. II and III

C. I, II, and III

D. II, and IV

36. The reaction: $(CH_3)_3CBr(aq) + OH^-(aq) \rightarrow$

$(CH_3)_3COH(aq) + Br^-(aq)$

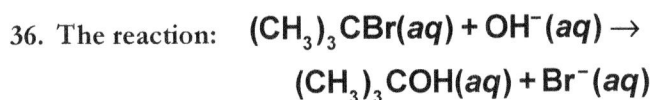

occurs in three elementary steps:

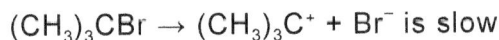

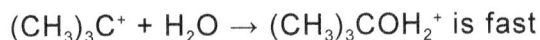

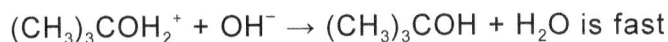

$(CH_3)_3CBr \rightarrow (CH_3)_3C^+ + Br^-$ is slow

$(CH_3)_3C^+ + H_2O \rightarrow (CH_3)_3COH_2^+$ is fast

$(CH_3)_3COH_2^+ + OH^- \rightarrow (CH_3)_3COH + H_2O$ is fast

What is the rate law for this reaction?

A. Rate $= k[(CH_3)_3 CBr]$

B. Rate $= k\left[OH^-\right]$

C. Rate $= k\left[(CH_3)_3CBr\right]\left[OH^-\right]$

D. Rate $= k\left[(CH_3)_3CBr\right]\left[OH^-\right]^2$

37. **Write the equilibrium expression K_{eq} for the reaction:**

$CO_2(g) + H_2(g) \leftrightarrow CO(g) + H_2O(l)$

A. $\dfrac{\left[CO\right]\left[H_2O\right]}{\left[CO_2\right]\left[H_2\right]^2}$

B. $\dfrac{\left[CO_2\right]\left[H_2\right]}{\left[CO\right]\left[H_2O\right]}$

C. $\dfrac{\left[CO\right]\left[H_2O\right]}{\left[CO_2\right]\left[H_2\right]}$

D. $\dfrac{\left[CO\right]}{\left[CO_2\right]\left[H_2\right]}$

38. **Which statements about reaction rates are not true?**

A. All reaction mechanisms are multi-step processes involving reaction intermediates

B. If one elementary reaction is the slowest reaction, it determines the overall reaction rate.

C. The slowest reaction in a series is called the rate-limiting step or rate determining step

D. Some reactions may not have a single rate-determining step.

39. The following represents the reaction for the destruction of ozone (O_3) in the upper atmosphere.

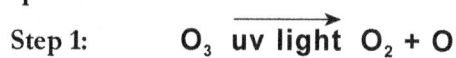

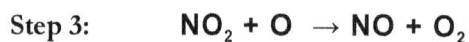

Step 1: $O_2 \xrightarrow{\text{uv light}} O_2 + O$

Step 2: $NO + O_3 \rightarrow NO_2 + O_2$

Step 3: $NO_2 + O \rightarrow NO + O_2$

The net overall reaction is:

A. $O_3(g) + NO(g) = NO_2(g) + O_2(g)$

B $NO_2(g) + O^-(\text{radical}) = NO(g) + O_2(g)$

C. $O_3(g) + O_2(g) = O_5(g)$

D. $2O_3(g) \rightarrow 3O_2(g)$

40. The energetic barrier to a reaction is

A.   determined by the Arrhenius Equation.

B.   related to the potential energy of the products

C.   not effected by a catalyst

D.   determines if a reaction is spontaneous.

41. Some typical bond energies are given in the following table.

C $=$ O	743 kJ/mol
H $-$ O	463 kJ/mol
C $\equiv$ C	829 kJ/mol
C $=$ C	614 kJ/mol

From this information we can conclude that

I.   The C $\equiv$ C bond is the shortest

II.   The H $-$ O bond is the shortest

III.   The C $=$ O is longer than the C$=$C

IV.   The C $\equiv$ C bond is the strongest

A.   I only

B.   I and III

C.   II and III

D.   I and IV

42. Doubling a reactant's concentration increases reaction rate by a factor of four indicates that

    A. the reaction is exothermic

    B. the reaction must have a catalyst

    C. the reaction is a second order reaction

    D. the reaction rate can be determined by a plot if [A] vs t.

43. **The melting of a solid is expected to have**

    A. $- \Delta H$ and a $+ \Delta S$

    B. $+ \Delta H$ and a $+ \Delta S$

    C. $-\Delta H$ and a $- \Delta S$

    D. $+\Delta H$ and a $- \Delta S$

44. **Which of the following has a standard state enthalpy of formation of zero?**

    A. CO_2 (g)

    B. HCL (g)

    C. Cl_2

    D. O_3

45. **At what temperature is the following process spontaneous?**

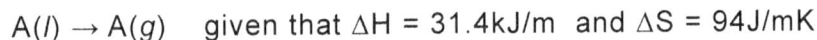

    $A(l) \rightarrow A(g)$ given that $\Delta H = 31.4 kJ/m$ and $\Delta S = 94 J/mK$

    A. $61°C$

    B. $58°C$

    C. $0.33°C$

    D. $3.3°C$

46. **Which of the following has a ΔS which is positive ?**

    A. Solid sugar is added to water to form a solution?

    B. Iodine vapor condenses on a cold surface to form crystals?

    C. $2NiS(s) + 3O_2(g) \rightarrow 2SO_2(g) + 2NiO(s)$

    D. A liquid cooled to a solid

47. Using a Styrofoam cup as a calorimeter, a student puts a piece of metal that weights 20 g at temperature of 100 °C into 50 g distilled water at temperature of 25 °C. He stirred the mixture and observed that the final temperature of the water and metal was 32.0 °C.

By ignoring heat gained or lost by the calorimeter, which of the following statement correctly describes the heat flow in calories?

A. The metal gained 1360 calories of heat and the water lost 350 calories of heat.

B. The metal gained 350 calories of heat and the water lost 350 calories of heat.

C. The metal lost 350 calories of heat and the water gained 350 calories of heat.

D. The metal lost 1360 calories of heat and the water gained 350 calories of heat.

48. What would occur if 1 mL of sulfuric acid is dissolved in a 100 mL beaker of water, indicating that the process is exothermic?

A. Gas will be evolved from the solution.

B. Decrease in temperature of the solution will be observed.

C. An increase in temperature of the solution will be noticed.

D. The solution changes in color.

49. Using the thermochemical equations below, which one of the following choices is correct for the calculation of the change in heat of the following reaction?

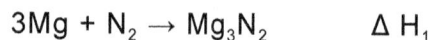

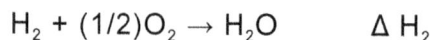

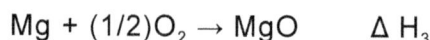

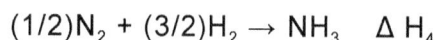

$$Mg_3N_2 + 3H_2O \rightarrow 3MgO + 2NH_3 \ \Delta H_{rxn}=?$$

$$3Mg + N_2 \rightarrow Mg_3N_2 \qquad \Delta H_1$$
$$H_2 + (1/2)O_2 \rightarrow H_2O \qquad \Delta H_2$$
$$Mg + (1/2)O_2 \rightarrow MgO \qquad \Delta H_3$$
$$(1/2)N_2 + (3/2)H_2 \rightarrow NH_3 \qquad \Delta H_4$$

A. $\Delta H_{rxn} = \Delta H_1 + \Delta H_2 + \Delta H_3 + \Delta H_4$

B. $\Delta H_{rxn} = \Delta H_1 + \Delta H_2 - \Delta H_3 - \Delta H_4$

C. $\Delta H_{rxn} = 2\Delta H_4 + 4\Delta H_3 - \Delta H_2 - 3\Delta H_1$

D. $\Delta H_{rxn} = 3\Delta H_3 + 2\Delta H_4 - \Delta H_1 - 3\Delta H_2$

50. The heat capacity of lead is 0.13 J/g° C. How many joules of heat would be required to raise the temperature of 15 g of lead from 22°C to 37°C?

A. 2.0 J

B. -0.13 J

C. 5.8 X10^{-4} J

D. 29 J

51. Consider the reaction: $2HCl + Ba(OH)_2 \rightarrow BaCl_2 + 2H_2O$

How many mL of a 0.7000 M HCl solution would just react with 8.000 g of $Ba(OH)_2$?

A. 0.35ml

B. 144ml

C. 133ml

D. 4.0ml

52. What are the pH and the pOH of 0.010 M HNO_3 (*aq*)?

A. pH = 1.0, pOH = 9.0

B. pH = 2.0, pOH = 12.0

C. pH = 2.0, pOH = 8.0

D. pH = 8.0, pOH = 6.0

53. For the reaction: $3 H_2(g) + N_2(g) \leftrightarrow 2 NH_3 (g)$
the addition of N_2 gas results in:

A. The amount of H_2 increasing,

B. The amount of H_2 decreasing

C. The amount of NH_3 decreasing

D. The system stays the same because it is in equilibrium.

54. Give the equation for Kc for the equation: $NaCl (s) \leftrightarrow Na^+(aq) + Cl^- (aq)$

A. Kc = $[Na^+][Cl^-]$ / $[NaCl]$

B. Kc = $[Na^+][Cl^-]$

C. Kc = $[Na^+][Cl^-]$ $[NaCl]$

D. Kc = $[NaCl]$ /$[Na^+][Cl^-]$

55. Which of the following will have a pH above 7 at the equivalence point:

A. strong base titrated with a weak acid

B. Weak base titrated with a strong acid

C. Strong base titrated with a strong acid

D. Strong acid titrated with a strong base

56. Which of the following will have a pH at 7 at the equivalence point:

A. Strong acid titrated with a strong base

B. Weak base titrated with a strong acid

C. Strong base titrated with a strong acid

D. Both A and C

57. Litmus is red when the pH is below 5 and blue when the pH is above 8. Methyl orange is red when the pH is below 3.7 and yellow-orange when the pH is above 4.4. Phenolphthalein is bright pink when the pH is above 9.8 and colorless when the pH is below pH 8.2. For the titration to the right, which of these indicators provides an endpoint that reflects the equivalence point?

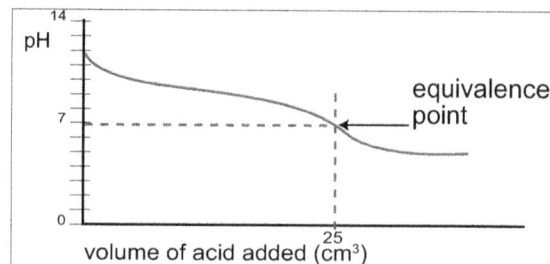

A. Phenolphthalein and litmus

B. Methyl orange only

C. Phenolphthalein only

D. Litmus only

58. Which way will the equilibrium shift if the system temperature goes up (heat is added):

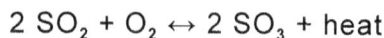

$$2 SO_2 + O_2 \leftrightarrow 2 SO_3 + heat$$

A. the reaction must shift to the left.

B. the reaction must shift to the right

C. the reaction will continue because adding heat will not effect the equilibrium.

D. Cannot determine from the information give.

59. You drop 10.0kg of solid NaOH in an Olympic-sized swimming pool full of pure water (volume = 2.5×10^6 L). What is the pH of the pool?

A. pH = 4

B. pH = 6

C. pH = 10

D. pH = 8

60. An indicator HInd produces a yellow colour in 0.1 M HCl solution and a red colour in 0.1 M HCN solution. Therefore, the following equilibrium:

$$HCN + I_{nd}^- \leftrightarrow HI_{nd} + CN^-$$

A. Products are favored and the stronger acid is HI_{nd}

B. Products are favored and the stronger acid is HCN

C. Reactants are favored and the stronger acid is H_{nd}

D. Reactants are favored and the stronger acid is HCN

Part II - Free Response Questions

1. **Use your knowledge of the Periodic Table to answer the following:**

   (a) Describe the 4 quantum numbers used to describe the electronic configuration of an atom, what they represent and their possible values.

   (b) Draw a diagram representing the electrons on Nitrogen, comment on what rules you used to arrive at this configuration.

   (c) Draw the Lewis dot structure for N_2.

   (d) Explain the trend of atomic radii from $Li^- > K$ and from $Li \rightarrow Ne$

   (e) Comment on the diagonal trends of Electron Affinity, Ionization Energy and Electronegativity in the Periodic Table.

2. **Use your knowledge of the Gas Laws to answer the following questions.**

   (a) What volume does 15.0g of H_2 gas occupy ?

   (b) A scuba diver's tank contains 0.29kg of O_2 compressed into a volume of 2.3L. What is the pressure inside the tank at 25° C?

   (c) The gas in a hair spray can at 27°C is compressed to a pressure of $30lbs/in^2$. The can will explode if the pressure reaches $90lbs/in^2$. At what temperature will the can explode.

   (d) Find the density of N at 0.326 atm and 18 °C

3. **Changes in Matter – Reactions**

   (a) Describe the 4 basic types of reactions and explain what happens in each one by giving an example.

   (b) Identify each reaction below as oxidation or reduction reaction

   i) $Pb \rightarrow Pb_2^+ + 2e^-$

   ii) $C_2H_2 + H_2 \rightarrow C_2H_4$

   iii) $Cr_3+ + 3e^- \rightarrow Cr$

   (c) We know that the following reaction can be used to generate an electric current in a voltaic cell.

$$Zn(s) + CuSO_4 (aq) \leftrightarrow Cu(s) + ZnSO_4 (aq)$$

   Sketch the voltaic cells made from the reaction, showing the key components of the two half-cells and indicating the cathode electrode and the anode electrode, the negative and positive electrodes, the direction of movement of the electrons in the wire between the electrodes, and the direction of movement of the ions in the system. Show a salt bridge and the movement of ions out of the salt bridge.

4. Rates of Chemical Reactions

(a) For the gaseous hydrogenation reaction below, what is the concentration for each substance at equilibrium?

$$C_2H_4(g) + H_2(g) \leftrightarrow C_2H_6(g)$$

with $Kc=0.98$ characterized from previous experiments and with the following initial concentrations:

$$[C_2H_4]_0 = 0.53 \qquad [H_2]_0 = 0.63$$

(b) For the following reaction:

$$A + B \rightarrow 2C$$

it is found that doubling the amount of A causes the reaction rate to double while doubling the amount of B causes the reaction rate to quadruple. What is the best rate law equation for this reaction?

(c) For the following reaction – comment on the effects of each of the changes below.

$$A(g) + B(g) \rightarrow AB(g)$$

i) increase in [A]

ii) increase in [AB]

iii) increase in pressure

iv) increase in the addition of the catalyst

(d) Draw a potential Energy Diagram that traces the PE vs Reaction for

i) an endothermic reaction

ii) an exothermic reaction

iii) an endothermic reaction with a catalyst.

5. Use your knowledge of Thermodynamics to answer the following

(a) Draw a diagram showing Temperature vs Time for the heating of ethanol from a solid to a gas. Label each of the parts.

(b) Calculate the boiling point of ethanol given the following data

Substance	ΔH_f° (kJ/mol)	ΔS° (J/K mol)
$C_2H_5OH(l)$	-277.7	160.6
$C_2H_5OH(g)$	-235.1	282.6

(c) Calculate ΔG (kJ) for a reaction at 53 °C given that $Keq = 4.33 \times 10\text{-}5''$

(d) If $\Delta H = 43kJ$ and ΔS is 49.1 J/K at what Temperature is the reaction spontaneous ?

6. **Equilibrium, Acids/Bases, Titrations etc**

   (a) Antacids use Aluminum Hydroxide to react with acid in the stomach. Assume this acid is predominately HCL:

   i. Write the balanced neutralization reaction

   ii. Calculate the number of grams of HCL that can react with 0.5g of Aluminum hydroxide.

   (b) Write a balanced equation that represent the oxidation of C_3H_8 to CO_2 and water.

   (c) Calculate the Kp when when 0.032 mol of NO react with 1.70 g of bromine to form NOBr at 25C. The concentration of nitrosyl bromide at equilibrium is 0.438 atm.

7. **Experimental**

   A summer intern is given the task of making buffer solutions for the lab. The first solution needed for an HPLC analysis is 1L of a 0.2 M solution of a TRIS buffer (Tris is tris(hydroxymethyl-amino-methane - $C_4H_6(OH)_3-NH_2$) at pH=6.9 Explain the steps you would take to make this solution; what materials and equipment you will need (include PPE).

Answers and Rationale to Practice Test 1 _____

1. **Answer C.**

   The molecular weight of Fe SO_4 is $(56 + 32 + 4(16)) = 152$ atomic mass units

   The number of moles of Fe SO_4 is 15.2 gr./152 = 0.1 moles

   The number of O atoms present in 15.2 gr. Fe SO_4

   $= 4$moles O/ mole Fe SO_4 * 0.1moles Fe SO_4 * $(6.02{\times}10^{23}) = 2.408 \times 10^{23}$

   *Note that you do not need to get an exact value for .4 * 6.02; you can get the right answer by eliminating the ones that are way off—this may be faster in certain cases when you are pressed for time.*

2. **Answer** C

   This diagram is a violation of Hund's rule, each p orbital should be filled with one electron before doubling up the electrons.

3. **Answer D**

   This diagram is a violation of the Pauli exclusion principle, two electrons occupy the same energy level (one arrow should point down).

4. **Answer A**

   This diagram shows the ground state orbital diagram for this element.

5. **Answer B**

   This diagram is for an excited state orbital because one of the electrons is in a higher orbital that it would be in, if it were in the ground state.

6. **Answer D.**

   Ge has 32 electrons and has an electron configuration of $1s^2\ 2s^2\ 2p^6\ 3s^2\ 3p^6\ 4s^2\ 3d^{10}\ 4p^2$

   When filling the orbitals remember that 4s fills before 3d, then 4s and 4p.

7. **Answer A.**

   This is the forth row on the periodic table. It starts by filling in the 4s orbitals then 3d, then 4p. So these elements represent the filling of the 3d orbitals.

8. **Answer A.**

   Figure A shows the proper filling for Ti which has 22 electrons. Recall that 4s fills before 3d.

9. **Answer D.**

   Arsenic (As atomic #33) is most similar to Sb (antimony #51) because it is in the same column and has the same number of valence electrons. It is therefore expected to have similar properties.

10. **Answer D.**

    The size (atomic radius) of the elements in the periodic table decrease from left to right for each period and increase from top to bottom for each column. In A the set of elements are located in columns so their sizes are increasing. In B and C the set of elements are located in the periods with the order of right to left so in B and C the sizes of atoms are increasing too. But in D the size of atom must decrease in the set of Al, Si, P, while it the atomic number is increased. Therefore D does not increase in size from left to right.

11. **Answer C.**

Avogadro's hypothesis states that equal volumes of different gases at the same temperature and pressure contain equal numbers of molecules. Therefore when 100 mL of SO_2 contains 2.2 x 1017 molecules, 100 mL of methane will also. Then 200mL of methane contains double amount of molecules or 4.4 x 1017

12. **Answer D.**

From the universal gas law $PV = nRT$ which can be rearranged $V=(nRT) / P$

At constant P, T is directly proportional to V. When T=0, V=0. Therefore the correct plot is D.

13. **Answer A.**

First calculate the molecular weight

C= 7 * 12 = 84

H = 7 * 1 =   7

O = 1 * 16 = 16

Cl = 1* 35 = 35

Total molecular weight = 142

%C = 84/142 = 59.15, or 59.2%

14. **Answer D.**

This statement is not true. Molecules have a distribution of velocities as shown by a Maxwell-Boltzman plot.

15. **Answer D.**

Using the Gas Law $PV = nrT$

If both P and V double ($P_2= 2P_1$ and $V_2= 2V_1$) then T must quadruple ($T_2=4T_1$). However, you must remember to use units of K. So the initial T= 10°C becomes 283.15 K. Quadrupling the temperature results in a t= 1132K.

16. **Answer B.**

The partial pressure of He can be found as a % of the total. If He is 30% by volume then :

pHe = 0.300 x 4.00 atm = 1.20 atm. pAr = 4.00 - 1.20

17. **Answer: A.**

I is true, double bonds are planar; single bonds are never planar. II is true sp2 hybridization is trigonal. Therefore only I and II are true.

18. **Answer A.**

Remember "grams to moles to moles to grams." Step 1 converts mass to moles for the known value. In this case, kg and kmol are used. Step 2 relates moles of the known value to moles of the unknown value by their stoichiometry coefficients. Step 3 converts moles of the unknown value to a mass.

$$15 \times 10^3 \text{ g C}_4\text{H}_8\text{O} \times \frac{1 \text{ mol C}_4\text{H}_8\text{O}}{60 \text{ g C}_4\text{H}_8\text{O}} \times \frac{6 \text{ mol CO}_2}{2 \text{ mol C}_4\text{H}_8\text{O}} \times \frac{44 \text{ g CO}_2}{1 \text{ mol CO}_2} = 33 \times 10^3 \text{ g CO}_2$$

$$= 33 \text{ kg CO}_2$$

The step labels above the equation read: step 1, step 2, step 3.

19. Answer D.

The Lewis structure for NH_3 is given to the right. This structure contains 4 electron pairs around the central atom, so the geometric arrangement is tetrahedral. However, the shape of a molecule is given by its atom locations, and there are only three atoms so choices B is not correct. Four electrons pairs with one unshared pair (3 bonds and one lone pair) give a trigonal pyramidal shape as shown to the left.

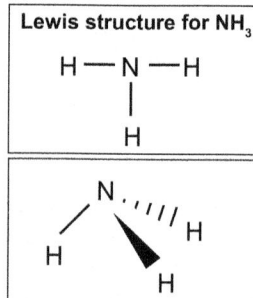

Lewis structure for NH_3

20. Answer C.

The equation of volume-concentration relation of solutions at constant temperature is

$M_1 \times V_1 = M_2 \times V_2$

$6 \times 150 = 0.6 \times V_2 \rightarrow V_2 = 6 \times 150/0.6 = 1500 \text{ mL}$

But this is the volume of the final solution after adding water to the original solution. So for answering the question of this problem we have to subtract 150mL from the final volume.

The amount of water needed to be added to the concentrated HNO_3 is $1500 - 150 = 1350 \text{ mL}$

21. Answer C.

$Ag^+ (aq) + e- \leftrightarrow Ag(s)$.
Therefore, in the cell described, reduction will occur at the Ag electrode and it will be the cathode. Using the equation:

$E^o_{cell} = E^o(\text{cathode}) \rightarrow E^o(\text{anode})$

$E^o_{cell} = 0.80V - (-2.37 \text{ V}) = 3.17V$ (Answer C)

Choice D results from the incorrect assumption that electrode potentials depend on the amount of material present.

22. Answer D.

The limiting reagent is found by dividing the number of moles of each reactant by its stoichiometric coefficient. The lowest result is the limiting reagent:

$$7 \text{ mol Fe} \times \frac{1 \text{ mol reaction}}{1 \text{mol Fe}} = 7 \text{ mol reaction if Fe is limiting}$$

$$10 \text{ mol HCl} \times \frac{1 \text{ mol reaction}}{2 \text{ mol HCl}} = 5 \text{ mol reaction if HCl is limiting}$$

Therefore, HCl is the limiting reagent (II is true) and Fe is the excess reagent. 5 moles of the reaction take place, so 5 moles of H_2 are produced, and of the 7 moles of Fe supplied, 5 are consumed, leaving 2 moles of the excess reagent (IV is true).

23. Answer D.

The oxidation state of atoms is altered in a redox process. During combustion (Choice A), the carbon atoms are oxidized from an oxidation number of -4 to +4. Oxygen atoms are reduced from an oxidation number of 0 to -2. All batteries (Choice B) generate electricity by forcing electrons from a redox process through a circuit. Li is oxidized from 0 in the metal to +1 in the $LiMnO_2$ salt. Mn is reduced from +4 in manganese (IV) oxide to +3 in lithium manganese (III) oxide salt. Corrosion (Choice C) is due to oxidation. Al is oxidized from 0 to +3. H is reduced from +1 to 0. Acid-base neutralization (Choice D) transfers a proton (an H atom with an oxidation state of +1) from an acid to a base. The oxidation state of all atoms remains unchanged (Ca at +2, C at +4, O at -2, H at +1, and Cl at -1), so D is correct. Note that Choices C and D both involve an acid. The availability of electrons in aluminum metal favors electron transfer but the availability of CO_3^{2-} as a proton acceptor favors proton transfer.

24. Answer C.

First the equation must be constructed:

$$2H_2 + O_2 \rightarrow 2H_2O$$

A fast and intuitive solution would be to recognize that:
1. One mole of H_2 is about 2.0 g, so about 16 moles of H_2 are present.
2. One mole of O_2 is 32.0 g, so one mole of is O_2 is present.
3. Imagine the 16 moles of H_2 reacting with one mole of O_2. 2 moles of H_2 will be consumed before the one mole of O_2 is gone. O_2 is limiting. (Eliminate choice A.)
4. 16 moles less 2 leaves 14 moles of H_2 or about 28 g. (Eliminate choice B.)
5. The reaction began with 64.0 g total. Conservation of mass for chemical reactions forces the total final mass to be 64.0 g also. (Eliminate choice D.)

A more standard solution is presented next. First, mass is converted to moles:

$$32.0 \text{ g } H_2 \times \frac{1 \text{ mol } H_2}{2.016 \text{ g } H_2} = 15.87 \text{ mol } H_2 \quad \text{and} \quad 32.0 \text{ g } O_2 \times \frac{1 \text{ mol } O_2}{32.00 \text{ g } O_2} = 1.000 \text{ mol } O_2$$

Dividing by stoichiometric coefficients gives:

$$15.87 \text{ mol } H_2 \times \frac{1 \text{ mol reaction}}{2 \text{ mol } H_2} = 7.935 \text{ mol reaction if } H_2 \text{ is limiting}$$

$$1.000 \text{ mol } O_2 \times \frac{1 \text{ mol reaction}}{1 \text{ mol } O_2} = 1.000 \text{ mol reaction if } O_2 \text{ is limiting.}$$

O_2 is the limiting reagent, so no O_2 will remain in the vessel.

25 Answer C.

We are given the unbalanced equation. Next we determine the number of atoms on each side. For reactants (left of the arrow): 5 H, 1 P, 6 O, and 1 Ca. For products: 2 H, 2 P, 9 O, and 3 Ca.

We assume that the molecule with the most atoms – $Ca_3(PO_4)_2$ – has a coefficient of one, and find the other coefficients required to have the same number of atoms on each side of the equation. Assuming $Ca_3(PO_4)_2$ has a coefficient of one means that there will be 3 Ca and 2 P on the right because H_2O has no Ca or P. A balanced equation would also have 3 Ca and 2 P on the left. This is achieved with a coefficient of 2 for H_3PO_4 and 3 for $Ca(OH)_2$. Now we have:

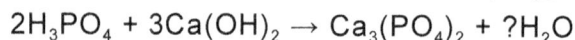
$$2H_3PO_4 + 3Ca(OH)_2 \rightarrow Ca_3(PO_4)_2 + ?H_2O$$

and then:

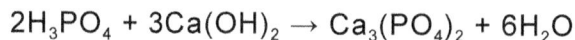
$$2H_3PO_4 + 3Ca(OH)_2 \rightarrow Ca_3(PO_4)_2 + 6H_2O$$

The coefficient for H_2O is found by balancing H or O. Whichever one is chosen, the other atom should be checked to confirm that a balance actually occurs. For H, there are 6 H from $2H_3PO_4$ and 6 from $3Ca(OH)_2$ for a total of 12 H on the left. There must be 12 H on the right for balance. None are accounted for by $Ca_3(PO_4)_2$, so all 12 H must be associated with H_2O. It has a coefficient of 6:

This is choice C, but if time is available, it is best to check that the remaining atoms are balanced. There are 8 O from $2H_3PO_4$ and 6 from $3Ca(OH)_2$ for a total of 14 on the left, and 8 O from $Ca_3(PO_4)_2$ and 6 from $6H_2O$ for a total of 14 on the right. The equation is balanced.

26. Answer D.

Explanation: The balanced chemical equation is $2CO\ (g) + O_2\ (g) \rightarrow 2CO_2\ (g)$. Coefficients are the numbers in front of the element symbols. Add 2 + 1 + 2 to get 5.

27. Answer A.

First examination of the overall reaction requires writing down the two half electrochemical reactions.
At the anode: $Ni(s) \rightarrow Ni^{2+}(aq) + 2e^-$
At the cathode: $MnO_2(s) + NH_4^+(aq) + 2e^- \rightarrow Ni\ (OH)_2(s) + NH_3(aq)$

By definition, oxidation occurs when electrons are lost and the ion becomes more positive with higher oxidation state. This fits with choice A.

28. Answer A.

First you need to write down and balance the complete chemical equation:

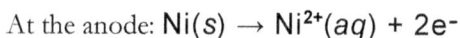
$$C_6H_{12}O_6 + 6\ O_2 = 6\ H_2O + \underline{\ \ 6\ \ }\ CO_2$$

If you balance correctly it should be 6 moles CO_2

The reaction starts with 18g of sugar – which must be converted to moles. To find the number of moles of sugar calculate the molecular weight

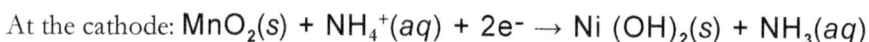
```
C =  6 * 12 = 72
H = 12 * 1 = 12
O = 6  * 16 =  96
```

MW glucose = 180.0g Therefore 18g of sugar is 0.1 mol of sugar (18/180) = 0.1

In the reaction 1mole sugar $\rightarrow$ 6 moles CO_2 0.1M sugar $\rightarrow$ **0.6M CO2**

[.6 moles CO_2 * gramCO_2/mol = gr CO_2 = 26g of CO_2 (but this was not the question asked).]

29. Answer B.

We are given 0.214 mol FeS and 0.664 mol HCl.

From the balanced equation we know for every 1 mole of FeS we require 2 moles of HCl.

Therefore for .214 mol FeS we require .214 x 2 = .428 mol HCl.

The amount we have is .664 mol which is more than enough making HCl the excess reagent

Therefore .664 - .428 = .236 mol HCl left over.

30. Answer B.

This is a decomposition reaction because the original compound is broken into it's component parts.

31. Answer: C.

A second order reaction plot will be a straight line for a plot of $1/[A]$ vs time

32. Answer B.

Knowing the equation of a straight line $y = mx + b$

we know that the intercept, b, is the concentration at time=0

Since the slope (m) is negative, that means the $[A]$ is decreasing.

A zero-order reaction would be independent of concentration, therefore this statement is false.

33. Answer A.

The energy of the products is greater than the energy of the reactants, thus the system absorbs this extra energy as heat from the surroundings. The system is therefore endothermic.

34. Answer is A. Calculate the reaction quotient at the actual conditions:

$$Q = \frac{[N_2][O_2]}{[NO]^2} = \frac{(0.090 \text{ M}) (0.020 \text{ M})}{(0.005 \text{ M})^2} = 72$$

This value is less than Keq ($72 < 2 \times 10^3$), therefore Q < Keq. To achieve equilibrium, Q must be equal to Keq. This occurs when products turn into reactants. Therefore NO will react to make more N_2 and O_2.

35. Answer B.

Catalysts provide an alternate mechanism in both directions, but do not alter equilibrium (I is false, II is true). The kinetic energy of molecules increases with temperature, so the energy of their collisions increases also (III is true). Catalytic converters contain a heterogeneous catalyst (IV is false).

36. Answer A.

The first step will be rate-limiting. It will determine the rate for the entire reaction because it is slower than the other steps. This step is a unimolecular process with the rate given by answer A. Choice C would be correct if the reaction as a whole were one elementary step instead of three, but the stoichiometry of a reaction composed of multiple elementary steps cannot be used to predict a rate law.

37. Answer D.

Product concentrations are multiplied together in the numerator and reactant concentrations in the denominator, eliminating choice B. The stoichiometric coefficient of H_2 is one, eliminating choice A. For heterogeneous reactions, concentrations of pure liquids or solids are absent from the expression because they are constant, eliminating choice C. D is correct.

38. Answer A.

A is incorrect: not all reactions are multi-step processes. Choices B and C are correct and define the rate-determining step. D is also correct, since some reactions may not have a single rate-determining step. If no particular step is significantly slower than the others, there is no one rate-determining step.

39. Answer D.

By adding the three equations and canceling out things that appear on both sides – the net result is

$$2O_3(g) \Rightarrow 3O_2(g)$$

40. Answer A.

The Arrhenius Equation is used to determine the activation energy of a reaction. It relates the activation energy to the rate constant.

41. Answer D.

The $C \equiv C$ (triple bond) has the most energy and is the strongest bond. It is therefore also the shortest. So I and IV are true.

42. Answer C.

If the rate doubles when the concentration of reactant is doubled, the reaction is the first-order. If the rate becomes 4-fold times (quadruples) when the concentration of the reactant is doubled, the reaction is the second-order .

43. Answer B.

Heat is added to the system to melt a solid therefore ΔH is positive and since going from a solid to a liquid increases the disorder of the system ΔS is + as well.

44. Answer C

The standard enthalpy of formation is defined as the change in enthalpy associated with the formation of a mole of a compound from its component elements in their most stable state. It is zero when the substance us already in its standard state (most stable form of the element). Cl exists most stably as Cl_2 and therefore has an enthalpy of formation equal to zero.

45. Answer A.

During a phase change, equilibrium exists between phases, so ΔG is zero, Therefore we can use the Gibbs Free energy equation to solve for T. (Remember to watch your units!)

$\Delta G = \Delta H - T \, \Delta S$

$0 = \Delta H - T \, \Delta S$

$T = \Delta H / \Delta S$

$= 31400 \, J/m \, / \, 94J/mK$

$= 334 \, K = 61C$

46. Answer A.

When sugar is added to water to form a solution the system becomes more disordered. In all the other cases the system is becoming more ordered. Therefore only in A is the entropy increasing and ΔS is positive.

47. Answer C.

Since the initial temperature of the water is colder than that of the metal, and heat flows from hot to cold, hence the metal will lose heat and the water gains heat. This eliminates answers A and B.

The amount of heat flow q could be estimated $(q = m \, DT)$, where m is the mass and DT is the difference between initial and final temperature of the system.

For water, $q = 50 \times 7 = 350$ calories of heat gained by the water.

Since heat loss must equal heat gain, hence the metal lost 350 cal of heat while the water gained 350 cal of heat.

> Note: The total heat loss by the metal is estimated to: $q = 20 \times 68 = 1360$ cal. Part of this heat should be transferred to the surrounding air since part of the metal is in contact with water and the other with air.

48. Answer C.

Exothermic reactions cause heat to be released. Sulfuric acid needs heat to dissolve and releases this heat energy into water. As the water receives the heat energy, its temperature will increase.

49. Answer D.

The overall $\Delta Hrxn$ is given by the sum of the products less the sum of the reactants. Remember to multiply by the stoichiometric coefficients of the balanced equation.

$\Delta H_{rxn} = 3\Delta H_3 + 2\Delta H_4 - \Delta H_1 - 3\Delta H_2$

50. Answer D.

$q = m^* \, \Delta T^* \, \text{heat-capacity}$

$= 15 * 15 * 0.13 \, J/g\,°C$

$= 29J$

51. Answer C.

First calculate how many moles of $Ba(OH)_2$ there are:

8.000 g / 171.3438 g/mol = 0.04669 mol

Then determine the moles of HCl required to neutralize:

The molar ratio between HCl and $Ba(OH)_2$ is 2:1

Therefore, 0.09338 moles of HCl is required to neutralize 0.04669 moles of $Ba(OH)_2$

then determine the volume of HCl required:

0.7000 mol/L = 0.09338 mol / x

x = 0.1334 L = 133.4 mL

52. Answer B.

HNO3 is a strong acid, so it completely dissociates:

$[H^+] = 0.010\ M = 1.0 \times 10^{-2}\ M$

$pH = \log_{10}[H+] = -\log_{10}(1.0 \times 10^{-2}) = 2.0$ (choices B or C)

From pH +pOH = 14: pOH = 12.0 (choice B)

53. Answer B.

The amount of H_2 decreases, because the reaction shifts to the right to decrease the amount of N_2 present therefore establishing a new equilibrium.

54. Answer B.

Kc is given by the concentration of the products divided by the [concentration] of the reactants. However solids are not included.

55. Answer A.

For a strong base you start out with an excess of $[OH^-]$ and therefore a high pH. As you add acid, the pH will decrease rapidly at first and then gradually as you approach the equivalence point,which is above pH=7, followed by a gradual slope to low pH.

56. Answer D.

57. Answer D.

The pH at the equivalence point is 7; of the three indicators, only litmus will change in the range of pH 7.

58. Answer A.

Even though heat is not a chemical substance, for the purposes of LeChatelier's Principle, you can treat it as if it has physical existence. Since heat is added, the reaction will shift to try and use up some of the added heat. In order to do this, the reaction must shift to the left.

59. Answer C.

The concentration can be found from :

Concentration = 10 Kg/2.5 x 106 L * 1000 g/1 Kg * 1 mol/40 g = .0001 M or 10^{-4} M

The concentration of $[OH^-] = 10^{-4}$M

pOH = 4 pH = 10

60. Answer C.

The indicator is yellow in its acidic form (HInd) and red in its basic form (Ind$^-$). HCl is a strong acid, while HCN is weak.

For the equilibrium: HCN + Ind$^-$ ↔ HInd + CN$^-$, the concentrations of the reactants will be much higher than the concentrations of the products. (The solution is red because the concentration of Ind- is much higher than the concentration of HInd). HInd is an acid and Ind- is its conjugate base. HCN is an acid and CN$^-$ is its conjugate base. For the reaction, HInd reacts with CN$^-$ to form HCN and Ind$^-$. If HCN was a stronger acid than HInd, it would react with Ind$^-$ to form HInd and CN$^-$. The answer to the problem is C: the reactants are favored, and HInd is stronger than HCN (the stronger acid dissociates, while the weaker does not).

1. a) The quantum numbers needed to describe the electronic configuration are

   n – principal quantum # $n =$

   Al – angular momentum quantum $l = 0 \to n-1$

   m – magnetic quantum # $= (m-l)\ldots 0 \ldots (m+e)$

   s – spin quantum # $= \frac{1}{2}, -\frac{1}{2}$

   b) N   [↑↓] [↑↓]   [↑] [↑] [↑]

   $n = 2 \quad e = 0, 1 \quad m = -1, 0, 1 \quad s = \frac{+1}{2}$

   c) $\ddot{N} \equiv \ddot{N}$

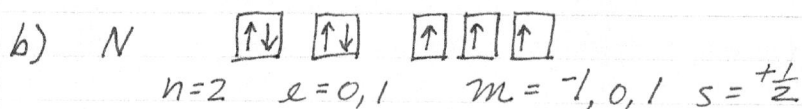

   d) – atomic radii increase from Li → K because e^- are added to higher energy levels

   – atomic radii decrease from Li → Ne across a row because in addition to more e^-, there are also more protons pulling electrons closer to the nucleus.

   e) Ionization energy increases from left to right along a row of the Periodic Table. It also decreases from top to bottom of a row.

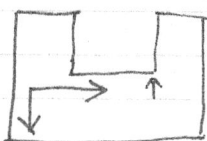

    Therefore the diagonal trend is to decrease from the upper right to the lower left.

   Electron Affinity – varies diagonally across the Periodic Table with F having the highest and Fe (lower left) having the lowest.

   Electronegativity – shows the same diagonal trend. F has the highest and $+r$ the lowest. Values increase across the row and decrease down a column.

2 a) $PV = nRT$

$(1atm)V = (7.5m)(.08206 L\text{-}°/_{mk})(273k)$ $(15gH_2)(\frac{1m}{2g}) = 7.5m$

$\quad = 168 L$

2 B.   Scuba tank $0.29 kg \; O_2$ $V = 2.3L$ $T = 25°C$ $P = ?$

$PV = nRT$ $\qquad\qquad T = 273 + 25 = 298K$

$P = (nRT)/V$ $\qquad\qquad n = (0.29 kg)(1000g)(\frac{1M}{32g})$
$\qquad\qquad\qquad\qquad\qquad\qquad\qquad\quad \overline{Kg}$

$\quad = \dfrac{(9.06)(0.08206)(298)}{2.3L}$ $\qquad = 9.06 \; M.$

$\quad = \dfrac{221.61}{2.3} = 96.3 \; L$

2 c)   Hair spray can 27°C $30 lbs/in^2$ will explode at $90 lbs/in^2$

$P_1 = 30 \; lbs/in^2$ $\qquad\qquad P_2 = 90 \; lbs/in^2$

$T_1 = 27° = 300K$ $\qquad\qquad T_2 = ?$

$V_1 = ?$ $\qquad\qquad\qquad V_2 = V_1$ (same size container)

$P_1 V_1 = n_1 R T_1$ $\qquad\qquad P_2 V_2 = n_2 R T_2$

$V_1 = \dfrac{n_1 R T_1}{P_1}$ $\qquad\qquad V_2 = \dfrac{n_2 R T_2}{P_2}$

$\dfrac{n_1 R T_1}{P_1} = \dfrac{n_2 R T_2}{P_2}$ $\qquad\qquad n_1 = n_2 \quad R = R.$

$\dfrac{T_1}{P_1} = \dfrac{T_2}{P_2}$ $\qquad \dfrac{300K}{30 \, lbs/in^2} = \dfrac{T_2}{90 \, lb/in^2}$

$\qquad\qquad\qquad\qquad\qquad T = 900 \, K = 627°C$

2d. density of N_2 at 0.326 atm & 18°C

$PV = nRT$

$P = 0.326$ atm

$T = 18°C = 291K$

$V = (nRT)(P)$ $\qquad$ $V = ?$ $\qquad$ $D = m/V$

$(M/n)/D = (nRT)/P$ $\qquad$ $n = ?$ $\qquad$ $= (M/n)/V$

$\qquad\qquad\qquad\qquad\qquad\qquad\qquad\qquad V = (M/n)/D$

$D = \dfrac{PM}{RT} = \dfrac{(0.326)(28 g/m)}{(0.08206)(291)}$ $\qquad$ $M_{N_2} = 28 g/m$

$= 0.3828$ g/L

3 (A)

① Synthesis nxws - 2 reactants combine
to form one product
$A(s) + B(g) \rightarrow AB(g)$

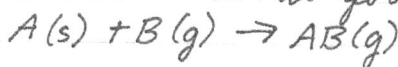

② decomposition nxw - a compound breaks
into component parts
$AB(s) \xrightarrow{\Delta} A(g) + B(g)$

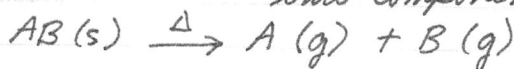

③ Single Displacement - one atom of a compound
is replaced by another atom
$A(s) + BC(aq) \rightarrow AC(aq) + B(s)$
- also called redox reaction

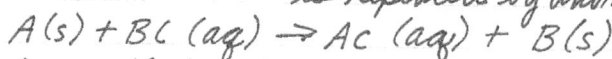

④ Double Displacement - elements from two
compounds displace each
other to form 2 new
compounds (also a
neutralize nxw)
$AB_2(aq) + 2CD(aq) \rightarrow AD(aq) + 2CB(s)$

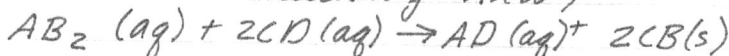

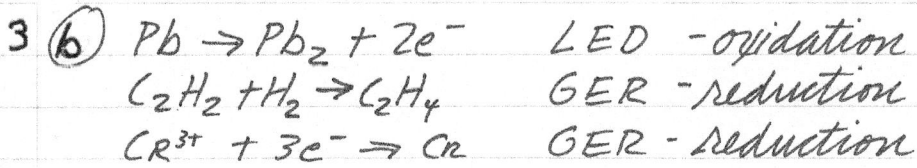

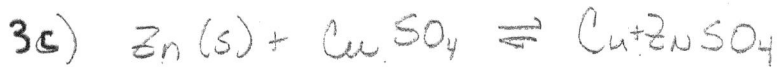

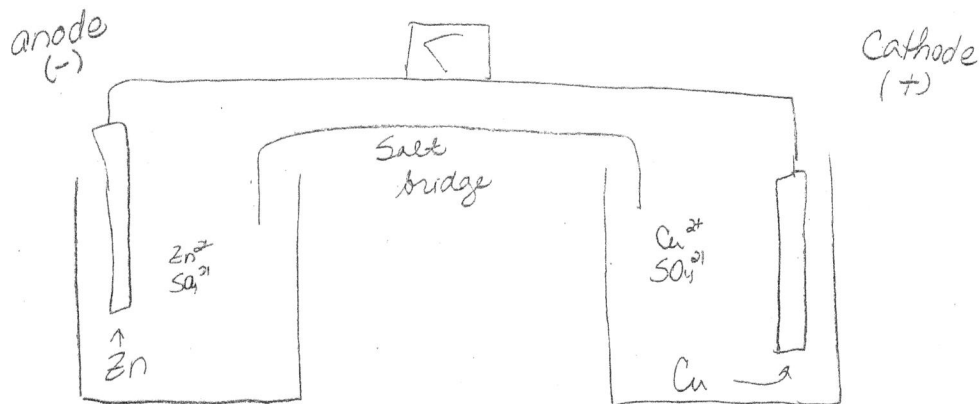

3 (b) $Pb \rightarrow Pb_2 + 2e^-$ LED - oxidation

$C_2H_2 + H_2 \rightarrow C_2H_4$ GER - reduction

$CR^{3+} + 3e^- \rightharpoonup Cr$ GER - reduction

3c) $Zn(s) + CuSO_4 \rightleftharpoons Cu + ZnSO_4$

anode (-)

Cathode (+)

Salt bridge

Zn^{2+} SO_4^{2-}

Cu^{2+} SO_4^{2-}

$\uparrow$ Zn

Cu $\rightarrow$

Zn is oxidized to Zn^{2+} at anode

Cu^{2+} is reduced to Cu(s) at cathode.

#4

$$C_2H_4\,(g) + H_2\,(g) \rightleftharpoons C_2H_6\,(g)$$

A.

$$K_c = \frac{[C_2H_6]}{[H_2][C_2H_4]}$$

ICE table

	C_2H_4	H_2	C_2H_6
I	0.53	0.63	0
C	$-x$	$-x$	x
E	$0.53-x$	$0.63-x$	x

$$K_c = \frac{x}{(0.53-x)(0.63-x)}$$

$$0.98 = \frac{x}{(0.3339 - .53x - 0.63x - x^2)}$$

$$0.98\,(0.3339 - 1.16x - x^2) = x$$

$$0.327 - 1.1368\,x - 0.98x^2 = x$$

$$0.327 - 2.1368x - 0.98x^2 = 0$$

need quadratic equation.

$x = 0.166$
$0.53 - x = 0.364$
$0.63 - x = 0.464$
check result.

$$x = -b \pm \sqrt{b^2 - 4ac} \;/\; 2a$$

$$= 2.1368 \pm \sqrt{4.56 - (4)(0.98)(0.327)} \;/\; 1.96$$

$$= 2.1368 \pm \sqrt{4.56 - 1.282} \;/\; 1.96$$

$$= 2.1368 \pm \sqrt{3.278} \;/\; 1.96$$

$$= (2.1368 \pm 1.811) \;/\; 1.96$$

$$x = 2.0 \quad \text{OR} \quad x = 0.166$$

Can't be 2.0 must be 0.166

4B) $A + B \rightarrow 2C$

if doubling A, causes rxn to double

" B " " to quadruple

∴ K is a second order rxn $k = [A][B]^2$

4c) $A(g) + B(g) \rightleftharpoons AB(g)$

i) ↑A will result in more products

ii) ↑AB " " in reverse rxn (more reactants)

iii) ↑ Pressure → ↑ products

iv) ↑ catalyst → no effect

4 d)

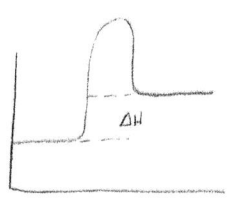
endothermic

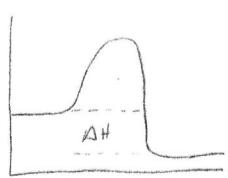
exothermic

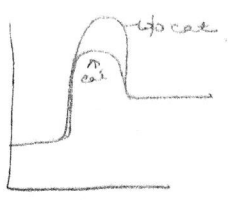

5A   Energy diagram

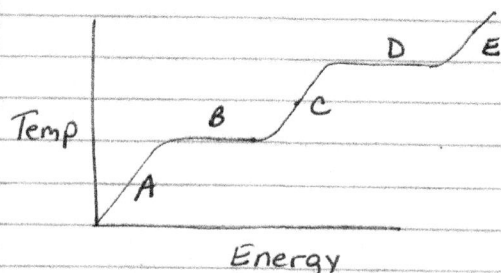

Energy

A – represents the heating of ethanol in the solid phase

B – represents the melting of the solid to a liquid. Although energy is added, the temperature does not change during a phase change.

C – this is heating of liquid ethanol

D – the vaporization of ethanol to a gas. Energy is added but the temp does not change

E – heating of ethanol in gas phase.

5 B) $\Delta G = \Delta H - T\Delta S$ $\Delta G = 0$ at equil

∴ $\dfrac{\Delta H}{\Delta S} = T$ $\Delta H = (-235.1 - ^-277.7)$

    $= 42.6$ KJ/m

$\dfrac{42.6 \text{ KJ/m}}{0.122 \text{ KJ/K·m}} = T$ $\Delta S = (282.6 - 160.6)$

    $= 122$ J/K·m $= 0.122 \dfrac{\text{KJ}}{\text{K·m}}$

$349.2 K = T$

$76.1°C = T$

5 C) $\Delta G = -2.303 \, RT \log K_{eq}$ $K_{eq} = 4.33 \times 10^3$

    $T = 53°C = 326K$

$\Delta G = (-2.303)(8.314 \dfrac{J}{K·m})(326K) \log K_{eq}$

    $= 27,238$ J $= 27.238$ KJ

5 d) $\Delta G = \Delta H - T\Delta S$

$(27.2 KJ) = 43 KJ - T(.0491 \text{ KJ/K})$

$-15.8 = -T(0.491)$

$T = 32 K$

#6 Equilibrium - Acids/Bases

A) i. $Al(OH)_3 (s) + 3HCl (aq) \rightleftharpoons AlCl_3 (aq) + 3H_2O$

a-ii. $0.5g (AlOH_3) \Rightarrow M$

$Al = 27.0$

$(OH) = 17 \times 3 = 51$

$\overline{Al(OH)_3 = 78}$

$(0.5g)\left(\dfrac{1M}{78g}\right) = 0.006 M$

$1 mol\ Al(OH_3) + 3mols\ HCl$ $MW_{(HCl)} = 36.4$

$0.006 M\ requires\ \ 0.006 \times 3$

$(0.0192 HCl)\left(\dfrac{36.4g}{M}\right) =$ $0.0192 M\ HCl$

<u>0.70g HCl</u>

B) $C_3H_8 + O_2 \rightarrow CO_2 + H_2O$

balances to $C_3H_8 + 5O_2 \rightarrow 3CO_2 + 4H_2O$

6C. Calc K_p when 0.032 mols of NO react w/ 1.70g Br_2 to form NOBr at 25°C. Given [NOBr] = 0.438 at equilibrium.

Write Balanced Eq.
$$2NO + Br_2 \rightleftharpoons 2NOBr$$

① Determine partial pressures of [NO] and [Br_2]

NO $PV = nRT$ n = 0.032 moles
$P = (0.032 \text{mol})(0.08206)(298)/V$ T = 25 + 273 = 298K
$P = 0.784$ L·atm /1L = 0.784 atm   V = 1L
                                   P = ?

Br_2 $P = nRT/V$

$\qquad\qquad\qquad\qquad\qquad\qquad n_{Br} = (1.70_g)(1/158.8)$
$\quad = (0.0106)(0.08206)(298)\qquad = 0.0106$
$\quad = 0.26013$ atm

② Determine Equil Pressure of NO & Br_2
   From Eq. NO:NOBr   ∴ if 0.438atm of NOBr

$\qquad\qquad\qquad\qquad\qquad$ must have used 0.438atm NO

$\quad$ initial → equil
$\quad 0.784 - 0.438$ atm = 0.3494 atm.

$\quad$ For Br_2 the ratio is 1:2 therefore $\frac{0.438}{2} = 0.219$ atm Br_2

$\quad 0.260135 - 0.219$ atm = 0.041135 atm.

③ $K_p = [P_{NOBr}]^2 / [P_{NO}]^2 [P_{Br_2}]$

$\quad = (0.438)^2 / ((0.34941)^2 (0.041135))$

$\quad = 38.2$

1. a) this experiment requires a few beakers, pepetes, a scale, a pH meter and of course the stock chemicals TRIS, and water

b) TRIS is $C_4H_6(OH)_3 \cdot NH_2$

$$MW = 121$$

$$(.2M)\left(\frac{121g}{1M}\right) = 24.2 \, g$$

- weigh out 24.2 g of the TRIS
- dissolve in about 500ml of destilled water
- add additional water to the 1 L level

c) measure pH.
   add HCl by drops to adjust the pH as needed

d) PPE include safety glasses, lab coat and gloves

SECTION IX:
Practice Test 2

**Plus Answers
and Rationale**

Practice Test 2

Part I - 60 Multiple Choice Questions

1. Rank the following bonds from greatest to least difference in electronegativity (ΔEN):

    C-C, C-F, C-N, Li-H, B-H.

    A. C-C > C-N > B-H > Li-H > C-F

    B. Li-H > B-H > C-C > C-N > C-F

    C. C-N > C-F > Li-H > B-H > C-C

    D. C-F > Li-H > C-N > B-H > C-C

2. When an Al atom combines with three Cl atoms to form $AlCl_3$, the Al atom will:

    A. lose electrons and reduce in size

    B. lose electrons and increase in size

    C. gain electrons and reduce in size

    D. gain electrons and increase in size

3. In which of the following set of elements is the order of reducing metallic characters of the elements is correct?

    A. Al > Ga > In > Tl

    B. In > Ga > Al > Ti

    C. Ti > Ga > In > Al

    D. Ti > In > Ga > Al

4. Which of the following is the electron configuration for an atom of antimony?

    A. 1s2 2s2 2p6 3s2 3p6 4s2 3d10 4p3

    B. 1s2 2s2 2p6 3s2 3p6 4s2 3d10 4p3

    C. 1s2 2s2 2p6 3s2 3p6 4s2 3d10 4p6 5s2 4d10 4p3

    D. 1s2 2s2 2p6 3s2 3p6 4s2 3d10 4p6 5s24d105p3

5. **Based on trends in the periodic table, which of the following properties would you expect to be greater for Ba than for Be?**

      I.   Atomic radius

      II.   Atomic number

      III.   Ionization energy

      IV.   Electronegativity

A.   I only

B.   I and II

C.   II and III

D.   I, III and IV

6. **In each set of the following elements, the size of the atom increases from left to right, except:**

A.   Li, Na, K

B.   As, Ge, Ga

C.   Br, Se, As

D.   Al, Si, P

7. **Given the elements Cl, Ge, and K and the values 418, 1255, and 784 kJ/mol of possible first ionization energies, match the atoms with their first ionization energies.**

A.   Cl: 784 kJ/mol; Ge: 1255 kJ/mol; and K:418 kJ/mol

B.   Cl: 418 kJ/mol; Ge: 1255 kJ/mol; and K:784 kJ/mol

C.   Cl: 1255 kJ/mol; Ge: 784 kJ/mol; and K:418 kJ/mol

D.   Cl: 1255 kJ/mol; Ge: 418 kJ/mol; and K:784 kJ/mol

8. **The graph on the right is of the potential energy between two atoms.**

Which of the following statements are TRUE:

      I. At C – the atoms are attracted to each other

      II. B is the optimal bonding distance

      III. A is the optimal bonding distance

      IV. D is the ideal resting state for these atoms.

A.   Only I is TRUE

B.   I and II are TRUE

C.   I and III are TRUE

D.   III and IV are TRUE

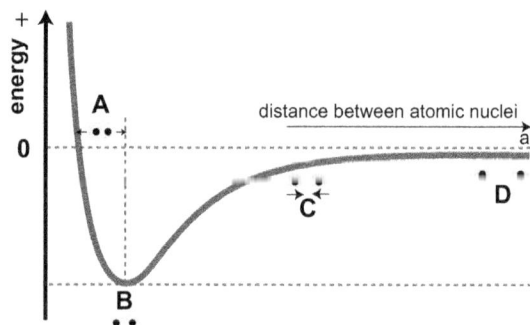

9. Which of the following shows the correct molecular orbital diagram for aluminum?

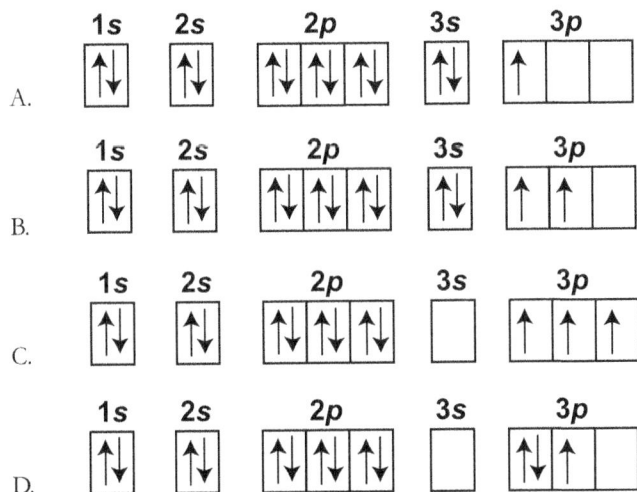

A.

1s	2s	2p	3s	3p
↑↓	↑↓	↑↓ ↑↓ ↑↓	↑↓	↑

B.

1s	2s	2p	3s	3p
↑↓	↑↓	↑↓ ↑↓ ↑↓	↑↓	↑ ↑

C.

1s	2s	2p	3s	3p
↑↓	↑↓	↑↓ ↑↓ ↑↓		↑ ↑ ↑

D.

1s	2s	2p	3s	3p
↑↓	↑↓	↑↓ ↑↓ ↑↓		↑↓ ↑

10. Which of the following represents the mass spec of Chlorine?

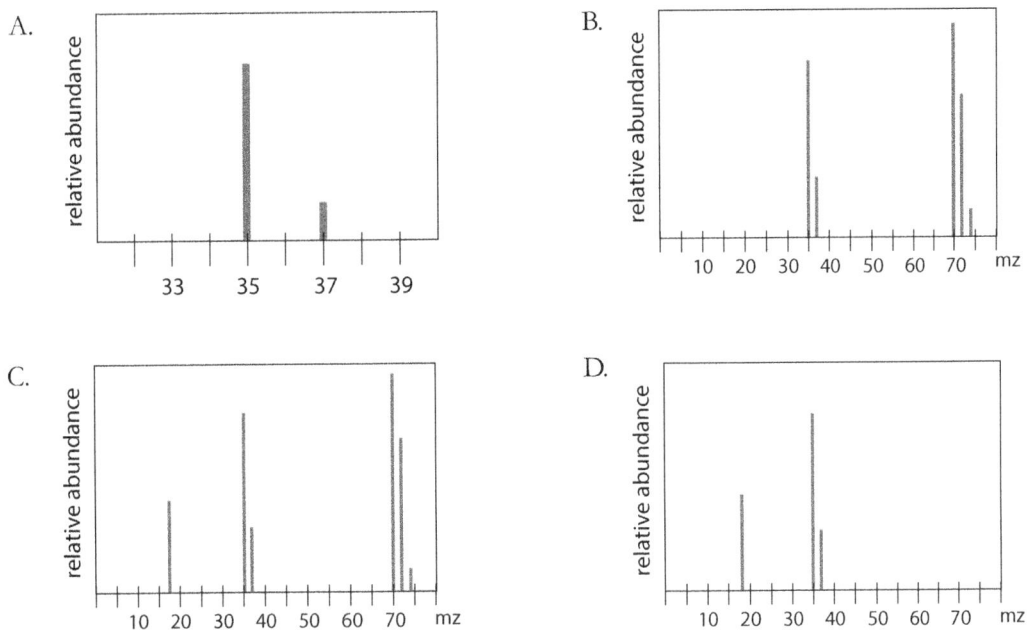

A.

B.

C.

D.

11. Which of the following statements regarding intermolecular forces (IMF) is/are true?

I Intermolecular forces result from attractive forces between regions of positive and negative charge density in neighboring molecules.

II The stronger the bonds within a molecule, the stronger the intermolecular forces will be.

III Only non-polar molecules have London forces.

A. I only correct

B. I and III

C. only III

D. I, II, and III

12. The following questions refer to the phase diagram below:

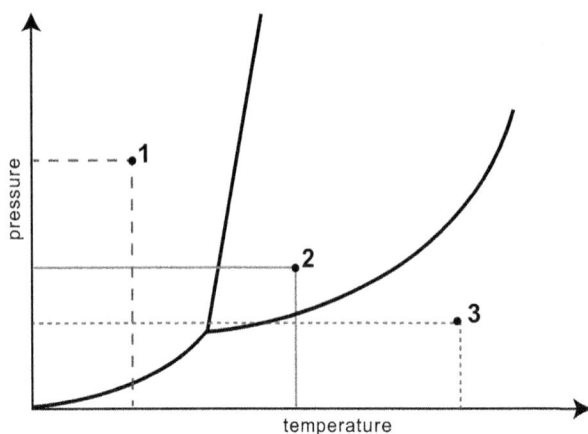

I.   As you move from point 1 to point 2 the solid melts

II. Samples in area 2 are solids.

III. Samples in area 3 are gases

IV. As you move from 2 to 3 the liquid becomes solid

A.   All of the above are TRUE

B.   II, III are TRUE

C.   I, III and IV are TRUE

D.   I and III are TRUE

13. According the VSEPR Theory, the molecular geometry of the central atom(s) above are:

Molecule	N_2	O_2	HF
Lewis Structure	:N≡N:	:O=O:	H—F:

A.   linear, linear, linear

B.   linear, bent, linear

C.   bent, bent, linear

D.   linear, bent, bent

14. SO_2 is a gas that reacts rapidly with water and is one of the major contributors to acid rain. What volume will 0.2 mol of this gas occupy at 760 torr and 541K?

A.   880 L

B.   8.88 L

C.   1.18 L

D.   118. L

15. Which of the following graphs represents the plot of pressure versus volume at constant temperature?

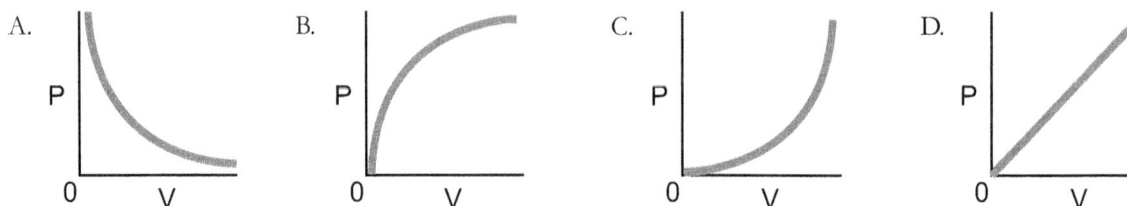

16. A mixture of nitrogen and neon gases contains equal moles of each gas and has a total mass of 10.0 g. What formula will give the density of this gas mixture at 500 K and 15.0 atm? (Assume ideal gas behavior.)

   A.   (15.0 x 10.0) / (0.08206 x 500)

   B.   (15.0 x 24.0) / (0.08206 x 500)

   C.   (15.0 x 1.0) / (0.08206 x 500)

   D.   (15.0) / (24.0 x 0.08206 8 500)

17. An ideal gas at 50.0° C and 3.00 atm is in a 300 cm³ cylinder. The cylinder volume changes by moving a piston until the gas is at 50.0° C and 1.00 atm. What is the final volume?

   A.   100. cm³

   B.   450. cm³

   C.   900. cm³

   D.   1.20 dm³

18. The boiling points of N_2, O_2 and Cl_2 are -196, -182, and -34 respectively. Cl_2 boils at a much higher temperature than expected. This is explained by:

   A.   Single bonds are easier to break than double or triple bonds

   B. Cl_2 has a longer bond length than the others

   C. Cl_2 has greater London Dispersion forces because it has more electrons.

   D.   Hydrogen bonding is stronger in Cl_2

19. What is the proper Lewis structure for HCOOH ?

   A.   :O: :: C : O : H
            ..
            H

   B.   O
               :
        H : C :: O : H

   C.   :O:
            ::
        H : C :: O
            H

   D.   :O: :: C : O : H :

20. In the lab, 32.0 g of hydrogen and 32.0 grams of oxygen react to form water until the limiting reagent is consumed. What is present in the vessel after the reaction is complete?

   A.   16.0 g O_2 and 48.0 g H_2O

   B.   24.0 g H_2 and 40.0 g H_2O

   C.   28.0 g H_2 and 36.0 g H_2O

   D.   28.0 g H_2 and 34.0 g H_2O

21. In an experiment, 0.200 g of H_2 and 1.600 g of O_2 were compressed into a 1.00 L bomb, which then was placed into a calorimeter with the capacity of 1.816×10^5 J/°C. The initial calorimeter temperature was measured to be 26° C and after the reaction take place, the calorimeter's final temperature is 26.155°C . Calculate the amount of heat given off in the reaction of H_2 and O_2 to form H_2O, expressed in kilojoules.

    A. 28.2 kJ

    B   0.6 kJ

    C. 92 kJ

    D. 282 kJ

22. Which of the following are true about Galvanic cells

    I   two half reactions take place in the two chambers

    II   oxidation takes place at the anode

    III   reduction takes place at the cathode

    IV   if the salt bridge is removed the voltage drops to zero

    V   Le Chatliers's principle can be applied to the system

    A. I , II, and III are true

    B. only II and III are true

    C. only I and IV are true

    D. All are true

23. Balance the equation for the neutralization reaction between phosphoric acid and calcium hydroxide by filling in the blank stoichiometric coefficients.

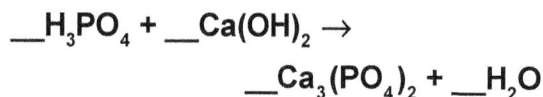

$$\_H_3PO_4 + \_Ca(OH)_2 \rightarrow$$
$$\_Ca_3(PO_4)_2 + \_H_2O$$

    A. 4, 3, 1, 4

    B. 2, 3, 1, 8

    C. 2, 3, 1, 6

    D. 2, 1, 1, 2

24. Household "chlorine bleach" is sodium hypochlorite. Which of the following best represents the production of sodium hypochlorite, sodium chloride, and water by bubbling chlorine gas through aqueous sodium hydroxide?

    A. $4CL(g) + 4NaOH(aq) \rightarrow NaClO_2(aq) + 3NaCl(aq) + 2H_2O(l)$

    B. $2Cl_2(g) + 4NaOH(aq) \rightarrow NaClO_2(aq) + 3\ NaCl(aq) + 2H_2O(l)$

    C. $2Cl(g) + 2NaOH(aq) \rightarrow NaClO(aq) + NaCl(aq) + H_2O(l)$

    D. $Cl_2(g) + 2NaOH(aq) \rightarrow NaClO(aq) + NaCl(aq) + H_2O(l)$

25. Table sugar is another word for sucrose. Sucrose is a disaccharide with the following structure:

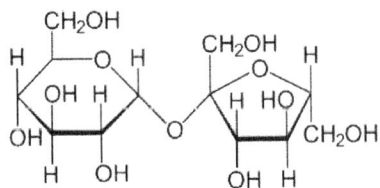

Under anaerobic conditions, yeast will ferment this compound into ethanol and CO_2. However, before they can do this, they must break this disaccharide into two monosaccharides, glucose and fructose, both of which will have the same empirical formula $(C_6H_{12}O_6)$. The unbalanced equation for the fermentation of glucose or fructose is:

$$C_6H_{12}O_6 \rightarrow C_2H_5OH + CO_2$$

If 36 grams of table sugar and an unlimited supply of water are given, how many moles of ethanol will be produced?

A. 0.5 moles of ethanol

B. 4 moles of ethanol

C. 0.4 moles of ethanol

D. 144 moles of ethanol

26. Ammonium nitrate is a good source of nitrogen that is very important in the agriculture industry. What percent of the mass of NH_4NO_3 is due to nitrogen?

A. 28%

B. 35%

C. 25%

D. 22%

27. After balancing the following redox reaction, what is the coefficient of H^+?

$$MnO_4^-(aq) + SO_2(g) + H_2O\ (1) \rightarrow Mn_2^+(aq) + SO_4^{2-}(aq) + H{+}(aq)$$

A. 1

B. 2

C. 4

D. 6

28. Which of the following are in the correct order of increasing bond length

A. $N{\equiv}N$, $N{=}N$, $N{-}N$

B. $N{-}N$, $N{=}N$, $N{\equiv}N$

C. $C{-}O$, $C{-}C$, $C{\equiv}C$

D. $C{=}O$, $C{=}C$, $N{=}N$

29. Approximately how long will it take to plate out copper in 200.0 mL of a 0.15 M Cu^{2+} solution using a current of 0.200 amps ?

    A. 29000 sec

    B. 5800 sec

    C. 20000sec

    D. 38000 sec

30. Will **Sn4+** oxidize aluminum?

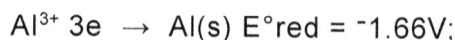

    $Sn^{4+}(aq) + 2e^- \rightarrow Sn^{2+}(aq)$ $E°_{red} = 0.15V$

    $Al^{3+} 3e \rightarrow Al(s)$ $E°red = ^-1.66V$;

    A. No, $E°cell = -1.81$ V

    B. Yes, $E°cell = 1.81$ V

    C. No, $E°cell = 1.55$ V

    D. Yes, $E°cell = -1.55$ V

31. At a certain temperature, **T**, the equilibrium constant for the reaction

    $$2NO\ (g) \leftrightarrow N_2\ (g) + O_2\ (g)$$

    is K_{eq} = **2 x 10³**. If a 1.0 L container at this temperature contains **90 mM N_2, 20 mM O_2,** and **5 mM NO**, what will occur?

    A. The reaction will make more N_2 and O_2.

    B. The reaction is at equilibrium.

    C. The reaction will make more **NO**.

    D. The temperature, T, is required to solve this problem.

32. Sodium chloride is used for melting ice in the winter months. This compound contains 1.54g Cl for every 1.00 g of sodium. Which of the following mixture would react to produce sodium chloride with no sodium or chlorine remaining?

    A. 7.11 g of sodium and 11.8g of chlorine

    B. 4.11g of sodium and 5.55g of chlorine

    C. 11.9g of sodium and 19.3g of chlorine

    D. 13.9 g of sodium and 21.4g of chlorine

33. The half-life for the first order decomposition of nitromethane, CH_3NO_2 , at 500K is 650seconds. If the initial concentration of CH_3NO_2 is 0.500M, what will it's concentration be (in M) after 1300 seconds have elapsed ?

    A. 0.125

    B. 0.140

    C. 0.250

    D. 0.450

34. **If a given reaction has K = 10, and presently has a Q = 5, what must happen in order for the reaction to reach equilibrium?**

    A. The reaction must proceed in the forward direction, producing more products until the value of Q is also 10.

    B. The reaction must proceed in the reverse direction, producing more reactants until the value of Q is also 10.

    C. The reaction is at equilibrium.

    D. Heat must be added to double the reaction until Q is doubled.

35. **For the reaction A(g) + B(g) → C(g) the following data are collected:**

Trial	[A]	[B]	rate
1	0.2	0.1	1×10^{-3}
2	0.2	0.2	2×10^{-3}
3	0.4	0.1	4.0×10^{-3}

    A. rate = k[A][B]/ [C]

    B. rate = k[A]^2[B]

    C. rate = k[A] [B]^2

    D. rate = k[A]^2[B]^2

36. **Two reactions have equilibrium constants of 1×10^{10} and 1.6×10^7 respectively**

    A + B ↔ C Kc = 1.0×10^{10}

    A + D ↔ E Kc = 1.6×10^7

    What is the equilibrium constant for the overall reaction

    C + D ↔ E + B

    A. Kc = 1.6×10^7

    B. Kc = 1.0×10^3

    C. Kc = 1.6×10^{-3}

    D. Kc = 1.6×10^{18}

37. **Which of the following is not the correct expression for Ksp**

    A. $Sn(OH)_2$(s) ↔ Sn_2+(aq) + 2OH$^-$(aq)   Ksp = [Sn^{2+}] [OH$^-$]^2

    B. Ag_2CrO_4(s) ↔ 2Ag+(aq)+CrO4$_2$$^-$(aq)   Ksp = [Ag$^+$]^2 [CrO$_4$$^{2-}$]

    C. $Fe(OH)_3$(s) ↔ Fe$_3$+(aq) + 3OH$^-$(aq)   Ksp = [Fe3$^+$] [OH$^-$]^3

    D. AgCl(s) ↔ Ag+(aq)+Cl$^-$(aq) Ksp = ([Ag+] [Cl$^-$]) / [AgCl]

38. The reaction **A + B → products** is found to be second order in **[A]** and first order in **[B]**. The rate equation would be:

A. $R = k[A] [B]$

B. $R = k[A]^2 [B]$

C. $R = k[A] [B]^2$

D. $R = k[B]$

39. A sample of sea water contains 0.53 M of Cl- ions and 8.4×10^{-4}M of Br-ions. What concentration of added Ag⁺ions would cause precipitation of AgCl and AgBr ? Which of these two halides would precipitate first? The Ksp for AgCl is $1.6 \times 10^{-3}$and the Ksp for AgBr is 6.5×10^{-13}.

A. AgCl will precipitate first at [Ag+] = 3.0×10^{-3}

B. AgCl will precipitate first at [Ag+] = 1.7×10^{-1}

C. AgBr will precipitate first at [Ag+] = 7.7×10^{-10}

D. Silver will not precipitate from this solution.

40. The entropy of the surroundings goes up when water vapor condenses. This is because:

A. Heat given off by the system increases the thermal motion of the surroundings.

B. The system is going to a more ordered state.

C. Heat absorbed by the system increases the thermal motion of the surroundings.

D. Heat given off by the system decreases the thermal motion of the surroundings.

41. Heat is added to a pure solid at its melting point until it all becomes liquid at its freezing point. Which of the following occur(s)?

A. Intermolecular attractions are weakened.

B. The freedom of the molecules to move about increases.

C. The temperature of the system remains constant.

D. All of the above

42. Assuming the apparatus itself absorbs no heat, what will be the change in temperature if 1000 mL of water are used as a heat sink for a reaction that releases 8 kJ of heat?

A. 1.91 K

B. 8.00 K

C. -265°C

D. -1.91 K

43. Which of the following reactions will have the largest change in Entropy?

A. $C_5H_{12}(l) + 8O_2(g) \rightarrow 6H_2O(g) + 5CO_2(g)$

B. $Na+(g) + Cl^-(g) \rightarrow NaCl(s)$

C. $S_3(g) + 9F_2(g) \rightarrow 3SFg(g)$

D. $2CH_4(g) + 2O_3(g) \rightarrow 4H_2O(g) + 2CHO(g)$

44. The enthalpy of fusion ($\Delta Hfus$) of tungsten is 35.23 kJ/mol, and its melting point is 3422°C. What is the entropy of fusion ($\Delta Sfus$) of tungsten?

A. 130.17 J/(mol·K)

B. 10.30 J/(mol·K)

C. 10.30 kJ/(mol·K)

D. 9.53 J/(mol·K)

45. Which of the following statements is TRUE:

A. A positive value for ΔH means that the system is absorbing heat, and the reaction is exothermic.

B. A negative value for ΔH means that the system is losing heat, and the reaction is exothermic.

C. A negative value for ΔS means that entropy is decreasing, and the reaction is exothermic.

D. A positive value for ΔS means that entropy is increasing, and the reaction is exothermic.

46. Use the following data to calculate the change in enthalpy. For the reaction:

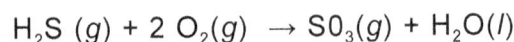

$$H_2S(g) + 2O_2(g) \rightarrow SO_3(g) + H_2O(l)$$

Given:

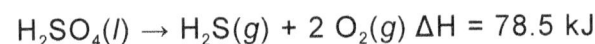

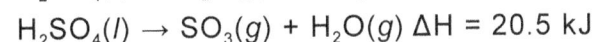

$$H_2SO_4(l) \rightarrow H_2S(g) + 2O_2(g) \quad \Delta H = 78.5 \text{ kJ}$$
$$H_2SO_4(l) \rightarrow SO_3(g) + H_2O(g) \quad \Delta H = 20.5 \text{ kJ}$$
$$H_2O(g) \rightarrow H_2O(l) \quad H = -11 \text{ kJ}$$

A. 88kJ

B. -1000kJ

C. -69kJ

D. 110.kJ

47. Determine the $\Delta Hrxn$ of the reaction using the provided bond energies:

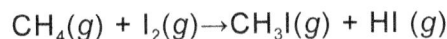

$$CH_4(g) + I_2(g) \rightarrow CH_3I(g) + HI(g)$$

Bond energies:

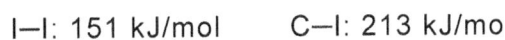

C–H: 416 kJ/mol   H–I: 299 kJ/mol

I–I: 151 kJ/mol   C–I: 213 kJ/mo

A. -1077kJ/mol

B. 55kJ/mol

C. -887kJ/mol

D. 1193kJ/mol

48. Determine the boiling point for an unknown metal given that:

ΔH_{vap}= 349.6 kJ/mol and ΔS_{vap}= 100.0 J/(mol·K)

A. T= 3496K

B. T = 35K

C. T = 280K

D. Cannot be determined.

49. What is true, considering the value of **ΔG°** in the following reaction?

$N_2(g) + 3H_2(g) \leftrightarrow 2NH_3$ $\Delta G°$= -32.96 kJ

A. ΔG° is free energy when all components of the reaction are present at standard-state conditions

B. The system is not in equilibrium at this temperature.

C. The reaction is spontaneous

D. All of the above

50. What does the below plot of Temperature vs Time tell us about the system ?

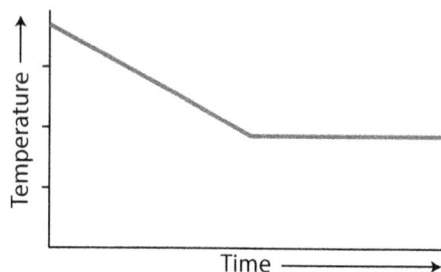

A. Heat is being removed from the system, then the system undergoes a phase change

B. Heat is added to the system throughout the time observed

C. The sample is melting and is then heated again

D. The sample is cooled to a solid.

51. For a given reaction, if Q is less than K, in what direction will the reaction proceed?

A. Forward, until Q = K

B. Reverse, until Q = K

C. Forward and reverse reactions will be equal, since the reaction is at equilibrium

D. Reverse if K is small and forward if K is large

52. Which of the following mixtures forms a buffer?

A. 1 M HCl and 1 M HCOOH

B. 1 M HCOOH and 1 M COOH-

C. 1 M HCOOH and 1 M NaOH

D. 1 M HCl and 0.5 M COOH

53. **Which way will the equilibrium shift if the system temperature goes up (heat is added):**

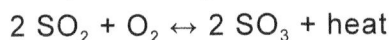

$$2\ SO_2 + O_2 \leftrightarrow 2\ SO_3 + heat$$

   A. The reaction will shift to the right

   B. The reaction will shift to the left

   C. The reaction is in equilibrium and will not change.

   D. There is not enough information.

54. **The container holding the following reaction (already at equilibrium) has its volume suddenly increased. Which way will the equilibrium shift to compensate?**

$$H_2 + Cl_2 \leftrightarrow 2HCl + heat$$

   A. The reaction will shift to the right

   B. The reaction will shift to the left

   C. The reaction is in equilibrium and will not change.

   D. There is not enough information.

55. **In the above reaction, what would adding heat do to the value of the K_{eq}?**

   A. The reaction is in equilibrium and will not change.

   B. The Keq decreases in value; heat is being added to an exothermic reaction.

   C. The Keq increased in value; heat is being added to an exothermic reaction.

   D. There is not enough information to determine.

56. **Determine the Ksp of silver bromide, given that its molar solubility is 5.71 x 10^{-7} moles per liter.**

   A. $Ksp = 5.71 \times 10^{-7}$ moles per liter

   B. $Ksp = 3.26 \times 10^{-13}$ moles per liter

   C. $Ksp = 5.71 \times 10^{-15}$ moles per liter

   D. $Ksp = 3.26 \times 10^{-14}$ moles per liter

57. **Given that the K_a for acetic acid is 1.8 x 10^{-5}, what is the K_b for the acetate ion?**

   A. 12.2

   B. 5.2

   C. 5.6×10^{-10}

   D. 12.2×10^{-10}

58. **Which of the following could be used to calculate the pH of a saturated solution of $Cu(OH)_2$?** (Given: Ksp = 1.6×10^{-19})

A. $Ksp = [Cu^{2+}] [OH^-]^2 = 1.6 \times 10^{-19} = 4s^3$

B. $Ksp = [Cu^{2+}] [OH^-] = 1.6 \times 10^{-19} = s^2$

C. $Ksp = [Cu^{2+}]2 [OH^-]^2 = 1.6 \times 10^{-19} = 16s^4$

D. not enough information is given.

59. **pH = pK_a when**

A. [conjugate acid] = [conjugate base]

B. when you are at the transition point of the titration

C. when you are at equilibrium

D. when the solution is neutral.

60. **Which of the following are true regarding the relationship between ΔG and K**

A. $\Delta G°$ cannot be related to K

B. If $\Delta G° < 0$ then K > 1, so products will be in greater concentration than reactants at equilibrium.

C. If $\Delta G° > 0$ then K > 1, so products will be in greater concentration than reactants at equilibrium.

D. If $\Delta G° < 0$ then K < 1, so products will be in greater concentration than reactants at equilibrium.

Part II. Free Response Questions

1. (a) Write the electronic configuration of Cobalt.

   (b) Draw the electronic structure of Cobalt

   (c) Co-60 was involved in the worst radioactive accident in recent history. If the half-life is 5.3 years, starting with 10mg of Co-60 how much is left after 10.6 years ?

   (d) The atomic masses of the five naturally occurring isotopes of germanium are germainium-70 (69.924u); germanium-72 (71.921u); germanium-73 (72.923u); germanium-74(73.92u) and germanium-76 (75.921u).

   Use these values and the data from the accompanying mass spectra to determine a weighted average atomic mass of germanium.

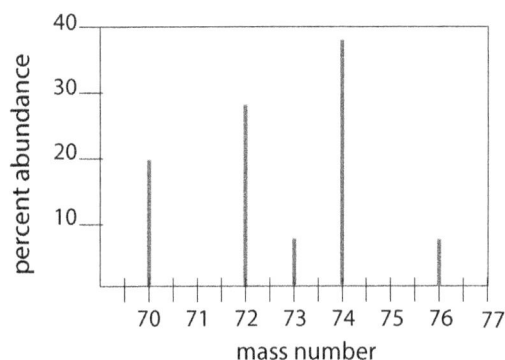

2. Consider the following reaction
   $$2NO(g) + O_2(g) \rightarrow 2NO_2(g)$$
   From the following data collected at 25° C, determine
   (a) the rate law (rate equation)
   (b) rate constant
   (c) rate of the reaction when [NO] 5×10^{-3} M and $[O_2] = 7 \times 10^{-3}$ M

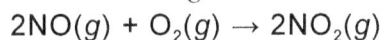

Experiment	[NO] (M)	$[O_2]$ (M)	Initial Rate (M/s)
1	2×10^{-3}	1×10^{-3}	1.25×10^{-4}
2	4×10^{-3}	1×10^{-3}	5.0×10^{-4}
3	4×10^{-3}	2×10^{-3}	10.0×10^{-4}

3. Calculate ΔH and ΔS for the following reaction:
   $$NH_4NO_3(s) + H_2O(l) \rightarrow NH_4^+ (aq) + NO_3^- (aq)$$
   Use the results of this calculation to determine the value of $\Delta G°$ for this reaction at 25° C, and explain why NH_4NO_3 spontaneously dissolves in water at room temperature.

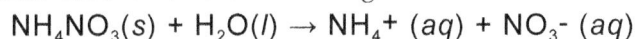

The following information can be found by using a standard-state enthalpy of formation and absolute entropy data table.

Compound	$\Delta H f°$(kJ/mol)	S (J/mol-K)
$NH_4NO_3(s)$	-365.56	151.08
NH_4^+ (aq)	-132.51	113.4
NO_3 (aq)	-205.0	146.4

4. One mole of acetic acid is dissolved in 1L of water following the reaction below. K for this process, known as the "acid dissociation constant" for acetic acid is about 21.8 x10^{-5}. Given that the pH of a solution is defined by pH = $^-$log[H_3O^+] what is the pH of this solution at equilibrium ?

$$HC_2H_3O_2 \ (aq) + H_2O \leftarrow \rightarrow C_2H_3O_{2-} \ ^- \ (aq) + H_3O^{+ \ (aq)}$$

5. (a) For the reaction below, when **0.5000 g of XI_3** reacts completely, **0.2360 g of XCl_3** is obtained. Calculate the atomic weight of element **X** and identify it.

$$2XI_3 + 3Cl_2 \ \rightarrow \ 2 \ x \ Cl_3 + 3I_2$$

   (b) If **0.520 grams of XCl3** are treated with iodine, **0.979 g of XI_3** are produced. What is the chemical symbol for this element?

$$2XCl_3 + 3I_2 \ \rightarrow \ 2XI_3 + 3Cl_2$$

6. (a) Phosphoric acid is usually obtained as an 87.0% phosphoric acid solution. If it is 13.0 M, what is the density of this solution?

   (b) How many grams of water must be used to dissolve 100.0 grams of sucrose ($C_{12}H_{22}O_{11}$) to prepare a 0.020 mole fraction of sucrose in the solution?

7. Experimental Observations of Electrochemical Cells

   I.   Consider the voltaic cell that contains standard $Co2^+/Co$ and $Au3^+/Au$ electrodes. The following experimental observations were noted:

   II. Metallic gold plates out on one electrode and the gold ion concentration decreases around that electrode, and

   III. The mass of the cobalt electrode decreases and the cobalt (II) ion concentration increases around that electrode.

   (a) Sketch the electrochemical cell and overall balanced redox reaction.

   (b) What is the standard reaction free energy for this cell?

   Given: $Au_3^+ + 3e^- \ \rightarrow \ Au + 1.50 \ V$

   $CO_2^+ + 2e^- \ \rightarrow \ CO \ -0.28 \ V$

   (c) Predict if this reaction is spontaneous by looking at the equilibrium constant for the overall redox reaction of this cell at 25° C. (Do not solve/calculate any equations)

   (d) Assume you use this cell as a battery to power one of your electrical devices that draws 1mA of current. If you ran the battery at a constant 1mA for 3000 hours, how many gr of CO would be consumed?

Answers and Rationale to Practice Test 2

1. **Answer D.**
   The polarity of a bond is given by the difference in electronegativity of the atoms, ΔEN. H has electronegativity 2.2, and the electronegativity for the second-row atoms starts with 1.0 at Li and increases by 0.5 as you go across the row until 4.0 at F. You don't need to know these exact values, though; you can solve this problem just by understanding the periodic table conceptually. First, clearly, C-C will have a difference of zero, so this will be the last bond, eliminating answers A and B. Between answers C and D, electronegativity increases as you go to the right in the periodic table, so C-F will clearly have a greater difference than C-N.

2. **Answer A.**
   Al will lose it's electrons to the chlorine atoms and the Al^{3+} ion will reduce in size. Also, the Al^{3+} ion's outer electrons are in the 2nd atomic orbital as opposed to the 3rd one (where an Al atom's outer electrons are). Therefore the ion will be smaller as it's outer electrons are closer to the nucleus.

3. **Answer A.**
   Metallic character increases as you move down a group because the atomic size is increasing. When the atomic size increases, the outer shells are farther away. The principle quantum number increases and average electron density moves farther from nucleus. The electrons of the valence shell have less attraction to the nucleus and, as a result, can lose electrons more readily. This causes an increase in metallic character.

4. **Answer D.**
   Antimony – Sb – has the atomic number 51. Therefore it has 51 electrons which are filled as
   $$1s^2\,2s^2\,2p^6\,3s^2\,3p^6\,4s^2\,3d^{10}\,4p^6\,5s^2 4d^{10} 5p^3$$
   Recall that for transition metals the 5s shell fills before the 4d shell.

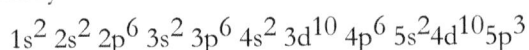

5. **Answer B.**
   Down a group, atomic radius increases. This the radius of Ba would be greater than Be. (I is TRUE) Atomic number increases across and down on the periodic table (II is TRUE) Ionization energies and electronegativites decrease down a column/group (so III and IV are false).

6. **Answer D.**
   The size (atomic radius) of the elements in the periodic table decrease from left to right for each period and increase from top to bottom for each column. In A the set of elements are located in columns so their sizes are increasing. In B and C the set of elements are located in the periods with the order of right to left so in B and C the sizes of atoms are increasing too. But in D the size of atom

7. **Answer C.**
   Cl is a Group 7 non-metal and will tend to form a ⁻1 ion, so it will be very reluctant to give up an electron and will have a very high first ionization energy. K is a group 1 metal and will readily lose an electron to form a +1 ion so the first ionization energy will be very low. Ge will form positive ions and be intermediate in its first ionization energy

8. **Answer B.**

At point B the optimal distance the two atoms are bonded. At Point A they are too close and there is a repulsion. At Point C there is a small attraction. At Point D the atoms are not close enough to interact.

9. **Answer A.**

This is the correct electron diagram for Al. The others either have too many electrons or violate one of the filling rules.

10. **Answer B.**

Chlorine will show peaks around 34 corresponding to the isotopes of Chlorine. However Cl also exists naturally at Cl_2 – so it will show another set of peaks at around 70.

11. **Answer A.**

Statement I is true - all IMF result from Coulombic attraction. Statements II and III are both false; the strength of the bonds within a molecule have no bearing on the strength of the bonds between molecules; all molecules have London forces.

12. **Answer D.**

Statements I amd II are TRUE
Area 1 represents the solid. Area 2 is the liquid. Area 3 is a gas.

13. **Answer A.**

According to VSEPR theory all are linear

14. **Answer B.**

When you are presented with questions like this one you should assume ideal gas behavior so you can use the ideal gas law: PV = nRT.

The next step is to notice that you can convert 760 torr to atm (atmospheres). In this case 1 atm = 760 torr so when you convert 760 torr to atm you get 1 atm (much easier to deal with)

Write down which values you are given in the problem and which ones are unknown. In this case

    R = 0.08206 L atm/K mol (taken from formula sheet)
    P − 760 torr = 1 atm
    n = 0.2 mol
    T = 541 K
    V = ?

First solve for the ideal gas law for V. Then plug in the numbers above

    PV= nRT
    V= (nRT) / P
    V= nRT/P = (0.2)(0.08206)(541)/(1) = 8.88 L

15. Answer A.

PV = nRT. As pressure increases the volume decreases as shown in graph A.

16. Answer B.

First let x = moles N_2 and y = moles Ne. The molecular weight of N_2 is 28, and the molecular weight of Ne is 20. Therefore, from the problem:

28x + 20y = 10

However x = y, therefore:

48x = 10

x = 0.208 moles N_2

y = 0.208 moles Ne

total moles in mixture = 0.416

"molar mass" of mixture = 10.0 g / 0.416 mol = 24.0 g/mol

Manipulate PV = nRT as follows:

Make sure to follow the units

PV = nRT   n = g/(g/mole) (that way the units are still moles here)

Molar Mass is the g/mol so you can replace n with g/MM

PV= RT (g/MM)

PV= RT g/MolarMass (rearrange)

P (MM) = RT g/V (substitue D for g/V)

D = g/V = P MM / RT - the answer comes out about 8 g/L (the units are correct for density)

units: g/L = (atm) (g/moles) / (L atm K_{-1}8722;1 mol_{-1}8722;1)(T)

= g/L

Insert values and solve:

(15.0 x 24.0) / (0.08206 x 500) = g/V

density = 8.8 g/L

NOTE: A common error would be:

15 atm * V = (0.416 moles * 24 g/1 mol) * R * 500K

15atm / (0.416mol * 24g/mol * (0.082 L-atm/mol-K * 500K) = g / V

here the moles cancel - which sounds right at first - but what you really

want is the moles to cancel with the R which has units of $(L\ atm\ K^{-1}mol^{-1})$

In this case if you look at just the units you are left with

atm / [(moles * g/mol * (L atm/K mol) * (K)

The atm cancel, the K cancel – but the moles do not.

17. Answer C.

A three-fold decrease in pressure of a constant quantity of gas at constant temperature will cause a three-fold increase in gas volume.

18. Answer C.

Liquid oxygen and liquid nitrogen and liquid chlorine are all non-polar substances that experience only London dispersion forces of attraction. These forces are greater for Cl_2 because it has more electrons, so it has the highest boiling point.

19. Answer A.

All elements have an octet.

$$\text{:}\overset{\displaystyle\cdot\cdot}{O}\text{::}C\text{:}\overset{\displaystyle\cdot\cdot}{\underset{\displaystyle\cdot\cdot}{O}}\text{:}H$$

$$H$$

The proper way to determine the Lewis structure, based on this example, is:

1.   Total valence electrons: **4 + 6 x 2 + 1 x 2 = 18**
2.   Total electrons: needed for octets/doublets: **8 x 3 + 2 x 2 =28**
3.   Total shared/bonding electrons: **28 - 18 = 10** (in other words, there are only 5 bonds)
4.   Total electrons in lone pairs: Step 1 - Step 3 = **18 − 10 = 8** (in other words, there are only four pairs of lone electrons; all for O)

20. Answer C.

First the equation must be constructed:
Error! Not a valid embedded object.

A fast and intuitive solution would be to recognize that:
One mole of H_2 is about 2.0 g, so about 16 moles of H_2 are present.
One mole of O_2 is 32.0 g, so one mole of is O_2 is present.
Imagine the 16 moles of H_2 reacting with one mole of O_2. 2 moles of H_2 will be consumed before the one mole of O_2 is gone. O_2 is limiting. (Eliminate choice A.)
16 moles less 2 leaves 14 moles of H_2 or about 28 g. (Eliminate choice B.)
The reaction began with 64.0 g total. Conservation of mass for chemical reactions forces the total final mass to be 64.0 g also. (Eliminate choice D.)
A more standard solution is presented next. First, mass is converted to moles:

$$32.0 \text{ g } H_2 \times \frac{1 \text{ mol } H_2}{2.016 \text{ g } H_2} = 15.87 \text{ mol } H_2 \quad \text{and} \quad 32.0 \text{ g } O_2 \times \frac{1 \text{ mol } O_2}{32.00 \text{ g } O_2} = 1.000 \text{ mol } O_2$$

Dividing by stoichiometric coefficients gives:

$$15.87 \text{ mol } H_2 \times \frac{1 \text{ mol reaction}}{2 \text{ mol } H_2} = 7.935 \text{ mol reaction if } H_2 \text{ is limiting}$$

$$1.000 \text{ mol } O_2 \times \frac{1 \text{ mol reaction}}{1 \text{ mol } O_2} = 1.000 \text{ mol reaction if } O_2 \text{ is limiting.}$$

O_2 is the limiting reagent, so no O_2 will remain in the vessel.

$$1.000 \text{ mol } O_2 \text{ consumed} \times \frac{2 \text{ mol } H_2O \text{ produced}}{1 \text{ mol } O_2} \times \frac{18.016 \text{ g } H_2O}{1 \text{ mol } H_2O} = 36.0 \text{ g } H_2O \text{ produced}$$

$$1.000 \text{ mol } O_2 \text{ consumed} \times \frac{2 \text{ mol } H_2 \text{ consumed}}{1 \text{ mol } O_2} \times \frac{2.016 \text{ g } H_2}{1 \text{ mol } H_2} = 4.03 \text{ g } H_2 \text{ consumed}$$

The remaining H_2 is found from: 32.0 g H_2 initially – 4.03 g H_2 consumed = 28.0 g H_2 remain

21. Answer A.

Solution:

To calculate the heat evolved we have to multiply the heat capacity (1.816 x 105 J/°C) by the temperature change in °C. The temperature change is 0.155. Therefore,

(1.816 x 105 J/1 °C) x 0.155 °C = 2.82 x 104J = 28.2 kJ.

So this is the amount of heat given off by the reactants.

22. Answer D.

All of the statements are true for Galvanic Cells.

23. Answer C.

We are given the unbalanced equation. Next we determine the number of atoms on each side. For reactants (left of the arrow): 5 H, 1 P, 6 O, and 1 Ca. For products: 2 H, 2 P, 9 O, and 3 Ca. We assume that the molecule with the most atoms – $Ca_3(PO_4)_2$ – has a coefficient of one, and find the other coefficients required to have the same number of atoms on each side of the equation. Assuming $Ca_3(PO_4)^2$ has a coefficient of one means that there will be 3 Ca and 2 P on the right because H_2O has no Ca or P. A balanced equation would also have 3 Ca and 2 P on the left. This is achieved with a coefficient of 2 for H_3PO_4 and 3 for $Ca(OH)_2$. Now we have:

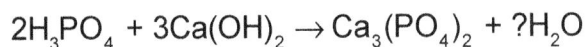

$$2H_3PO_4 + 3Ca(OH)_2 \rightarrow Ca_3(PO_4)_2 + ?H_2O$$

The coefficient for H_2O is found by balancing H or O. Whichever one is chosen, the other atom should be checked to confirm that a balance actually occurs. For H, there are 6 H from $2H_3PO_4$ and 6 from $3Ca(OH)_2$ for a total of 12 H on the left. There must be 12 H on the right for balance. None are accounted for by $Ca_3(PO_4)_2$, so all 12 H must be associated with H_2O. It has a coefficient of 6:

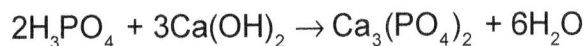

$$2H_3PO_4 + 3Ca(OH)_2 \rightarrow Ca_3(PO_4)_2 + 6H_2O$$

This is choice C, but if time is available, it is best to check that the remaining atoms are balanced. There are 8 O from $2H_3PO_4$ and 6 from $3Ca(OH)_2$ for a total of 14 on the left, and 8 O from $Ca_3(PO_4)_2$ and 6 from $6H_2O$ for a total of 14 on the right. The equation is balanced.

24. Answer D.

Chlorine gas is a diatomic molecule, eliminating choices A and C. The hypochlorite ion is ClO– eliminating choices A and B. All of the equations are properly balanced.

25. Answer C.

First you need to determine and balance the overall reaction from the two step reaction.

The empirical formula for sucrose is $C_{12}H_{22}O_{11}$ In the first step, sucrose needs to undergo decomposition to glucose and fructose:

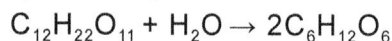

$$C_{12}H_{22}O_{11} + H_2O \rightarrow 2C_6H_{12}O_6$$

Then, each monosaccharide needs to be fermented. The balanced equation for glucose/fructose fermentation is:

$$C_6H_{12}O_6 \rightarrow 2C_2H_5OH + 2CO_2$$

The overall balanced equation is:

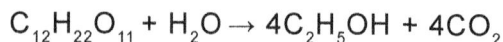

$$C_{12}H_{22}O_{11} + H_2O \rightarrow 4C_2H_5OH + 4CO_2$$

From the periodic table we have Carbon = 12 g/mol; Hydrogen = 1 g/mol; Oxygen = 16 g/mol

Therefore the molecular mass of glucose is 180.16 g/mol

(6 * 12 g/mol) + (12 * 1 g/mol) + (6 * 16 g/mol) = 180.

The molecular mass of sucrose is twice that of glucose, or 360 g/mol.

So, 36 grams of sucrose are

36 grams sucrose * 1 mol/360 g sucrose → 0.1 mol sucrose

Plugging this into the equation:

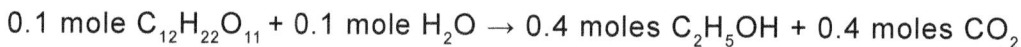

0.1 mole $C_{12}H_{22}O_{11}$ + 0.1 mole H_2O → 0.4 moles C_2H_5OH + 0.4 moles CO_2

answer = 0.4 moles of ethanol

26. Answer B.

Determine the molar mass of NH_4NO_3

From Periodic Table : 14.0 g N, 1.01 g H, 16.0 g O

Molar Mass = (2mol N * 14g N/mol N) + (4 mol H * 1g H/mol H) + (3 mol O * 16gO/mol O)

= 28.0 g N + 4.0 g H + 48.0 g O

= 80g NH_4NO_3

80.0 g of ammonium nitrate contains 28.0 g of nitrogen, so the percent nitrogen is

28/80 * 100% = 35.0%

27. Answer C.

$MnO_4^-(aq) + SO_2(g) + H_2O(l) \rightarrow Mn^{2+}(aq) + SO_4^{2-}(aq) + H^+(aq)$

Since we are told that this is a redox reaction, the easiest way to balance a complex equation like this is to break out the half reactions.

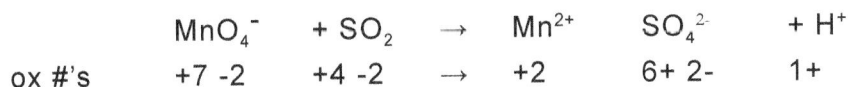

	MnO_4^-	$+ SO_2$	$\rightarrow$	Mn^{2+}	SO_4^{2-}	$+ H^+$
ox #'s	+7 -2	+4 -2	$\rightarrow$	+2	6+ 2-	1+

oxidation

$S^{4+} \rightarrow S^{6+} + 2e-$

add in the oxygens

$SO_2 \rightarrow SO_4^{2-} + 2e-$

balance H's and Oxygens

$SO_2 + 2H_2O \rightarrow SO_4^{2-} + 2e- +4H^+$

(multiply by 5)

$5SO_2 + 10 H_2O \rightarrow$

$16 H^+ + 2MnO_4^- + 10e- \rightarrow$

reduction

$Mn^{7+} + 5e- \rightarrow Mn^{2+}$

$MnO_4^- + 5e- \rightarrow Mn^{2+}$

$8 H^+ + MnO_4^- + 5e- \rightarrow Mn^{2+} +4H_2O$

(multiply by 2)

$5SO_4^{2-} + 10e- +20H^+$

$2Mn_2^+ +8H_2O$

add equations

$5SO_2 + 10H_2O+ 16 H^+ + 2MnO_4^- + 10e- \rightarrow \quad 5SO_4^{2-} + 10e- +20H^+ + 2 Mn_2^+ + 8H_2O$

cancel like terms – and left with

$5SO_2 + 2H_2O + 2MnO_4^- \rightarrow \quad 5SO_4^{2-} + 4H^+ + 2Mn_2^+$

Therefore the coefficient of the H+ is 4.

28. Answer A.

$N\equiv N$, $N=N$, $N-N$ a triple bond is the shortest, followed by the double bond and finally a single bond is the longest.

29. Answer A.

First determine moles of copper to be plated out:

0.2000 L x 0.1500 mol/L = 0.03000 mol

then determine moles of electrons required:

$Cu^{2+} + 2e^- \rightarrow Cu$

Every mole of Cu plated out requires two moles of electrons.

0.03000 mol x 2 = 0.06000 mol e⁻ required

Convert moles of electrons to Coulombs of charge: (Faraday's Constant is given in the Formula Sheet)

0.06000 mol e⁻ x 96,485.309 C/mol = 5789.12 C

Convert to seconds required to deliver the Coulombs determined in step 3 (remember, 1 A = 1 C/sec):

5789.12 C divided by 0.200 C/sec = 28945.6 sec (approximately 29,000 seconds, answer A)

30. Answer B.

Yes $E°$cell $= 1.81$ V

Sn^{4+} is a stronger oxidant than Al^{3+}. Balancing the equation and the electrons gives:

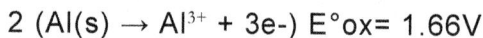

2 $(Al(s) \rightarrow Al^{3+} + 3e-)$ $E°$ox$= 1.66$V

3 $(Sn^{4+}$ (aq) + 2e- $\rightarrow$ Sn^{2+}(aq)) $E°$red$= 0.15$V

add the two equations and recall that electrode potentials do not depend on the amount of material present.

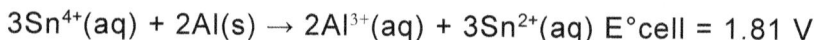

$3Sn^{4+}$(aq) + $2Al(s)$ $\rightarrow$ $2Al^{3+}$(aq) + $3Sn^{2+}$(aq) $E°$cell $= 1.81$ V

31. Answer A.

Calculate the reaction quotient (product concentrations/reactant concentrations)at the actual conditions:

$$Q = \frac{[N_2][O_2]}{[NO]^2} = \frac{(0.090 \text{ M})(0.020 \text{ M})}{(0.005 \text{ M})^2} = 72$$

This value is less than K_{eq} ($72 < 2 \times 10^3$), therefore $Q < K_{eq}$. To achieve equilibrium, Q must equal K_{eq}, and in this case this occurs when enough reactants turn into products. Therefore NO will react to make more N_2 and O_2. You can check this by the "alligator quick check" method: spelling "Quick checK", substitute the middle letters with the $<$ sign in between Q and K, and imagine it as an open alligator's mouth. The alligator is moving forward, and that is the direction in which the reaction will go until the $<$ turns into an $=$.

32. Answer D.

The equation is $2Na + Cl_2 \rightarrow 2NaCl$

the mole ratio of sodium to chlorine is 1:1 therefore, the ratio stays the same with the atomic masses. so you can leave in g (or use MW) (MW(Na) = 22.9, MW(Cl_2) = 70.8 MW NaCl = 58.3)

1.54g Cl : 1 g Na :: 11.8g Cl : x x = 7.66 g Na   not the correct choice

1.54g Cl : 1 g Na :: 5.55g Cl : x x = 3.60 g Na   not the correct choice

1.54 g Cl : 1 g Na :: 19.3 g Cl : x x = 12.53 g Na   not the correct choice

1.54 g Cl : 1 g Na :: 21.4 g Cl : x x = 13.89 g Na   This is the correct choice because the ratio **matches the choice**

33. Answer A.

The half life is 650sec. The initial concentration is 0.5M, after 650sec the remaining concentration is 0.25M. After another 650sec, the remaining concentration is half of that – or 0.125M.

34. Answer A.

The reaction must proceed in the forward direction, producing more products until the value of Q is also 10.

35. Answer B.

In trials 1 and 2 [A] is constant, [B] doubles and the rate doubles. Therefore the reaction is 1st order for [B]. In trials 1 and 3 the [B] is constant, the [A] doubles and the rate quadruples. Therefore the reaction is second order with respect to [A]. The overall reaction rate then is given by B.

rate = $k[A]^2[B]$

36. Answer C.

The overall reaction is the sum of the two reactions

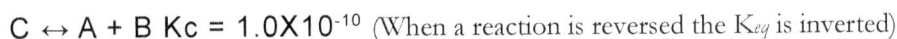

C $\leftrightarrow$ A + B Kc = 1.0×10^{-10} (When a reaction is reversed the K_{eq} is inverted)

A + D $\leftrightarrow$ E Kc = 1.6×10^7

net: C + D $\leftrightarrow$ B + E When reactions are added K_{eq} are multiplied. Thus:

Koverall = $1.6 \times 10^7 * 1.0 \times 10^{-10}$

     = 1.6×10^{-3}

37. Answer D.

The solid is never included in the Ksp.

38. Answer B.

Second order means the [A] is squared, first order means only [B] is used.

Thus the correct answer is choice B.

39. Answer C.

Since the Ksp for AgBr is smaller than the Ksp for AgCl, AgBr will precipitate first.

To calculate concentration write the equilibrium reaction and the equation for K:

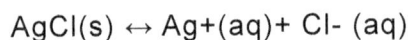

AgCl(s) $\leftrightarrow$ Ag+(aq)+ Cl- (aq)

Ksp = [Ag+][Cl-]

1.6×10^{-3} = [Ag+][0.53]

[Ag+] = 3.0×10^{-3} M

For Br

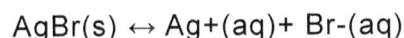

AgBr(s) $\leftrightarrow$ Ag+(aq)+ Br-(aq)

Ksp = [Ag+][Br-]

6.5×10^{-13} = [Ag+][8.4×10^{-4}]

[Ag+] = 7.7×10^{-10} M

Therefore AgBr will precipitate first at 7.7 x 10-10M concentration of [Ag+]

40. Answer A.

Heat given off by the system increases the thermal motion of the surroundings.

41. Answer D.

Intermolecular attractions are lessened during melting. This permits molecules to move about more freely, but there is no change in the kinetic energy of the molecules because the temperature has remained the same.

42. Answer A.

1,000 ml of water has a mass of 1,000 g. The specific heat of water is 4.184 J/g*K (from your sheet of constants). Using these numbers and those supplied in the problem:

$$q = m \times c_{\Delta}T$$

$$8000\,J = (1000\,g) \times (4.184\,\frac{J}{g \times K}) \times_{\Delta}T$$

$$_{\Delta}T = 1.91K$$

43. Answer A.

The reaction with the greatest positive value for Δn_{gas} will have the greatest value of ΔS_{rxn}

44. Answer D.

Phase changes are equilibrium processes, and $\Delta G_{fus} = \Delta H_{fus} - T_{fus}\,\Delta S_{fus}$.

At equilibrium $\Delta G_{fus} = 0$.

Thus: $\Delta S fus = \Delta H_{fus}\,/T_{fus}$

$$= (35{,}230\,J/mol)\,/\,(3695\,K) = 9.53 J/molK$$

45. Answer B.

Only statement B is TRUE.

46. Answer C.

The overall reaction is equal to the reverse of reaction 1 plus the forwards reactions of reactions 2 and 3.

$$\Delta H_{total} = -\Delta H_1 + \Delta H_2 + \Delta H_3$$
$$= -78.5\,kJ + 20.5\,kJ \quad 11\,kJ = -09\,kJ$$

47. Answer C.

ΔH_{rxn} = sum of bond energies of reactants - sum of bond energies of products

ΔH_{rxn} = [4(C-H) + 1(I-I)] - [3(C-H) + 1(C-I) + 1(H-I)]

ΔH_{rxn} = [4(416 kJ/mol) + (151 kJ/mol)] - [3(416 kJ/mol) + (213 kJ/mol) + (299 kJ/mol)]

ΔH_{rxn} = +55 kJ/mo

48. Answer A.

Phase changes are equilibrium processes, hence $\Delta G = 0$ and $\Delta H = T\Delta S$

$T = \Delta H / \Delta S = (349.6 \text{ kJ/mol}) / (.100 \text{ kJ}/(\text{mol·K})) = 3496 \text{ K}$

Note: make sure you put entropy and enthalpy both in either kJ or J

49. Answer D.

All of the statements are true based on the given data.

50. Answer A.

The heat is being removed from the system, so the temperature is dropping, then the system undergoes a phase change during which there is no temperature change.

51. Answer A.

When Q is less than K, the reaction will go forward. To check this, use the "alligator quick check" method: spelling "Quick checK", substitute the middle letters with the < sign in between Q and K, and imagine it as an open alligator's mouth. The alligator is moving forward, and that is the direction in which the reaction will go until the < turns into an =.

52. Answer B.

1 M HCOOH and 1 M COOH-

This is a classic buffer. Buffers need to include both a weak acid and weak base. A weak acid in balance with its conjugate base is a great way to do this. [H+] = Ka(Ca/Cb) = 10-5 (1/1) = 10-5 pH = 5

The others are: A. this is a strong and a weak acid; C. this is a weak base pH= 9.5; D. This is a strong acid/weak acid, not a buffer pH = 0.3

53. Answer B.

For the purposes of LeChatelier's Principle, you can treat it as if it has physical existence. Since heat is added, the reaction will shift to try and use up some of the added heat. In order to do this, the reaction must shift to the left.

54. Answer C.

Neither side is favored over the other since both sides have the same number of total molecules (two). No matter which way the reaction shift, the total number of molecules would remain unchanged.

In cases like this, where there is an equal number of molecules on each side, the equilibrium would remain unchanged by the change in pressure (in either direction).

55. Answer B.

Answer = make it smaller. To see this, write the Keq expression for the reation:

$$K_{eq} = [HCl]^2 / ([H_2][Cl_2])$$

As the equilibrium shifts to the left, the [HCl] goes down and both the $[H_2]$ and $[Cl_2]$ increase. This makes the numerator smaller and the denominator larger. The K_{eq} decreases in value as heat is added to an exothermic reaction.

56. Answer B.

When AgBr dissolves, it dissociates like this:

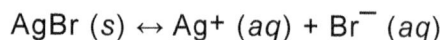

$$AgBr\ (s) \leftrightarrow Ag^+\ (aq) + Br^-\ (aq)$$

The Ksp expression is:

$$Ksp = [Ag^+][Br^-]$$

There is a 1:1 molar ratio between the AgBr that dissolves and Ag^+ that is in solution. In like manner, there is a 1:1 molar ratio between dissolved AgBr and Br^- in solution. This means that, when 5.71×10^{-7} mole per liter of AgBr dissolves, it produces 5.71×10^{-7} mole per liter of Ag^+ and 5.71×10^{-7} mole per liter of Br^- in solution.

Putting the values into the Ksp expression, we obtain:

$$Ksp = (5.71 \times 10^{-7})(5.71 \times 10^{-7}) = 3.26 \times 10^{-13}$$

57. Answer C.

Recall $Kw = 1.0 \times 10^{-14}$

$$K_w = K_a \times K_b$$

$$1 \times 10^{-14} = (1.8 \times 10^{-5}) \times K_b$$

$$\frac{1 \times 10^{-14}}{1.8 \times 10^{-5}} = K_b$$

$$5.6 \times 10^{-10} = K_b$$

58. Answer A.

Balance the equation $Cu(OH)^2 \leftrightarrow Cu^{2+} + 2\ OH^-$

$$Ksp = [Cu^{2+}][OH^-]^2$$

Solve for the [concentrations] $1.6 \times 10^{-19} = (s)(2s)^2 = 4s^3$

$$1.6 \times 10^{-19} = 4s^3$$

$$4.0 \times 10^{-19} = s^3$$

After dividing by 4 and then taking the cube root (which you are not expected to do but recognise that this is how you solve the problem):

$$s = 3.42 \times 10^{-7} \text{ M}$$

This can be used to determine the pH.

This is an important point: what we have calculated is 's' and it is NOT the $[OH^-]$. That value is '2s.'

$$[OH^-] = 6.84 \times 10^{-7} \text{ M}$$
$$pOH = -\log 6.84 \times 10^{-7} = 6.165$$
$$pH = 14 - 6.165 = 7.835$$

This can be used to determine the pH.

59. Answer A.

[conjugate acid] = [conjugate base]

60. Answer B.

If $\Delta G° < 0$ then $K > 1$, so products will be in greater concentration than reactants at equilibrium.

$$K = e^{-\Delta G°/RT}$$

#1 Cobalt has 27 electrons.

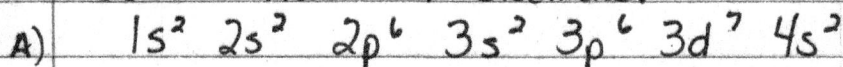

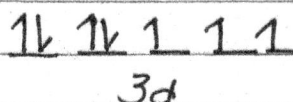

A) $1s^2\ 2s^2\ 2p^6\ 3s^2\ 3p^6\ 3d^7\ 4s^2$

B)   ⇅   ⇅   ⇅⇅⇅   ⇅   ⇅⇅⇅   ⇅
     1s   2s   2p   3s   3p   4s

     ⇅ ⇅ ↑ ↑ ↑
           3d

c) Half-life is the time for ½ the amount
    to be left.   Half-life = 5.3 years
    Start w/10g in 5.3 years ⇒ 5g.
    in 10.6y (another ½ life) ⇒ 5/2 ⇒ 2.5g.

d   Average atomic · mass
       read %-age & # off graphs

    Ave = (20%)(70) + (27%)(72) + (7.5%)(73)
              + (38)(74) + (7.5%)(76)

OR use given #'s & % from graph
    Ave= (20%)(69.9), (27%)(71.9) + (7.5%)(72.9)
              + (38%)(73.9) + (7.5%)(75.9)
         = 72.7

#2   The rate equation can be determined by looking at the data. If you compare experiments 2 & 3 (where [NO] is constant, you see that when [O_2] doubles, the rate doubles. This is what happens when a reaction is first order. Therefore the reaction is 1st order for [O_2]

If you compare experiments 1 & 2 where [O_2] is constant, you see that [NO] doubles, and the rate quadruples This means the reaction is 2nd order for [NO].

$$Rate = k \, [NO]^2 \, [O_2]$$

Alternate Method   (Mathematically)

$$Rate \, 1 = k \, [NO]^x \, [O_2]^y$$
$$Rate \, 2 = k \, [NO]^x \, [O_2]^y$$
$$Rate \, 3 = k \, [NO]^x \, [O_2]^y$$

when [NO] is constant then

$$\frac{Rate \, 3}{Rate \, 2} = \frac{k \, [4 \times 10^{-3} M]^x \, [2 \times 10^{-3} M]^y}{k \, [4 \times 10^{-3} M]^x \, [1 \times 10^{-3} M]^y}$$

k can cancel, [NO] cancels.

$$\frac{[10 \times 10^{-4} \text{ M/s}]}{[5 \times 10^{-4} \text{M/s}]} = \frac{[2 \times 10^{-3}]^y}{[1 \times 10^{-3}]^y}$$

$$2 = 2^y \quad \therefore \boxed{y = 1}$$

Similarly: for experiments 1 & 2

$$\frac{\text{Rate 2}}{\text{Rate 1}} = \frac{k'[4.0 \times 10^{-3}]^x}{k'[2.0 \times 10^{-3}]^x} \cdot \frac{[1 \times 10^{-3}]^y}{[1 \times 10^{-3}]^y} \qquad [O_2] \text{ cancel}$$

$$\frac{[5 \times 10^{-4}]}{[1.25 \times 10^{-4}]} = \frac{[4.0 \times 10^{-3}]^x}{[2.0 \times 10^{-3}]^x}$$

$$4 = 2^x \qquad \therefore x = 2 \qquad \boxed{x = 2}$$

Rate Law $\Rightarrow$ $k\,[NO]^2\,[O_2]$

B) The rate constant can be found from the data and the rate law

$$\text{Rate} = k\,[NO]^2\,[O_2]$$

$$k = \frac{\text{Rate}}{[NO]^2\,[O_2]} \qquad \text{use data from any experiment}$$

$$= \frac{5 \times 10^{-4}}{[4 \times 10^{-3}]^2\,[1 \times 10^{-3}]} =$$

$$= \frac{[5 \times 10^{-4}]}{[1.6 \times 10^{-5}][1 \times 10^{-3}]} = 3.1 \times 10^4$$

$$k = 3.1 \times 10^4$$

c) Calculate the reaction when $[NO] = \boxed{5 \times 10^{-3}\text{M}}$

$ [O_2] = \boxed{7 \times 10^{-3}\text{M}}$

$$Rate = k[NO]^2 [7 \times 10^{-3}\text{M}]$$

$$= (3.1 \times 10^4 \times 5 \times 10^{-3})^2 (7 \times 10^{-3}\text{M})$$

$$= 0.00547$$

#3 It's an endothermic reaction
the enthalpy of reaction is unfavorable:

$$\Delta H° = \Sigma H f° (products) - \Sigma H f° (reactants)$$

$$= [1\text{ mol } NH_4 \times 132.51 \text{ kJ/mol} + 1 \text{ mol } NO_3 \times \overline{205.0} \text{ kJ/mol}]$$

$$- [1 \text{ mol } NH_4NO_3 \times 365.56 \text{ kJ/mol}]$$

$$= 28.05 \text{ kJ}$$

Because the reaction leads to a significant increase in the disorder of the system, it is favored by the entropy of reaction:

$$\Delta S° = \Sigma S°(products) - \Sigma S°(reactants)$$

$$= [1\text{ mol } NH_4 \times 113.4 \text{ J/mol-K} + \text{mol } NO_3^- \times 146.4 \text{ J/mol-K}]$$

$$- [1 \text{ mol } NH_4NO_3 \times 151.08 \text{ J/mol-K}]$$

$$= 108.7 \text{ J/K}$$

Compare the $\Delta H°$ and $T\Delta S°$ to see which is larger and to decide if NH_4NO_3 should dissolve in water at 25°C.
First, convert the temp from °C to Kelvin:

$$TK = 25°C + 273.15 = 298.15 \text{ K}$$

$\Delta H°$ for this reaction are in kilojoules ⎫ convert to
$\Delta S°$ units are in joules per kelvin ⎬ consistent set
of units

convert $\Delta°$ to joules and multiply the entropy term by the absolute temp and subtract this quantity from the enthalpy term:

$$\Delta G° = \Delta H° - T\Delta S°$$

$$= 28,050 \text{ J} - (298.15 \text{ K} \times 108.7 \text{ J/K})$$

$$= 28,050 \text{ J} - 32,410 \text{ J}$$

$$= -4360 \text{ J}$$

Because the entropy term at 25°C is larger than the enthalpy term, the standard-state free energy for this reaction is negative:

$$\Delta G° = -4.4 \, kJ$$

This reaction is spontaneous at room temp.

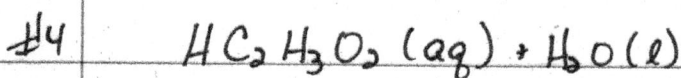

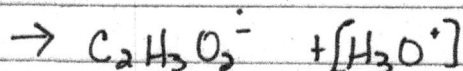

#4 $HC_2H_3O_2 \, (aq) + H_2O \, (l)$

$$\rightarrow C_2H_3O_2^- + [H_3O^+]$$

$K = 1.8 \times 10^{-5}$

need to find $[H_3O^+]$ at equilibrium.
Make an ICE table

	$[H_2O]$	$[HC_2H_3O_2]$	$[C_2H_3O_2^-]$	$[H_3O^+]$
Initial.	—	1M	0	0
Change	—	$-x$	x	x
Equil.	—	$1-x$	x	x

assume $[H_2O]$ essentially unchanged

$$K = \frac{[C_2H_3O_2^-][H_3O^+]}{[HC_2H_3O_2]} = \frac{[x][x]}{(1-x)} = \frac{[x]^2}{(1-x)}$$

$$1.8 \times 10^{-5} = \frac{(x^2)}{(1-x)}$$
$$(1-x)(1.8 \times 10^{-5}) = x^2$$
$$1.8 \times 10^{-5} - 1.8 \times 10^{-5}x - x^2 = 0$$
$$0 = x^2 + 1.8 \times 10^{-5}x - 1.8 \times 10^{-5}$$

need to use quadratic formula

$$X = \frac{-b \pm \sqrt{b^2 - 4ac}}{2a}$$

$a = 1$

$b = 1.8 \times 10^{-5}$

$c = -1.8 \times 10^{-5}$

$$X = \frac{-(1.8 \times 10^{-5}) \pm \sqrt{(1.8 \times 10^{-5})^2 - (4)(1)(-1.8 \times 10^{-5})}}{(2)(1)}$$

$$= \frac{(-1.8 \times 10^{-5}) \pm \sqrt{3.2 \times 10^{-10} + 0.000072}}{2}$$

$$= \frac{(-1.8 \times 10^{-5}) \pm (0.0084)}{2}$$

$X = -0.0042 \qquad X = 0.0043$

a negative value is not real so use

$X = 0.0043$ as $[H_3O^+]$

$pH = -\log [0.0043] = \underline{\underline{2.37}}$

#5 $2XI_3 + 3Cl_2 \rightarrow 2XCl_3 + 3I_2$

$2mols\ XI_3 \rightarrow 2moles\ XCl_3$

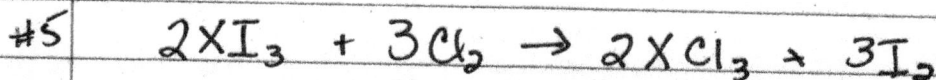

$gr \rightarrow moles$

$MW\ XI_3 = X + 3(I) = X + 381$

$XCl_3 = X + 3(Cl) = X + 106.5$

$\dfrac{0.5g\ XI_3}{X + 381} = moles\ XI_3$

$\dfrac{0.236g}{X + 106.5} = moles\ XCl_3$

$moles\ XI_3 = moles\ XCl_3$

therefore.

$\dfrac{0.5g}{(X + 381)} = \dfrac{0.236}{X + 106.5}$

$(X + 106.5)(0.5) = (0.236)(X + 381)$

$0.5X + 53.25 = 0.236X + 89.9$

$0.264X = 36.65$

$X = 138.8$

mw of X $\approx$ 138.8

This must be Lanthanum.

B/   FROM the balanced equation we
know   moles XCl_3 = moles $X I_3$ produced.

$$\frac{0.520g}{(x+106.5)} = \frac{0.979}{(x+381 g/m)} \qquad Cl_3 = 106.5 \qquad I_3 = 381$$

Solve for x   X = 204.5

the element is Tl, thallium.

#6

A   Phosphoric Acid is 87% phosph
acid soln. If it is 13M, what is
density?

① Determine # moles in 100 gr of 87% sln.
   87.0 g of 100g is H_3PO_4

$$MW \; H_3PO_4 = \begin{array}{l} H = (1)(3) = 3 \\ P = (31)(1) = 31 \\ O = \underline{(16)(4) = 64} \\ \qquad\qquad\quad 98 \end{array}$$

$$\frac{87 g}{98 g/m} = 0.887 \text{ moles}$$

②   DETERMINE Volume
   13M is 13 moles/L

$$13M = 6.887 \text{ mole}/x$$
$$X = 0.0683 L$$

③ Determine Density $D = m/v = \dfrac{100g}{68.3 ml}$

$$= 1.46 \; g/ml.$$

7a

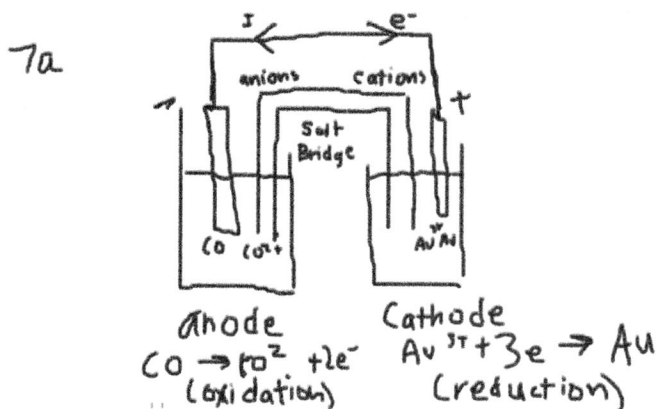

anode
$CO \rightarrow CO^{2+} + 2e^-$
(oxidation)

Cathode
$Au^{3+} + 3e \rightarrow Au$
(reduction)

7B   Std Free Energy

$$\Delta G = -nFE°$$

$$E°_{cell} = E°_{cath} - E_{anode} = 1.50V - (-0.28V) = 1.78V$$

$$\Delta G = -nFE°$$
$$= -(6)(96485)(1.78 \text{ J/c})$$
$$= -1030.46$$

6 moles of
e^-

C   Is the Reaction Spontaneous.

$$\log K = \frac{n \cdot E°}{RT} \approx \frac{(6)(1.78)}{0.0591} = 180.7$$

$$\log K = 180.7$$

therefore K is very large and the
process is spontaneous

7D Battery - 1 mA current   3000 hours

Convert

charge = current · t
= (0.001 C/s) (3000 hrs) (60m/h) (60s/m)
= 10800 C

Convert charge → moles
$10800 C * \dfrac{1\ mole^-}{96485 C} = 0.1119\ mole\ e^-$

moles of e^- → moles of Co $0.1119\ mole\ e^- \times \dfrac{1\ mol\ Co}{2\ mole}$

= 0.0560 mol Co

moles Co → gr Co
$0.0560 \times 58.9 g\ Co/mol\ C = 3.30 g\ Co$

<u>3.30 g</u> Co would be consumed

AP

The Advanced Placement® program is designed to offer students college credit while still in high school. The more than 30 AP courses culminate in an intensive final exam given every year in May.

Successful completion of a course and a passing score on the exam not only provides students with a deep sense of accomplishment, but also gives them a jumpstart on their college careers. AP credit is almost universally accepted by post-secondary schools, however each school has different guidelines as to what scores they will accept.

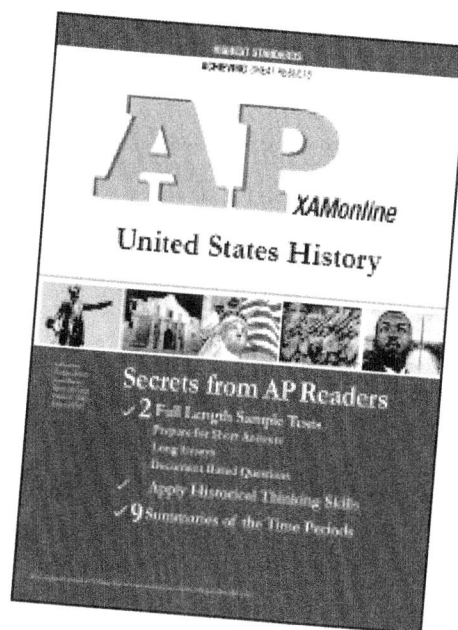

AP US History
ISBN 978-1-60787-552-9 $21.99

AP US Government and Politics
ISBN 978-1-60787-601-4 $21.99

AP Biology
ISBN 978-1-60787-553-6 $21.99

AP Calculus
ISBN 978-1-60787-555-0 $21.99

AP Chemistry
ISBN 978-1-60787-554-3 $21.99

AP Psychology
ISBN 978-1-60787-556-7 $21.99

AP English
ISBN 978-1-60787-557-4 $21.99

AP Spanish
ISBN 978-1-60787-558-1 $21.99

AP Macroeconomics/Microeconomics
ISBN 978-1-60787-585-7 $21.99

TO ORDER

XAMonline.com

or **amazon** or **BARNES & NOBLE** BOOKSELLERS

CPSIA information can be obtained
at www.ICGtesting.com
Printed in the USA
BVOW10s1735110817
491837BV00016B/328/P